Animal Cell Biotechnology

Volume 1

Contributors

J. R. Birch

S. G. Garland

J. B. Griffiths

G. J. Harper

J. L. Harris

H. Katinger

K. J. Lambert

Gian Franco Panina

P. A. Riley

W. Scheirer

R. E. Spier

G. Threlfall

Michael G. Tovey

Anton L. van Wezel

Animal Cell Biotechnology

Volume 1

Edited by

R. E. SPIER

Department of Microbiology
University of Surrey
Guildford, Surrey
United Kingdom

J. B. GRIFFITHS

Vaccine Research and Production Laboratory
Public Health Laboratory Service
Centre for Applied Microbiology and Research
Salisbury, Wiltshire
United Kingdom

1985

ACADEMIC PRESS

(Harcourt Brace Jovanovich, Publishers)

London   Orlando   San Diego   New York
Toronto   Montreal   Sydney   Tokyo

7297-9070

CHEMISTRY

ACADEMIC PRESS INC. (LONDON) LTD.
24–28 Oval Road
LONDON NW1 7DX

United States Edition published by
ACADEMIC PRESS, INC.
Orlando, Florida 32887

British Library Cataloguing in Publication Data

Animal cell biotechnology.
  1. Cell culture
  I. Spier, R. E.   II. Griffiths, J. B.
  591'.0724   QH585

Library of Congress Cataloging in Publication Data
Main entry under title:

Animal cell biotechnology.

  Includes index.
  1. Cell culture.   2. Biotechnology.   I. Spier, R. E.
II. Griffiths, J. B.
QH585.A58   1984   615'.36   84-18567
ISBN 0–12–657551–7 (v. 1)

PRINTED IN THE UNITED STATES OF AMERICA

85 86 87 88   9 8 7 6 5 4 3 2 1

Contents

3   Cell Biology: Experimental Aspects

J. B. GRIFFITHS

4   Cell Growth Media

K. J. LAMBERT AND J. R. BIRCH

Contributors

Numbers in parentheses indicate the pages on which the authors' contributions begin.

J. R. Birch (85), Celltech Ltd., Slough, Berkshire SL1 4DY, United Kingdom
S. G. Garland (123), Celltech Ltd., Slough, Berkshire SL1 4DY, United Kingdom
J. B. Griffiths (3, 17, 49), Vaccine Research and Production Laboratory, Public Health Laboratory Service, Centre for Applied Microbiology and Research, Salisbury, Wiltshire SP4 0JG, United Kingdom
G. J. Harper[1] (141), Experimental Microbiology and Safety Reference Laboratory, Public Health Laboratory Service, Centre for Applied Microbiology and Research, Salisbury, Wiltshire SP4 0JG, United Kingdom
J. L. Harris (283, 321), L. H. Fermentation, Stoke Poges, Buckinghamshire SL2 4EG, United Kingdom
H. Katinger (167), Institute of Applied Microbiology, University of Agriculture, A-1190 Vienna, Austria
K. J. Lambert (85), Celltech Ltd., Slough, Berkshire SL1 4DY, United Kingdom
Gian Franco Panina (211), Istituto Zooprofilattico Sperimentale, 25100 Brescia, Italy
P. A. Riley (17), Department of Biochemical Pathology, University College School of Medicine, London WC1E 6JJ, United Kingdom
W. Scheirer (167), Sandoz Research Institute, A-1235 Vienna, Austria
R. E. Spier (3, 243, 283, 321), Department of Microbiology, University of Surrey, Guildford, Surrey GU2 5XH, United Kingdom
G. Threlfall (123), Celltech Ltd., Slough, Berkshire SL1 4DY, United Kingdom
Michael G. Tovey (195), Laboratory of Viral Oncology, Institut de Recherches Scientifiques sur le Cancer, 94802 Villejuif, France
Anton L. van Wezel (265), Rijksinstituut voor Volksgezondheid en Milieuhygiëne, NL 3720 BA Bilthoven, The Netherlands

[1]Present address: "S'Argamassa", Quidhampton, Salisbury, Wiltshire SP2 9AR, United Kingdom.

Preface

"Animal Cell Biotechnology" is directed towards two groups of people. Those entering the field for the first time will find many details about how to set up and operate cell cultures in a variety of ways and at scales ranging between 0.001 and 10,000 litres. Others who are already engaged in the area will find that the comprehensive and detailed coverage of the selected topics written by experts in the various subjects will provide them with a well-referenced and well-indexed state-of-the-art report. This will enable them to expand their horizons further and to appreciate more fully those aspects of the subject with which they have not yet become deeply involved. It will also provide team leaders with the overview necessary to better coordinate and direct the various unit operations of manufacturing a product from animal cells in culture.

The contributing authors have been asked to assume that the reader will have been exposed at university or polytechnic level to two or more basic science subjects (mathematics, physics, chemistry and biology), although he/she may have qualified in the specialist disciplines of microbiology, chemical engineering, biochemistry, genetics, immunology or even biotechnology. In addition the writers have stressed, where appropriate, the underlying principles of the subject area and have provided many illustrations, diagrams, graphs and tables to present information which would otherwise take up much text and make for laborious reading. A balanced, thoughtful and fair assessment of alternative methods to achieve particular ends has been requested, yet in this burgeoning field the reader will have to contend with the clear enthusiasm of one or another of the various contributors. As editors, we are confident that there is a sufficient wealth of views for readers to have little difficulty in developing an appreciation of the strengths and weaknesses of alternative technologies and thereby be fortified in the choice of system for their specific application.

These volumes survey a new and as yet uncharted facet of biotechnology. They are designed both as an introduction and as an in-depth survey of the present situation. In view of the exciting scientific achievements but disappointing practical results derived from trying to express mammalian genes in prokaryotic organisms and the new capabilities in the area of genetic engineering of animal cells, we can anticipate that animal cell biotechnology is likely to retain its place at the cutting edge of biotechnology for some time to come. We trust that these volumes will help with the realisation of that future.

R. E. Spier
J. B. Griffiths

Contents of Volume 2

PART I

AN INTRODUCTION

1

Introduction

R. E. SPIER
Department of Microbiology
University of Surrey
Guildford, Surrey, United Kingdom

J. B. GRIFFITHS
Vaccine Research and Production Laboratory
Public Health Laboratory Service
Centre for Applied Microbiology and Research
Salisbury, Wiltshire, United Kingdom

1. THE AIMS AND SCOPE OF THIS BOOK

A technological revolution in the way in which bacteria were used for generating useful products occurred over a period of some 30 years. This time period began with Weizman's pioneering acetone–butanol fermentation in 1914 and culminated in the development of unit processes based on stirred tank reactors of 20,000 litres to generate penicillin in the mid-to-late 1940s. A similar revolution is in progress with regard to the way in which animal cells in culture are grown and exploited for the generation of products. Its beginnings may be discerned in the late 1940s with Enders' work on the production of poliomyelitis virus from human embryonic tissues. This gave a direct impetus for the large-scale production, in tissue culture, of sizeable

Animal Cell Biotechnology, Vol. 1

quantities of immunogenic material to make polio vaccines. The transition from bench-scale cultures to full-scale industrial apparatus whose maximum development at present stands at 8000-litre cultures for cells which do not require a supporting solid substratum (suspension cells) and 1000-litre cultures for cells which require a substratum is a process which is still in its dynamic phase.

It is the purpose of this book to enable the reader to become knowledgeable about the current status of this rapidly advancing technology and to appreciate that much exciting work yet remains to be done both in the engineering of animal cell lines to meet specific industrial and academic requirements and in the innovation, design and development of new ways to cultivate animal cells on the large scale. This infers collaboration between experts in many diverse fields, including molecular biologists, biochemists, immunologists, cell biologists and engineers. An additional aim of this book is to enable scientists of these different disciplines to understand each other's aims and problems, and to be able to communicate with each other in order to gain a better collaboration. It has also necessitated the multiauthor approach that has been used so that the extensive scope of animal cell biotechnology can be expertly presented.

We have presented the material in such a way that the practical aspects of the subject have been emphasised. This has been coupled with a requirement of the contributors to provide well-referenced texts so that the reader may become acquainted with the finer details by such a comprehensive introduction to the growing body of published work. The level at which we have pitched the contributions is such that a person who is being, or has been, educated at a university/polytechnic in one of the basic areas of the biosciences would be able to grasp the principles and practices of this new area of activity. Also, the text has been so organised as to serve as a broad and extensive survey of the present state of the art for those who are currently engaged in advancing knowledge, understanding and capabilities in the area of the biotechnology of cultured animal cells.

The flow of subject matter follows naturally from the kinds of activities which proceed when generating products from animal cells. The early chapters describe the cells, the media they grow in and the unit operations which enable the cultured cells to be grown in equipment free of viable contaminants. This is followed by a section which deals, in some detail, with the different kinds of technologies which are currently in use and under development for the production of the two basically different sorts of animal cells: the suspension cells and the substratum-adherent cells, often referred to as "monolayer" or "anchorage-dependent" cells. Having grown the cells, attention is transferred to the kinds of products which are currently made from such biological substrates or raw materials. In this section emphasis is given to the two areas which dominate this field, these being the viruses for use in vaccines and the antibodies, whose uses are presently expanding rapidly but already include applications for purifying immunoactive materials, diagnostic systems and detailed epidemiology and molecular analysis of disease-causing organisms. To have produced a material which forms the major and critical ingredient of a product is insufficient in itself. Such materials have then to be rendered in a suitable form to generate the maximum effect in the safest configuration. This is achieved by the series of downstream processing operations which result in more concentrated, less contaminated and safe to use materials. These are then formulated to potentiate their effects, prolong their shelf life and make them easy to use in the field. The bioproducts which result from these processes have to meet those requirements of safety and efficacy as determined by regulatory authorities whose function is to protect society at large against the dangers inherent in the applications of any prophylactic, therapeutic or other product which has been manufactured for direct bodily application. As such requirements are stringent, the many tests involved in the process of product acceptance have been given a separate section. It is important to realise that such tests may account for two-thirds of the total manufacturing costs (before packaging, storage overheads, sales, profits etc.). The agreement by a regulating agency to a particular testing regime (and the expected or acceptable results) also constitutes a major portion of the development costs of a product (typically in the range of £5–15 million) so that "regulatory agency requirements" impinge markedly on the design and development of the upstream equipment, processes and biocomponents.

In setting the scene for the depiction of the present state of animal cell biotechnology it was mentioned that this area is presently in the throes of its revolution. The final section, therefore, attempts to indicate the direction in which future changes in product profiles and technological developments can be expected. This area does not yet constitute an open and closed package in the history of technology. It is the view of the authors that we are but at the beginning of a journey of exploration and exploitation of the biochemical potential inherent in the animal cell in culture. Where we shall be in the decades ahead may well depend on you, the reader. We, as editors and authors, will be well satisfied if we have helped in some way to extend the scope of the applications and increase the rate of progress in this rapidly moving area of endeavour.

2. THE HISTORY OF THE USE OF LARGE-SCALE CULTURES OF ANIMAL CELLS

Acceptance of the idea that a cell belonging to a higher animal could be removed from that animal and thence made to grow and replicate in an *in vitro* situation did not occur until the early decades of the 20th century. Once it became apparent that such procedures were possible a second phase of activity began with the demonstration that the filterable infectious agents, viruses, could be grown and replicated in the extra-corporeal environment in such cells. A third phase may be delineated from the time of the demonstration of the production of useful amounts of virus for vaccine applications up to the time when (1) it became possible to insert and express particular exogenously derived genes in animal cells and (2) it was shown that it is possible to grow in culture a population of cells derived from a single cell. When such populations are derived from an antibody-excreting cell, the antibody molecules of the culture supernatant are all alike. The implications and consequences of these latter two capabilities are currently under intensive investigation. This marks the beginning of a fourth phase of activity and brings this historical survey up to the present. These phases will be considered in more detail below.

2.1. Growing the Cells of Higher Animals in Culture (29, 34, 35)

The demonstration of the capability of animal cells to grow and divide in culture required that a number of techniques were available. These may be summarised as

1. A means of obtaining some animal cells free of exogenous prokaryotes and fungi.
2. The development of a medium in which such excised cells could thrive.
3. A methodology to view such cells and observe their developments.
4. A means of continuously propagating such *in vitro* cultures and of keeping them free of other biological agents.

The intellectual background to these developments was provided by the development of the concept of a cell as the fundamental unit of living organisms of both the animal and plant kingdoms. This idea arose from Hooke's 1665 observation that the vesicles and cavities observable in cork were like the "cells" which form part of a monastery. Such observations were extended by Malpighi (1674) and Greur (1682), who observed plant cells filled with fluid and bounded by cell walls. Later (1806) Treviranus noted, as did Hugo von Mohl in 1830, that meristem cells elongated and divided. It was left to Schleiden to formulate a definitive theory of the cellularity of plants (1838), in which he included an important role for the nucleus, previously discovered by Brown in 1831. While comparisons between animal and plant tissues had been made by such as Muller (1835) and Henle and Purkinge (1837), it was Theodor Schwann, having met Schleiden in 1837 and discussed the existence of nuclei in both animal and plant cells, who published in 1839 a series of papers with the title "Mikroskopische Untersuchungen uber die Ubereinstimmung in der Structur und dem Wachstum der Tiere und Pflanzen" and thereby established the universality of the cellular basis for all the then known forms of animal and plant life. For these reasons our modern concept of the cell is said to have arisen from these two pioneers—Schleiden and Schwann (31).

A second idea, that of Claude Bernard (1813–1878), held that it is a characteristic of living things to preserve the status of their internal conditions despite changes in their external environment. His concept related to the organism as a whole, yet it can be equally well applied to a cell growing in culture outside the body. It followed from this approach that a cell outside the normal environment of an animal would seek to maintain its internal conditions and that it would be most likely to grow and divide when the difference between its internal environment and its external environment was minimal. Such thinking led to the development of fluids capable of sustaining and promoting cellular life outside the body.

At about the same time a number of investigators were showing that living tissue could be maintained outside the body. W. Roux (1885) retained the viability of a medullary plate of a chick embryo in warm saline and became an active publicist of this *in vitro* embryology. In 1887, Arnold demonstrated that the cells which had colonised a piece of alder pith when inserted into frogs could be shown to leave their new home when it was transplanted into an *in vitro* situation. Loeb (1897) demonstrated the survival of the cells of the blood and connective tissues in tubes of serum and plasma, and Ljunggren (1898) showed that he could keep an explant of human skin sufficiently viable in ascitic fluid for its successful reimplantation.

Additional experiments were performed by Jolly (1903), who observed cell division in hanging drops containing salamander leucocytes, and Beebe and Ewing (1906), who observed the same phenomena in an explanted canine lymphosarcoma.

Following Roux's activities, Ross Harrison became interested in such work and developed the "hanging drop" technique further (15), using small pieces of excised frog medullary tube embedded in a clot of frog lymph and held in this structure on the underside of a coverslip perched over a hollow in a glass slide. In 1907, by using such chambers, he was able to observe the growth of nerve cells over periods of several weeks. He quotes growth rates for such fibres as up to 20 μm in 25 min.

While the Harrison experiments were organised to answer questions relating to the physiology of frog nerve cells, the techniques which he used were applied to other cells from the tissues of warm-blooded animals by Burrows, who used a fowl plasma clot in place of a lymph clot (1910). In the same year Alexis Carrel entered the field. In 1913 he used a plasma supplemented with an embryo extract which had a strong growth-promoting effect. Such work had a much better chance of success than that of Lewis (1911) and Reed (in 1908 she prepared cultures of guinea-pig bone marrow which yielded cells), who jointly attempted to grow explants on chemically defined media.

Carrel attracted much attention to his work as he publicised it under the banner of cultivating "immortal" cells. This chick embryo heart cell preparation was initiated on 17 January 1912 and was passaged by Ebling for, it is claimed, 34 years. As Carrel was a surgeon and well versed in aseptic technique, he was well equipped to make a contribution to the *in vitro* cultivation of animal cells. His approach was ponderous. Assistants wore full-length black rubber gowns with attached whole head covering. The procedures were long-winded and burdened with much detail. His work was difficult to repeat by his contemporaries and as a result of his requirement for elaborate precautions to prevent contamination he created about the subject a mystique and exclusivity which retarded progress rather than aided it. Nonetheless, his achievements were many: he did succeed, in the absence of antibiotics, in the passaging of cells using surgical techniques to cut up the colonies and transfer them, he demonstrated to his colleagues the scientific value of the observations which could be made on such cultures and in 1923 he developed the Carrel flask (36). His work was carried on in a more or less unaltered form by Strangeways and Walton and in a simplified form by Drew.

During this work developments were made in media via the modification of Ringer's solution by Tyrode and the use of fibrin coagulum in addition to the fowl serum and embryo extract. The use of cinephotomicrography was also developed to observe dividing animal cells [Canti (1928)].

One further, and essential, piece of the jig-saw was developed during this period. It was the use of trypsin to free cells from the tissue matrix in which they were embedded (30), a technique which did not come into its own until 1937, when Simms and Stillman used it in passaging cells between plasma cultures. Further developments were initiated by Willmer and Medawar but it was the Mosconas who in 1952 defined the techniques for the trypsin-aided passaging of cell cultures that we use today (25). It also allowed the use of individual cells in culture rather than tissues, i.e. cell culture as opposed to tissue culture. The phenomenon of generating single cells from a tissue or cell sheet is a very powerful stimulus for cell proliferation.

Until the human diploid cell line (HDC) WI-38 was derived in 1961 by Hayflick and Moorhead (16) it was assumed that a cell line, once established, had an unlimited life-span. However, the WI-38 line was shown to have a limited life-span in culture of approximately 50 population doublings and, before dying out, showed senescence or ageing phenomena. This mortality was limited to the fact that the WI-38 cell remained diploid and showed no malignant characteristics. Immortality is a characteristic of cells derived from tumours or transformed in culture and is correlated with heteroploidy. Diploid cell lines are therefore analogous to normal somatic tissue and heteroploid cells to transplantable tumours. A further impact of HDC cells in cell culture history is that they were found to be genetically stable and free of all known latent and oncogenic viruses; they, therefore, became the permitted cell line to generate products for human use. This dogma still applies today, although recent discoveries have clearly demonstrated the presence of putative oncogenes, identical with those found in such known oncogenic viruses as the Rous sarcoma virus or Moloney murine sarcoma virus, in cells derived from normal tissues (7). Furthermore, the transforming protein of a simian sarcoma virus has been shown to be related to a material produced in healthy animals which is a platelet-derived growth factor (33). The effects of these most recent findings have yet to have an influence on the cells that are used to produce materials for injection into humans.

2.2. Virus Growth in Animal Cells in Culture
(29, 34, 35)

In the general area of science, cell culture has had most impact on the subject of virology. The capability of producing and assaying viruses *in vitro* led to the emergence of animal cell biotechnology. Although cell culture has expanded to include the generation of other products, notably immunobiologicals, it is the link with viruses that is important in tracing the history of cell culture and biotechnology.

The first discovery of a filterable infectious agent which would induce foot-and-mouth disease when administered to cattle occurred in 1897 (21). Such agents became known as viruses, a word colloquially used to describe poisons or noxious fluids. The difficulty in maintaining or augmenting the infectivity of such fluids outside whole animals was resolved, for the most part, by the infection of cultured animal cells. The history of these discoveries constitutes the history of the second phase of the use of cultured animal cells.

In 1913 two groups of workers are on record as attempting to grow viruses in cultured cells: Levaditi (polio and rabies) and Steinhardt, Israeli and Lambert used explanted cornea for vaccinia maintenance. Later, in 1917, Harde found some proliferation of vaccinia in the type of culture system used by Steinhardt *et al.* Better results were obtained by Parker (1925) when he used cultures of rabbit testis in rabbit plasma, and 2 years later Carrel and Rivers used flask culture to grow vaccinia virus. They speculated that such a technology would supplant the calf as a source of vaccine. (The last official vaccine to be made were made in sheep and bovines in the late 1970s!) An adaptation of this technique was used effectively by the Maitlands (1928), who grew vaccinia virus in minceates of hen kidneys held in Carrel flasks and treated them as if they were cell cultures (24). Other viruses were grown in such systems. A latent herpes virus of rabbits was grown in a kidney minceate (Andrewes, 1929) and in 1931 Hallauer attempted to grow fowl plaque virus in pulped chick embryo tissue. This naturally led to the use of embryonated eggs as a culture system for viruses (37), a system still in use for influenza, yellow fever and Newcastle disease viruses.

Although there was a swing away from cell culture to embryonated eggs in 1949, this trend was reversed by the finding of Enders, Weller and Robbins that the Lansing strain of polio would grow in cultivated cells (10). This work also demonstrated that cytopathogenicity occurred in the cultured cells, a phenomenon which could henceforth be used as a marker for virus growth.

One additional capability remained in order to fully utilise cell cultures for virus growth systems, and that required the development of plaque assay systems similar to those developed in 1921 for bacteriophage by d'Herelle. This was effected by Dulbecco in 1952 (8) and further developed by Cooper in 1961, using an improved efficiency plating technique for poliovirus.

Once it became possible to produce poliovirus in culture the scale-up of such cultures to produce commercially useful quantities of vaccine became imperative. The development of suitable technologies and the extension of those technologies to other vaccines are the subjects of the next section.

2.3. Animal Cells in Culture Used for Commercial Virus Vaccines

The Carrel flasks of 1923 are static culture systems. However, other work-ers branched out into tubes and Strangeways and Fell used a slope tube-like system in 1926. This was preceded by the demonstration of a roller tube culture system in 1925 by Lowenstadt (22). Later, in 1933, Gey published his version of this system (12). His system, when fully exploited, would operate at 4700 cm^2 of surface for cell growth (24 Roux bottle equivalents). This apparatus leads directly to the multiroller tube racks commercially available today. From such beginnings it is simple to perceive the origin of the multiple processes used in the production of monolayer cells for vaccine production (see Chapter 9). Vaccines made by such processes include those for mumps (3), measles (18), adenovirus (17) and rubella (4). They can each use avian cells as the biological substratum for virus growth, although new cell lines based on human diploid cells WI-38 (19) and MRC-5 (16) are also used.

As the human vaccines were first made from whole-animal cultures [originally man was used for vaccinia (1798) before calves and sheep were used] veterinary vaccines were also made in whole animals (1). Furthermore, foot-and-mouth disease vaccines were and still are made in tongue tissue explants after a process begun by Frenkel in 1947 (11). Cell culture vaccines for foot-and-mouth disease began with the use of trypsin-dispersed bovine kidney cells in 1960 (28) and this was followed by the use of a line of cells which could be passaged indefinitely [BHK cells (32)] for FMD virus growth (26). Such cell culture systems were based on cells which grew on the surface of stationary or rotated glass bottles; a dramatic change was introduced when it became possible to exploit cultured cells which could grow and replicate without being associated with a surface.

The first cultures of animal cells in suspension were generally based on cells derived from tumourous tissues, such as HeLa cells derived from a human cervical carcinoma by Gey (13) and Earle *et al.* (9) and mouse L cell fibroblasts (2, 6, 24). Later (1962) systems were adapted to use cells of "normal" tissues and the cells of such "normal" tissues were adapted to grow in suspension (5). Following the demonstration of such capabilities, the scale-up of the 0.2-litre system of Capstick *et al.* (5) to the 8000-litre systems of today proceeded with little change in technology.

Having described the first three phases of the history of the use of cultured animal cells, it remains for the following chapters to describe many of the other significant discoveries that are now part of the history of biotechnology. An example is the pioneering work of Okada and Harris in the 1960s in their development of somatic cell hybridisation techniques (14, 27), which laid the foundation for Kohler and Milstein's (20) development of hybridomas that secrete monoclonal antibodies. The following chapters also consider in detail the unit processes involved in that technology, to point out problems and shortcomings of such processes and to give a clear indication of the way ahead.

REFERENCES

1. Belin, M., Houillon, G. (1928). Vaccination antiaphteuse avec les complexes vaccino-aphteux sterilises. *C. R. Seances Soc. Biol. Ses Fil.* **99**, 1194–1196.
2. Bryant, J. C., Schiling, E. L., and Earle, W. R. (1958). Massive fluid suspension cultures of certain mammalian tissue cells. *J. Natl. Cancer Inst. (U.S.)* **21**, 331.
3. Buynack, E. B., and Hilleman, M. R. (1966). Live attenuated mumps virus vaccine. 1. Vaccine development. *Proc. Soc. Exp. Biol. Med.* **123**, 768–774.
4. Buynack, E. B., Hilleman, M. R., Weibel, R. E., and Stokes, J. (1968). Live attenuated rubella virus vaccines prepared from duck embryo cell culture. *JAMA, J. Am. Med. Assoc.* **204**, 103–108.
5. Capstick, P. B., Telling, R. C., Chapman, W. G., and Stewart, D. L. (1962). Growth of a

cloned strain of hamster kidney cells in suspended cultures and their susceptibility to the virus of foot-and-mouth disease. *Nature (London)* **195**, 1163–1166.

6. Danes, B. S. (1956). A glass stirrer apparatus for the cultivation of cell suspension. *Exp. Cell Res.* **12**, 169.

7. Duesberg, P. H. (1983). Retroviral transforming genes in normal cells? *Nature (London)* **304**, 219–226.

8. Dulbecco, R. (1952). Production of plaques in monolayer tissue cultures by single particles of an animal virus. *Proc. Natl. Acad. Sci. U.S.A.* **38**, 747–752.

9. Earle, W. R., Bryant, J. C., Schilling, E. L., and Evans, V. S. (1956). Growth of cell suspensions in tissue culture. *Ann. N.Y. Acad. Sci.* **63**, 666.

10. Enders, J. F., Weller, T. H., and Robbins, F. C. (1949). Cultivation of the Lansing strain of poliomyelitis virus in cultures of various human embryonic tissues. *Science* **109**, 85–87.

11. Frenkel, H. S. (1947). La culture du virus de la fierre aphteuse sur l'epithelium de la langue des bovides. *Bull. Off. Int. Epizoot.* **28**, 155–162.

12. Gey, G. O. (1933). An improved technic for massive tissue culture. *Am. J. Cancer* **17**, 752–756.

13. Gey, G. O. (1953). *J. Exp. Med.* **97**, 695.

14. Harris, H., and Watkins, J. F. (1965). Hybrid cells derived from mouse and man: Artificial heterokaryons of mammalian cells from different species. *Nature (London)* **205**, 640–646.

15. Harrison, R. G. (1907). Observations on the living developing nerve fibre. *Proc. Soc. Exp. Biol. Med.* **4**, 140–143.

16. Hayflick, L., and Moorhead, P. S. (1961). The serial cultivation of human diploid cell strains. *Exp. Cell Res.* **25**, 585–621.

17. Hilleman, M. R. (1958). Efficacy of and indications for the use of adenovirus vaccine. *Am. J. Public Health*, 153–158.

18. Hilleman, M. R., Buynack, E. B., Weibel, R. E., Stokes, J., Whitman, J. E., and Leagus, M. B. (1968). Development and evaluation of the Moraten measles virus vaccine. *JAMA, J. Am. Med. Assoc.* **206**, 587–589.

19. Jacobs, J. P. (1970). Characteristics of a human cell designated MRC-5. *Nature (London)* **227**, 168–170.

20. Kohler, G., and Milstein, C. (1975). Continuous cultures of fused cells secreting antibody of predefined specificity. *Nature (London)* **256**, 495–497.

21. Loeffler, F., and Frosch, P. (1897). Summarischer Bericht uber die Engebnisse der Unterschungen zur Enforschung der Maul-und Klausensenche. *Zentralbl. Bakteriol., Parasitenkd. Infektionskr., Abt. 1* **22**, 257.

22. Lowenstadt, H. L. (1925). Einige neue Hilfsmittel zur Anlegung von Gewebekulturen. I. *Arch. Exp. Zellforsch. Besonders Gewebezuecht.* **1**, 251–256.

23. McLimans, W. F., Fiardinello, F. E., Davis, E. V., Mucera, C. J., and Rake, C. W. (1957). Submerged culture of mammalian cells: The 5 L fermenter. *J. Bacteriol.* **74**, 768.

24. Maitland, H. B., and Maitland, M. C. (1928). Cultivation of vaccinia virus without tissue culture. *Lancet* **215**, 596–597.

25. Moscona, A. A., and Moscona, M. H. (1952). The dissociation and aggregation of cells from organ rudiments of the early chick embryo. *J. Anat.* **86**, 287.

26. Mowat, G. W., and Chapman, W. G. (1962). Growth of foot-and-mouth disease virus in a fibroblastic cell line derived from hamster kidneys. *Nature (London)* **194**, 253–255.

27. Okada, Y., and Murayama, F. (1965). Multinucleated giant cell formation by fusion between cells of two different strains. *Exp. Cell Res.* **40**, 154–158.

28. Patty, R. E., Bachrach, H. L., and Hess, W. R. (1960). Growth of foot-and-mouth disease virus in bovine kidney cell suspensions. *Am. J. Vet. Res.*, 144–149.

29. Paul, J. (1975). "Cell and Tissue Culture," 5th ed., pp. 1–8. Churchill-Livingstone, Edinburgh and London.

30. Rous, P., and Jones, F. S. (1916). A method for obtaining suspension of living cells from the fixed tissues for the plating out of individual cells. *J. Exp. Med.* **23,** 546.
31. Schleiden, M. J., and Schwann, T. (1910). "The Encyclopedia Britannica," 11th ed., Vol. 7, p. 710. Encyclopedia Britannica, Inc., New York.
32. Stoker, M., and Macpherson, I. (1964). Syrian hamster fibroblast cell line BHK21 and its derivatives. *Nature (London)* **203,** 1355–1357.
33. Waterfield, M. D., Scrace, G. T., Whittle, N., Stroobant, P., Johnsson, A., Wasteson, A., Westmark, B., Heldin, C.-H., Huang, J. S., and Deuel, T. F. (1983). Platelet-derived growth factor is structurally related to the putative transforming protein Ps8 of simian sarcoma virus. *Nature (London)* **304,** 35–39.
34. Waterson, A. P., and Wilkinson, L. (1978). "An Introduction to the History of Virology," pp. 67–77, 210–234. Cambridge Univ. Press, London and New York.
35. Willmer, E. N., ed. (1965). "Cells and Tissues in Culture," Vol. 1, pp. 1–17. Academic Press, London.
36. Witkowski, J. A. (1979). Alexis Carrel and the mysticism of tissue culture. *Med. Hist.* **23** (3), 270–296.
37. Woodruff, A. M., and Goodpasture, E. W. (1931). The susceptibility of the chorioallantoic membrane of chick embryos to infection with the fowl-pox virus. *Am. J. Pathol.* **7,** 209–222.

PART II

BASIC COMPONENTS OF CELL CULTURE SYSTEMS

This part provides the interface between the science of cell biology/tissue culture and biotechnology (i.e. the exploitation of the cell culture for applied purposes). This is a crucial interface because the use of cell cultures in an industrial process means that the process may be controlled by an engineer, or a biochemist or, because of their long history in biotechnology, a microbiologist. Many of the idiosyncrasies of cells in culture are often beyond the comprehension of the most dedicated cell biologist. An awareness that cells do not behave like bacteria or other microorganisms has to be highlighted and the extra conditions this imposes on the production process have to be understood. These include such mundane considerations as the need for high-quality distilled water (it is surprising how many culture problems can be cured by improving the quality of the water), meeting fastidious and complex nutrient requirements and the fact that a slow growth rate makes the cell culture very vulnerable to microbial contamination.

This part on the ingredients of the cell culture system is an attempt to draw attention to the unique requirements of animal cells. Thus, the structure, function and behaviour of cells are described, so that one is aware that the cultured cell is not normally a free-living organism, but is uniquely adapted (selected) as a gregarious component of a multicellular organism (Chapter 2). The part on methods and techniques describes how this transfer to culture conditions is made, maintained and then manipulated so that the required cellular function is expressed (Chapter 3). This needs complex media and physiological fluids to keep the cell alive, to encourage it to increase in number and to carry out cellular functions (Chapter 4), many of which normally occur only when the cell is in an organised *in vivo* environment. To maintain the culture for long periods of time it has to be safeguarded from microbial contamination by organisms which can grow at rates up to 100 times faster than a cell, and can rapidly take over the culture and destroy the cells. The safeguards against this are (1) successful freeze-preservation of a reserve stock and (2) keeping the cell at all times in a micro-biologically-free environment. This is achieved by using special laminar-flow cabinets supplying filtered air to the area where cell handling occurs (Chapter 6). A valuable aid to contamination control is the high-efficiency filter. These are used to sterilise all liquid components that come into contact with the cell at any time, whether it is for growing the cell or for making up the mixture used for freeze-preservation. Filters also allow an interchange between the culture and the outside environment—a requirement for pH and oxygen control. Finally, all equipment used for growing and handling cells or the components of the culture has to be sterile (Chapter 5).

Bacterial and fungal contamination is one of the main problems in the scale-up of cell cultures. Using small cultures, such as a litre flask, meticulous experimental technique minimises the risk of accidental contamination to a very low level. If the culture is lost then, assuming it is not the sole sample of its kind, it is not a great loss in time and effort. However, when the scale is tens or hundreds of litres, not only is the risk of contamination much higher because the culture system is more complex, and there is not so much time for ensuring that a good aseptic technique is maintained, but the consequent loss in resources can be huge. Attention to the experimental details, techniques and equipment described in this section, together with good experimental technique, is essential to obtain reproducible data and production systems from cell cultures.

2

Cell Biology: Basic Concepts

J. B. Griffiths
Vaccine Research and Production Laboratory
Public Health Laboratory Service
Centre for Applied Microbiology and Research
Salisbury, Wiltshire
United Kingdom

P. A. Riley
Department of Biochemical Pathology
University College School of Medicine
London, United Kingdom

1. INTRODUCTION

The basis of animal cell biotechnology is the fact that individual cells, released from complex tissues or organs, can be maintained in artificial

Animal Cell Biotechnology, Vol. 1

conditions and can be treated as discrete organisms *in vitro*. Such cells have been removed from a milieu in which they were specialised (differentiated) and often non-dividing, and under the control of chemical, neural and spatial growth-regulatory mechanisms. Transfer from these controlled conditions in a tissue to those existing in culture has two important consequences. Firstly, the process is highly selective since it is those cells which can adapt and grow quickly that predominate in the culture. Secondly, even if the cells with the desired property are obtained, it is difficult to retain highly specialised properties. Cultured cells have a tendency to undergo changes such as dedifferentiation, degeneration or transformation. To some extent these changes are the consequence of altered architecture. If a cell population can be retained in a histologically organised fashion which simulates proper tissue architecture (e.g. as tissue slices or in artificial capillary or sponge matrices), then it can often retain specialised functions. This is the correct meaning of the term *tissue culture*. In *cell culture*, the tissues are disassociated into discrete cells and often this loss of histological organisation results in the loss of specialised functions. However, cell separation is an essential stimulus to growth.

The purpose of animal cell technology is to use cultivated cells and by manipulating their environment fulfil a particular requirement. The environmental change needed may be simple, such as using particular agents to induce a response, an example being the induction of interferon synthesis by exposure to a virus. On the other hand, the cell may have to be manipulated genetically by hybridisation to carry out the required function. These techniques are discussed in Chapters 3 and 16, Volume 2. The aim of this chapter is to provide fundamental information on cell structure, function and behaviour so that the reader is aware of the restrictions imposed by the cell itself when exploited for biotechnology. To cover such a vast subject in one chapter has meant a highly selective treatment and preference has been given to the factors most relevant to biotechnology. In general, the bibliography consists of reviews rather than citations of original papers. A detailed treatment of this subject may be found in several books (e.g. *10, 23, 31*).

2. CELL STRUCTURE

2.1. Composition of the Animal Cell

The chemical composition of animal cells is shown in Table I. The range in values is given because not only is there variation between different cell types but individual cells vary depending upon their physiological state and

TABLE I Average Values for the Chemical Constituents of a Cell

	pg/cell	Range	Percentage
Wet weight	3500	3000–6000	—
Dry weight	600	300–1200	—
Protein	250	200–300	10–20
Carbohydrate	150	40–200	1–5
Lipid	120	100–200	1–2
DNA	10	8–17	0.3
RNA	25	20–40	0.7
Water	—	—	80–85
Volume	4×10^{-9} cm^3	—	—

their current phase of growth. Functionally, the cell is a collection of organelles delimited by a plasma membrane (Fig. 1) through which all communication with the outside environment is modulated. An understanding of the structure and function of the membrane is, therefore, vital because of its involvement in growth, division, attachment, movement, nutrient uptake,

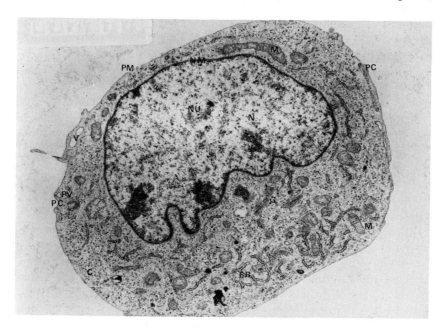

Fig. 1. An electron micrograph ($\times$14,500) of a Chinese hamster ovary (CHO) cell. C, cytoplasm; ER, endoplasmic reticulum; G, Golgi apparatus; M, mitochondrion; N, nucleus; NU, nucleolus; NM, nuclear membrane; PC, pinocytosis; PM, plasma membrane; PV, pinocytotic vesicle; V, vacuole.

antigenicity and viral infection. The following sections describe the functions of the various cell organelles, but they are, of necessity, brief and specialised. More detailed information can be obtained from many text-books (e.g. 37).

2.1.1. Nucleus—Structure and DNA Replication

The nucleus is bound by a double membrane structure (the two membranes being separated by a space) which is continuous around the nuclear pores. Within the nucleus is the double helical DNA packed in proteins (positively charged histones and negatively charged non-histone nucleoproteins). The DNA strands are supercoiled to less than 1/50 their length between nuclear division, but during nuclear division are condensed even further to less than 1/100 their length.

The mammalian genome contains about 3×10^9 DNA nucleotides, but only a small fraction of these are thought to code for essential proteins. The function of the remainder is unknown, but some may be structural elements involved in forming the eukaryotic chromosomes. Chromosomes are probably composed of single long DNA molecules that are organised into a series of domains. Despite the fact that they contain tens of thousands of base pairs, each of the looped domains may contain only one or a few protein coding sequences. In eukaryotic cells the DNA is tightly bound to basic proteins called histones, forming a string of small beadlike DNA–protein particles called nucleosomes. Nucleosome structure appears to be interrupted at intervals by short stretches of free DNA that may be binding sites for regulatory proteins. Nucleosomes are packed together to form condensed structures which exist in a variety of modified forms and are associated with different non-histone proteins. The complex of DNA and its associated proteins is known as chromatin, and different regions of DNA are differently packed in the chromatin. There seems to be some functional significance in this packing pattern and the structural units of chromatin seem to correspond to "banding" in the mitotic chromosomes and also correspond with replication "units".

DNA replication occurs from a series of origins and is achieved by a process in which two forks of DNA replication form and move in opposite directions away from one another. The chromatin structure is re-formed after the replication fork passes and appears to incorporate the old histones which are bound to the nucleosomes of one daughter strand, while new histones are assembled into nucleosomes on the other daughter strand. Replications are activated more or less simultaneously in adjacent loop domains of chromatin during the S phase (DNA synthesis) of the cell cycle. The replication forks move at about 50 nucleotides per second, so that it would take less than an hour to complete DNA synthesis in each cluster of origins.

Throughout a typical 6–8-hr S phase, different replication units are activated in a sequence which is determined by the chromatin structure, the most condensed regions of chromatin being replicated last. The significance of this sequence of replication in relation to the maintenance of the differentiated characteristics of the progenitor cell may be regarded of fundamental importance.

DNA replication takes place on both strands of the unwound helix in a 5′–3′ direction. Since the two strands of the helix are antiparallel, the synthesis can take place continuously on only one of the strands, the "leading" strand. The other "lagging" strand gives rise to short DNA fragments which are made by a "back-stitching" process. Because the DNA polymerase cannot start a new chain, the lagging strand DNA fragments require series of short RNA primer molecules, which are subsequently degraded and replaced with DNA. DNA replication thus requires the cooperation of many proteins. These include: DNA polymerase; RNA polymerase (primase) to catalyse DNA "helicases" and helix de-stabilising proteins to help separate the strands of the helical region to be copied; DNA ligase; an enzyme that degrades RNA primers so that the discontinuously synthesised lagging strand of DNA fragments can be joined together; and DNA topoisomerases, which are thought to reduce the helical winding problems of the newly synthesised double strands. In addition there are initiation proteins, as yet poorly characterised, which are necessary for the generation of replication forks at the replication origins.

2.1.2. Mitochondria

One of the most significant organelles in eukaryotic cells is the mitochondrion. The major function of mitochondria, although not the only one, is the synthesis of ATP from ADP and inorganic phosphate by a process of oxidative phosphorylation. The evolution of mitochondria, which has had the effect of increasing the cell's capacity for ATP synthesis, can be regarded as one of the most crucial factors permitting the evolution of larger cells and multicellular organisms. The number of mitochondria in eukaryotic cells varies but in rat liver cells, which have been used for many of the studies on mitochondria, there are an estimated 1000 per cell. The overall shape of mitochondria is variable (Fig. 1) and, in some cases, they appear to have a regular shape varying from spherical to cylindrical. The diameters of these organelles are in the region of 0.5–5 μm. Cylindrical mitochondria may be more than 10 μm in length. Mitochondria consist of a double membrane structure, the two membranes defining the intermembrane space and the internal matrix. The outer membrane is relatively featureless, although high-resolution electron micrographs imply the presence of pits approximately 30 Å in diameter. The inner membrane, which follows the contour of the outer membrane, in places invaginates to produce structures known as cristae. Negatively stained preparations of mitochondrial inner membrane show that the inward-facing surface contains an array of regularly stacked particles, which are thought to be related to the mechanism for electron transport. Mitochondria contain DNA, but the amount of genetic information available in mitochondrial DNA is less than that needed to code for all the mitochondrial protein and a major contribution to the coding of mitochondrial proteins is made by nuclear DNA. Mitochondria also contain the protein-synthesising machinery. The matrix contains ribosomes which are physically distinguishable from those of the endoplasmic reticulum (the site of synthesis of surface membrane proteins) and cytosol. They seem to be largely bound to the matrix-facing surface of the inner membrane. There are important structural differences between the ribosomes of mitochondria and those of the cytosol, and also differences in sensitivity to a number of inhibitors of prokaryotic protein synthesis such as erythromycin and chloramphenicol.

The mitochondria play a central role in metabolism in that the tricarboxylic acid cycle, which is central to the carbohydrate metabolism in eukaryotic cells, takes place in mitochondria with the production of ATP. In addition, there are the so-called anaplerotic reactions, which take place in mitochondria to replenish any of the tricarboxylic acid cycle intermediates that are deficient. Also, part of the urea cycle occurs in the mitochondria (the conversion of ornithine to citrulline) and beta oxidation of fatty acids is also a mitochondrial process.

2.1.3. Lysosomes

The recognition of enzyme-containing subcellular organelles as lysosomes was made by de Duve et al. (12). Lysosomes are heterogeneous organelles, differing greatly in size, density and structure. The diverse physical characteristics have made it impossible to separate completely the various components of the lysosomal system in a pure state. However, studies suggest the existence of two main subcellular routes of production of the so-called primary lysosomes. The first is from the rough endoplasmic reticulum through the cisternae of the Golgi apparatus. The alternative pathway is by the Golgi-associated endoplasmic reticulum (the so-called GERL), which is a specialised portion of the smooth endoplasmic reticulum. By whichever route these organelles arise, they contain hydrolytic enzymes which have an acid optimum pH. These primary lysosomes fuse with pinocytotic or phagocytic vacuoles to form secondary lysosomes, in which the material ingested by plasma membrane invagination is degraded by the hydrolases. A proportion of lysosomes appear to act as secretory vesicles but this is a much smaller proportion of the total number than would be expected by chance. It has been suggested that primary lysosomes have no mechanism of attachment to the internal surface of the plasma membrane. In addition to their function as digestive vacuoles for material ingested from the exterior of the cell, the lysosomes are also involved in autophagic degradation of cellular constituents. Recently, some distinction has been made between primary lysosomes of different kinds which contain enzymes with different pH optima. Thus, neutral proteases appear to be secreted in separate primary lysosomal bodies. These latter organelles may be regarded in the same category as secretory vacuoles in exocrine glands and are of importance in cellular biotechnology, since secretory products are often of great significance.

2.1.4. The Cytoskeleton

Within the cytoplasm of animal cells are three types of cytoplasmic fibre (Table II). They are: the microtubules, which are 25 nm in diameter and hollow, the intermediate filaments (10 nm in diameter) and the microfilaments (6 nm in diameter). This latter group is closely associated with the plasma membrane and is part of the cell's contractile system. The densely interwoven network of these various cytoplasmic fibres forms the cytoskeleton, or cytoskeletal framework. Functionally, however, they are more than just a mechanical support and are involved in cell division, membrane and cytoplasmic transport and movement. It is only the intermediate filaments that are truly skeletal. The microtubules are associated with cilia and flagellae, the mitotic spindle during mitosis and with maintenance of cell polarity. Of significance is their role in cytoplasmic streaming in endocytosis, exocytosis and in blebbing. These functions are of particular importance as they partially offset the lower diffusion capability of eukaryotic cells, which are bigger and have a much lower surface-area-to-volume ratio than most prokaryotes.

The cytoskeleton is not a fixed structure but is highly dynamic, capable of

TABLE II Elements of the Cytoskeleton

	Microfilaments	Microtubules
Diameter (nm)	7	25
Protein	Actin	Tubulin
Molecular weight	42,000	120,000
Disassociation agents	Cytochalasin B	Colchicine, colcemid, vinblastine, vincristine
Association agents	ATP	GTP
Functions	Cytokinesis, cytoplasmic streaming, pino-, exo- and endocytosis	Mitotic spindle and centriole, flagellae, cilia, exocytosis

dissociation and reassembly (Table II). It is this ability that permits the cytoskeleton to have regulatory properties on cell growth (25), locomotion and mitosis.

2.1.5. The Plasma Membrane

The widely accepted model of membrane structure is the fluid mosaic model proposed by Singer and Nicolson (38). The basis of this model is that the membrane is plastic and that certain membrane components (glycoproteins, glycolipids and lipids) can diffuse laterally in the bilayer of phospholipids (Fig. 2). Some proteins penetrate the bilayer, usually with glycoprotein chains on the outside and with communication to the cytoskeleton on the inside, and this modifies their lateral mobility. Mobility of proteins in the membrane is demonstrated by the fact that after fusion of two different cell types, specific receptors from each cell are intermixed on the surface of the hybrid cell. The importance of this phenomenon is that surface receptor sites can move and form cross-linked complexes and this provides an explanation for certain aspects of the behaviour of cells during adhesion to a substratum, the formation of intercellular junctions, fusion between cells and the uptake of materials from outside the cell (endocytosis and phagocytosis).

The chemical composition of membranes is given in Table III. Membrane proteins include enzymes, receptors and structural proteins. Membrane carbohydrate is either attached to exterior protein (forming glycoproteins) or to the polar ends of phospholipids (forming glycolipids). The oligosaccharide

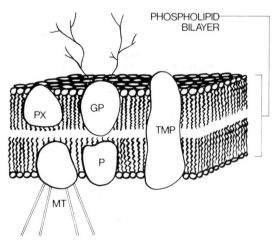

Fig. 2.   Diagrammatic representation of the cell membrane based on the fluid mosaic model (38). GP, glycoprotein complex; MT, microtubules; P, protein; PX, protein complexes; TMP, transmembrane protein.

TABLE III Composition of the Cell Membrane

	Percentage	μmoles $\times$ 10^{-10}
Protein	46–58	
Lipid	42–54	
Carbohydrate	2–10	
Sialic acid	6	8.9
Glucosamine	36	58.3
Fucose	4	6.3
Mannose	33	52.8
Galactose	21	32.8

chains of the membrane are formed from the sugars N-acetyl-D-glucosamine and galactosamine, L-fucose, D-galactose, D-mannose and sialic acid (N-acetylneuraminic acid) and are fundamental to the antigenic properties of the cell.

3. MEMBRANE-ASSOCIATED FUNCTIONS

3.1. Adhesion

Although cells can be treated as individual organisms, most of them are derived from tissues in which there is contact and direct intercellular communication by means of a flow of ions and small molecules. Not all intercellular contacts permit the exchange of metabolites and several types of contact can be distinguished. The initial, or primary, contact made between cells, or between cells and a substrate, is called a site of adhesion and has a morphological pattern known as a tight junction (Fig. 3). These points of contact may develop through intermediate stages to form a physical bridge between cells (gap junctions) (27). Other forms of adhesive specialisation include desmosomes and cross-linking by certain molecules (e.g. antibodies). It is the short-range (50 Å) junctions (gap and tight) which allow equilibrium, co-ordination and metabolic interchange (metabolic co-operation) between cells, and which also play a role in growth regulation.

The basis of many membrane-related cellular functions is the ability of the membrane to form electron-dense caps (plaques) of glycoprotein complexes as a result of the lateral movement of glycoproteins in the membrane (Fig. 4). These plaques form in response to antibodies, agglutinating agents (lectins) or neighbouring cells which cross-link with the membrane receptors. During cell adhesion, the substrate acts as a multivalent antibody, and the plaques formed are known as adhesion patches. These patches, rich in "ad-

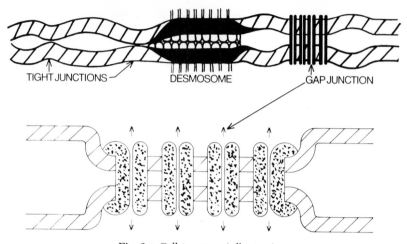

Fig. 3.   Cell junctions (adhesions).

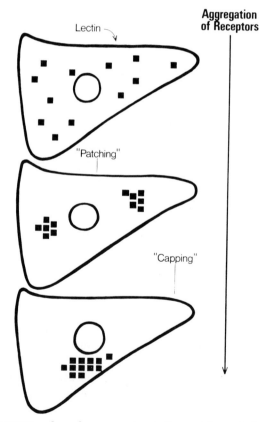

Fig. 4.   Aggregation of membrane receptors to form patches and plaques or caps.

hesion proteins", always radiate cytoskeletal elements which anchor glyco-proteins, and this has the effect of decreasing membrane fluidity and pre-venting the cell from rounding up. The formation of adhesion patches and fibre bundles is the successful outcome of the search by cellular protrusions for a suitable substrate on which to attach, spread and move. Cells are negatively charged and can, therefore, form an adhesion to a positively charged substrate by electrostatic forces (see Chapter 9, this volume). If the substrate is negative, then divalent ions are needed as well as the forces of protein-mediated adhesion. A suitable substrate for many cells, particularly fibroblasts and non-malignant cells, is fibronectin (or cold-insoluble globulin) [see review by Ruoslahti (36)]. Fibronectins are high-molecular-weight (200,000–250,000) adhesive glycoproteins and are found on the surface of fibroblast and glial cells (cellular fibronectin) or in plasma, spinal and am-niotic fluids (plasma fibronectin). The concentration of fibronectin on the external surface of cells is variable, being highest in confluent (non-dividing or serum-starved) cells and lowest in dividing cells, especially during mitosis. The various properties of fibronectin are to promote cell–substrate and cell–cell adhesion and cell spreading, to increase cell motility, and to alter the cell surface morphology and microfilament organisation. Fibronec-tins bind to collagen, fibrin and glycoproteins, including certain surface glycoproteins that specifically recognise and bind to fibronectin, thus form-ing the cross-linking which leads to capping, fixing of the cytoskeletal ele-ments and flattening of the cells. A glycoprotein that has similar cell attach-ment properties, but is more effective for epithelial and malignant cells, is laminin (40). During cell movement across the substrate, adhesion patches are formed in discrete foci (about 1 μm^2) situated mainly near the cell periphery behind areas of active membrane ruffling. Thus, the actual area of contact between cell and substrate is very small. Movement is achieved when the end of the cell contracts and pushes the cytoplasm forward. An adhesion patch remains stationary as the cytoplasm flows past and so, eventually, it has a position at the end of the cell. This is an area of cytoskele-tal depolymerisation and the adhesion patches will become disassembled, allowing the cell to progress forward. To summarise, a moving cell will show cytoskeleton polymerisation, formation of patches and immobilisation of the membrane at the leading end, and the complete reverse at the rear end.

3.2. Contact Inhibition Phenomena

Cells move over a substrate by extending cytoplasmic extrusions (called variously pseudopodia or lamellipodia according to their shape) which form adhesions to the substrate. This process is called membrane ruffling. Contact between lamellipodia results in the inhibition of movement, and the cells then extend lamellipodia in alternative directions. This phenomenon was called contact inhibition by Abercrombie and Heaysman (2). When a culture becomes confluent, lamellipod activity and movement cease. The degree to which contact inhibition is exhibited depends upon the cell type and the environmental conditions. Normal fibroblasts are more sensitive than most other cells and, as a consequence, exhibit a reduced potential for overlapping and multilayering in dense cultures. Contact inhibition of movement is usually accompanied by inhibition of growth, and some confusion in terminology arose when the term "contact inhibition" was used indiscriminately for both functions. To define cessation of growth due to cell crowding, the terms density-dependent inhibition and topoinhibition are commonly used. That density-dependent inhibition is reversible was shown by the classic wounding experiments—i.e. that scraping off a column of cells across a confluent monolayer causes the remaining cells to move into the space, divide and make good the damage (15). All cell lines show a typical saturation density cell count, at which stage multiplication (but not necessarily increase in cell mass) ceases. However, this cell density can be increased, and inhibited cells induced to further replicative cycles, by environmental stimuli, e.g. addition of fresh serum.

3.3. Cell Fusion

Although fusion occurs naturally between certain cells (e.g. macrophages, myoblasts and sarcoblasts) to form syncytia or multi-nucleate cells, it is generally rare. It occurs with very low frequency in mixed populations of cells *in vitro* (somatic hybridisation). The ability of dissimilar cells to fuse and form hybrids is a useful research tool for the study of gene expression, and for transferring certain properties of specialised non-dividing cells to dividing cells. The frequency of fusion can be artificially increased with many RNA viruses (particularly inactivated Sendai virus), with some DNA viruses (herpes group especially) and also with chemicals such as lysolecithin and polyethylene glycol. The massive adsorption of virus onto the plasma membrane of adjacent cells permits the phospholipids to fuse and allows cytoplasmic bridges to be formed between adjacent cells in a micellar structure. As the bridge enlarges, the two cells become integrated. The probability of generating hybrid cells with the required characteristic is low, and some sort of selection procedure has to be used. An example is to utilise cells lacking in hypoxanthine phosphoribosyltransferase (HPRT) as one of the fusion partners. These cells will not grow in medium containing hypoxanthine, aminopterin and thymidine (HAT medium) unless fusion has taken place with the other donor cell which supplies the HPRT enzyme. For further information, the review by Baserga *et al.* (4) is recommended. This technique is of special importance because it forms the basis of hybridoma technology for the production of monoclonal antibodies (see Chapter 3, Volume 2).

The product of fusion is a cell with nuclei from both parent cells (a hetero-karyon). If this multi-nucleate cell proliferates, it forms a hybrid cell line. The usefulness of this technique for studying gene expression is that chromosomes are randomly lost during subsequent divisions. Thus a population of cells with different combinations of chromosomes is derived, and properties of these cells can be compared, and related, to the chromosomes still present. The expression of malignancy has been studied using these hybridisation techniques and has shown that the ability to form tumours can be suppressed by normal cells, except when "malignancy" has been induced by viruses. Fortunately, the myeloma cell line, used for hybridoma production, has virus receptor sites indicating a viral aetiology—otherwise, if its malignant characteristics were suppressed, it could not be induced to secrete monoclonal antibody on fusion with a B lymphocyte.

3.4. Membrane Transport

The plasma membrane forms the interface between the cell and its exterior environment. It is vital that water, ions and nutritional and regulatory metabolites can pass through the membrane. The mechanisms for this are as follows:

1. Passive movement (osmosis for water and diffusion for solutes) occurs as a result of the physical laws of concentration gradients in establishing an equilibrium. Equilibrium imbalances can occur if solutes once inside the cell become insoluble, or are metabolised.

2. Facilitated diffusion is the means by which sugars and amino acids pass through the membrane by forming a solute complex with a membrane carrier. This system is specific, pH-dependent and facilitates transport but cannot work against a concentration gradient.

3. Active transport requires the transported substance to be combined with a specific carrier molecule in the membrane, analogous to an enzyme–substrate complex. Sodium ions are actively pumped out of the cell, which leaves the intracellular level lower than the extracellular level. This causes an inward gradient of Na^+ ions and provides the driving force for the inward transport of K^+, amino acids and glucose through the plasma membrane against a concentration gradient. This process consumes significant quantities of ATP, particularly in epithelial cells, which are very active in absorption and secretion processes. A high intracellular level of K^+ (100–150 mM) is needed to maintain ribosomal protein synthesis, and pyruvate kinase activity for glycolsis. The Na^+ and K^+ gradients across the membrane are primarily responsible for the transmembrane potential difference. This process is summarised in Fig. 5.

4. Bulk transport through the membrane occurs by the invagination of the

(a)

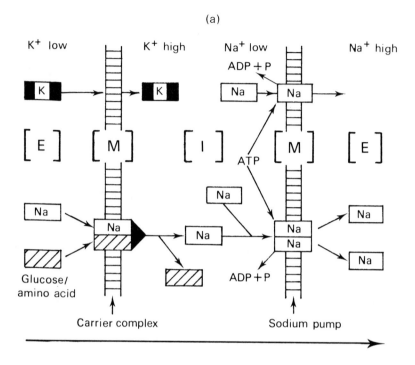

(b)

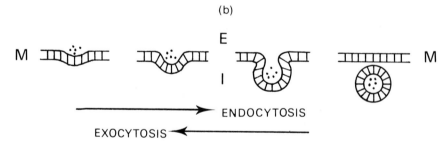

Fig. 5.   Diagrammatic representation of (a) active transport by means of the sodium pump and (b) bulk transport by invagination (endocytosis) and exocytosis. M, membrane; I, intracellular; E, extracellular.

plasma membrane to form a sac. The sac is eventually fully enclosed by the membrane to form an intracellular vesicle (Fig. 5). This process is known as endocytosis, but is referred to as pinocytosis, when the vesicles are small, and phagocytosis, when the vesicles are large and contain particular matter. Often these events are initiated by the stimulation of specific receptor sites on the cell membrane and this is termed receptor-mediated endocytosis. The secretion of material (exocytosis) enclosed within a cell vacuole is effected by fusion of the vacuole and the plasma membrane, followed by membrane breakdown and the release of the vacuole contents to the extracellular medium [see review by Plattner (28)].

4. CELL GROWTH

Growth is an increase in biomass (cell number × mean cell mass) and can be achieved by an increase in mean cell mass (hypertrophy) or in cell number (hyperplasia). Regulatory factors keep the cell mass within strongly defined limits and, therefore, in practical terminology "growth" mainly refers to proliferation, i.e. an increase in numbers. Growth is one of the fundamental and most studied aspects of cell biology. An adult human has more than 10^{14} cells all derived from a single cell (the fertilised ovum) and, although the greatest proportion are non-dividing, there is an enormous replacement rate—in the order of 20 million divisions per second for maintenance purposes. It is clear that sophisticated regulation is needed to control each tissue, to balance growth to loss and to respond to trauma by wound healing. This is achieved by a combination of signals such as hormones, chalones and metabolic products, by neural signals and by proximal contacts. The growth of cells outside these regulatory controls in the formation of tumours is also an important area of study. Much of the data and understanding of growth kinetics, and regulation of growth, has come about by the comparison of normal and abnormal cells in the study of neoplasia (Table IV), and this has been extensively reviewed by Pardee (26).

TABLE IV Altered Properties of Transformed Cells

1. Increased cell population density—multilayering
2. Increased life-span
3. Loss of anchorage dependence—growth in suspension
4. Decreased serum growth factor requirements
5. Increased glucose transport and glycolysis
6. Changes in metabolite concentrations, e.g. cAMP
7. Increased lectin agglutination
8. Increased mobility of surface proteins
9. Absence of 250,000-dalton surface protein
10. Decrease in ganglioside content of lipids
11. Alterations in cytoskeletal elements
12. Changes in surface antigens
    Expression of foetal antigens
    Disappearance of virus-specific
      transplantation antigens

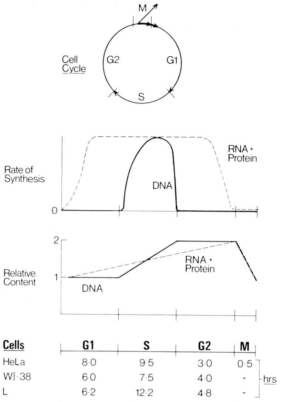

Fig. 6. The phases of the cell cycle correlated with macromolecular synthesis and concentration in the cell.

Cells	G1	S	G2	M
HeLa	8·0	9·5	3·0	0·5
WI-38	6·0	7·5	4·0	-
L	6·2	12·2	4·8	-

hrs

4.1. Cell Cycle

The cell cycle, or events between successive mitotic divisions, is a sequential series of steps and has been divided into four main phases (Fig. 6). Two of these phases are easily recognised: mitosis (M phase) and the period in which DNA synthesis occurs (S phase). The two periods in between were designated as gap 1 (G1) and gap 2 (G2). However, since this simplistic model was proposed (18), much more has become known of the biochemical and genetic sequence of events during the cell cycle (3). Cells traverse the proliferative cycle until a signal is received that conditions are sub-optimal for further growth. This could be due to nutrient deficiency, spatial restriction or drugs, etc. Sub-optimal conditions cause the cells to be arrested in G1 and, for convenience, a resting phase (G0) has been proposed for non-cycling cells which can re-enter the cycle if conditions change. This led to the concept of a switch-point (R) in G1, operated by labile proteins, which allows cells to either leave or enter the cycle (restriction point theory) (26). It has also been proposed that this part of the G1 phase is missing in malignant cells and, with no G0 phase to escape to, they are thus committed to go on dividing, or die.

This model of the cell cycle is not totally accepted and the continuum model has been proposed as an alternative or modification (11). In the continuum model, it is suggested that there are no G1 specific events regulating cell division, but that an initiator of DNA synthesis is continuously synthesised, and that a threshold amount of this initiator is needed for DNA synthesis. If synthesis of the initiator is impaired, then cells become arrested after mitosis.

Biochemical data can be used to support either model, but the continuum model does reconcile eukaryote and prokaryote cells into a common type of cell cycle. The important point is the recognition that a cell between divisions does go through a sequential order of biochemical events, and that its properties and susceptibility to the environment (e.g. drugs, metabolites) will vary. A population of cells is, therefore, never completely homogeneous unless synchronised (Chapter 3), and this must be taken into account in many biotechnological processes, e.g. induction of metabolites, harvesting of products, and in cytotoxicity and other quality control studies.

4.2. Regulation of Growth

Regulation of growth has received much attention because of its relevance to cancer. As a result, a number of differences between cells before and after transformation have been catalogued (Section 6). The dominant theme has been differences in the cell membrane [see review by Baserga (3)] and the effect of external agents (growth factors) on the membrane and on cell growth [see review by Pollack and Hough (29)]. This led to the view that control of proliferation is the consequence of a series of sequential steps, e.g. (1) the effect of extracellular agents on the cell membrane receptors, (2) the cytoplasmic transmission of information from the membrane and (3) the action of the signal on the nucleus to initiate replication. Probably a more balanced view is that regulation of growth is, in fact, an interplay of many factors which have different priorities depending upon the state of the cell and the environment. A descriptive model, which takes into account the modulation between different regulatory systems, is the hierarchical control theory (34), which classifies the regulatory agents as follows:

1. Intrinsic—intracellular regulation to decide whether mitosis will take place (e.g. cell mass and conformation)

2. Intercellular—controls that operate within a population (e.g. space, nutritional factors)

3. Interpopulation—controls that act between populations (e.g. hormones, growth promotors and inhibitors)

The aim of the following discussion is to describe briefly the many facets and hypotheses of growth control and then to summarise and integrate them. A full review of the subject, together with summaries of different theories, and chemical and physical control signals, is given by Leffert and Koch (21).

4.2.1. Growth Control Mechanisms

1. Cell mass. Cells divide because otherwise they would become too big, with too small a surface area for their volume. It has been thought for a long time that cells maintain a constant relationship between the mass/volume of the cytoplasm and the mass of chromatin that they contain [see review by Riley (34)], but the mechanism for implementing this control is unknown.

2. Cell conformation (shape and cell–substrate interaction). The ability of transformed cells to grow rounded up in suspension, and the tendency for normal cells to grow only when attached and flattened out on a substrate, is a very consistent correlation. Wittelsberger and Folkman (42) were able to study cells in different degrees of "flattening" by treating the culture surface with polyHEMA. They concluded that the growth of normal cells was related to their degree of flattening on the substrate. Cells in confluent monolayers are pushed together and have a sub-optimal height-to-mass ratio (34). One consequence of this is the reduced surface area available for nutrient uptake. Transformation appears to result in a loss of sensitivity to conformationally-related growth impedance.

3. Growth factors. The initial observation was that the addition of serum allowed density-inhibited cells to continue to grow and normal cells to multilayer (41). Transformed cells either would grow without serum or were far less serum-dependent. Subsequently, this effect was found to be due to growth factors in serum, particularly insulin and epidermal growth factor (EGF). The effect of serum can be duplicated by the action of proteases (e.g. trypsin at 3 μg/ml), mitogens, raised concentration of nutrients (13), hormone-like polypeptides (e.g. somatedins) and viral transformation.

The reverse effect has been found by the addition of chalones [an intracellular secretion for controlling, by inhibition, the mitotic activity of that cell or tissue (20)]. Early models of growth regulation assumed that cAMP was the modulator because all the above-listed stimulators bring about a fall in cAMP levels, followed by a rise in the concentrations of cGMP and Ca^{2+}

(9). Low cAMP, and high cGMP and Ca^{2+}, levels are features of malignant cells and also of normal cells during mitosis, but not of normal interphase cells. Dibutryl cAMP, which raises the cAMP level within the cell, has the effect of bringing about a morphological change in that transformed cells become more consistent with normal cells in morphology and behaviour. The influence of cAMP in the regulation of mammalian cell division is reviewed by Abell and Monahan (1). As cAMP is so ubiquitous in cell metabolism, it was not believed to be the primary division signal. This role was proposed for Ca^{2+} because it is modulated by cAMP and it plays a vital role in embryogenesis, lymphocyte transformation and neoplasia [see review by Berridge (5)].

In conclusion, a group of hormone-like growth factors [see review by Bradshaw and Rubin (7)] react with specific membrane-bound receptors and trigger a number of different events. These all start with a change in the properties of the membrane and lead to the stimulation of mitosis. This regulatory mechanism can be suppressed, or by-passed, temporarily by environmental (phenotypic) changes, or permanently by viral transformation causing genotypic changes. Recently, the interest in growth factors has been renewed by the realisation that cells which previously had been thought to be incapable of *in vitro* growth could be cultured if the correct factors are present (see Section 8).

4. Membrane in growth regulation. Changes in membrane structure and function have for a long time been correlated with the transformation of normal to transformed cells. Differences in enzymes, glycolipids, glycoproteins, gangliosides, plasminogen activator and fibronectin are reviewed by Baserga (3) and Hakomoris *et al.* (16). The role of surface component proteins in transformation is demonstrated in chick cells infected with the temperature-sensitive mutant of Rous sarcoma virus when they are transferred between 36° and 41°C. At the lower temperature the cells express transformed characteristics, but at 41°C they revert to normal cell behaviour [see review by Pollock and Hough (29)].

Interest in the role of the membrane in transformation was stimulated by the observation that transformed cells were agglutinated by certain lectins (proteins that react specifically with an amino sugar), such as concanavalin A and wheat germ agglutinin, whereas non-transformed cells were not agglutinable, except during mitosis (8). This was not due to receptor sites on the normal cells being covered, or fewer in number, but to the fact that their lateral mobility was inhibited by the cytoskeleton. The movement of membrane components, followed by grouping of receptor sites (into plaques, caps and patches) which then become anchored by the microtubule system, has been described as the functional result of membrane interaction with specific growth factors, neighbouring cells, substrate (fibronectin), lectins, antibodies and viruses. The key is the availability of surface receptor sites to form these complexes, followed by receptor-mediated endocytosis, which enables the cell to sample the environment and somehow to modulate the growth-regulatory system. This may be mediated through the cytoskeleton, as an intact array of cytoplasmic microtubules exerts a restrictive effect on initiation of DNA synthesis. The disruption of microtubules by colchicine both enhances the effect of growth factors and shortens the lag phase of growth [see review by Otto (25)].

4.2.2. *Interrelationship of Cellular Functions with Growth Control*

In describing the factors involved in growth regulation, it is obvious that these are not cellular functions in isolation, but part of the overall control of cell function and behaviour. Many aspects of the metabolic changes brought about by the membrane are well understood, for instance modulation of the levels of cAMP and cGMP, calcium, protein kinase, adenylate cyclase, phosphodiesterase and phosphatase. The actual trigger for proliferation is still unknown, but Pardee (26) suggests that assembly of enzymes and other proteins into a complex on the DNA may be an activating step. The role of F1 histone phosphorylation has also been suggested as a possible control step (6).

In order to indicate that growth regulation and other cellular functions described in this chapter are interdependent, a diagrammatic representation of the interplay between these factors is given (Fig. 7). This is not meant to represent a complete model of cellular behaviour but a summary of the features described, many of which are common to descriptions of growth, movement, adhesion etc. The aim is to show that the cell reacts to certain signals and that, under certain physiological conditions, it is in a cultural sense "dynamic" (i.e. dividing and moving) or, alternatively, "stationary or resting" (which allows adhesions, differentiation etc.). Cells in culture are selected for their growth potential. When growth is restricted they either differentiate, become transformed or become senescent.

5. DIFFERENTIATION

One of the important features of metazoans is the extent of specialisation in the functions of the various cells that comprise them. By contrast with prokaryotic and unicellular organisms, the metazoan cells become specialised as the organism grows. This progressive specialisation, which results in individual tissues with particular functions, is known as differentiation. It is clear that the major distinguishing feature of the cells comprising different

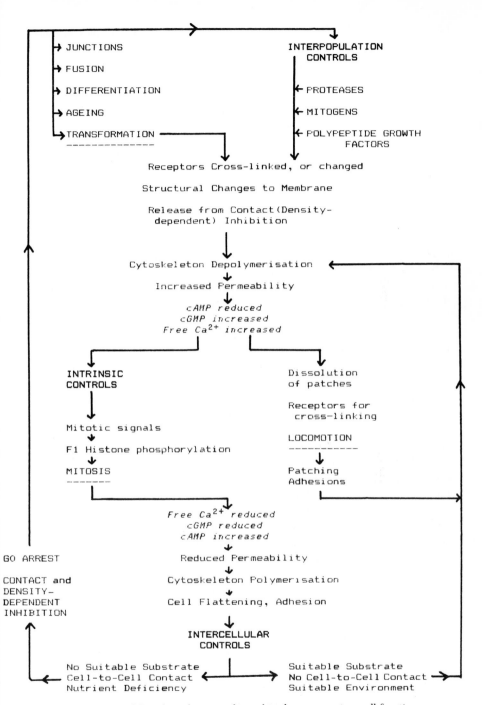

Fig. 7. Model to show the interrelationships between various cell functions.

specialised tissues is the difference in the phenotype; i.e. the sets of genes which are expressed in different specialised (differentiated) cells differ in certain important respects. Thus, while all cells manifest certain fundamental genetically determined traits (i.e. those controlling general functions such as respiration, energy production, motility etc.) certain genes, which control important specialised functions, are expressed differentially in different cell types. The mechanisms by which this is achieved are still far from clear, despite our considerable knowledge of the nature of genetic instructions. It is now thought likely that differentiation involves fairly large-scale changes to the structure of the genome. Nevertheless, there are important experiments [see review by Gurdon (14)] which imply that the process of differentiation does not involve any loss of genetic information, but is merely a differential expression of such information in different cells.

In the case of micro-organisms, powerful selective pressures can be exerted in order to retain in culture those cells that exhibit the property of interest. In the case of specialised eukaryotic cells, in addition to the problems of limited growth, there is the difficulty of applying a selective pressure to the population that will result in the retention of the differentiated property. In general terms, this is because in micro-organisms it is possible to make use of genes which confer resistance to antibiotics and, by linking them with genes which are of interest, exert an antibiotic selection pressure for the property desired. In the case of differentiated cells, no such possibility exists at present. It is, therefore, necessary to rely on the continued expression of the differentiated phenotype in the culture. This seems rarely to be the case, since cells committed to continued proliferation tend to reduce the number of specialised functions which are expressed. This does not necessarily mean that there is any loss of specialisation in a genetic sense (i.e. so-called dedifferentiation), since in many instances it is possible to cause the cells to re-express their differentiated characteristics by slowing or stopping the growth of the culture. Such "stop–go" tactics can sometimes be used for harvesting certain specialised products. On the other hand, some cells tend to express the specialised function better during the growth phase (e.g. in the production of a proteolytic enzyme from keratocytes—see Fig. 8).

Marked genetic heterogeneity appears to be one of the important features of cells that will grow for prolonged periods in culture conditions, and is frequently manifested as extreme chromosomal variability. These changes in genetic arrangement frequently result in alterations in the cells' properties and, in the absence of selective pressures as mentioned above, may result in the loss of the specialised property that it is desired to exploit.

Thus, there are three major problems facing the potential user of animal cells for the production of specialised differentiated products: (1) genetic instability, (2) variable phenotypic expression and (3) often a limited life-span

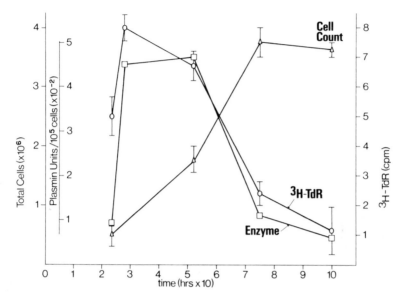

Fig. 8. Production of a fibrinolytic enzyme by keratocyte cells. At each time point the medium was replaced with serum-free medium (plus TdR) for 30 min. This medium was assayed for the enzyme using fibrin plates and a fibrin standard. The data show that the enzyme production is correlated to growing, not stationary, cells.

(see Section 7). To date, the most successful strategy to overcome these limitations has been the use of hybrid cells. Of these the most notable system is the monoclonal antibody system (see Chapter 3, Volume 2). This system is specialised because the cells with the required property (i.e. production of an antibody) can be selected by using the antigen which initially induced this antibody. In other words, the system is somewhat unique in selecting for a complementarity.

The possible use of cell cultures instead of animals for screening procedures, for example in toxicology, has focussed a considerable amount of interest on the possibility of growing cultures of cells capable of certain specific metabolic activities. One such system would be the use of hepatocytes in metabolising certain compounds, either to more, or to less, active products. One of the difficulties of such cultures is that, as described above, the cells tend to lose their differentiated characteristics as they proliferate, so that the manifestation of particular metabolic properties is lost. Thus, for example, the production of albumin, which is a normal function of hepatocytes, is reduced to negligible amounts when the cells have been cultivated for a relatively short time. Rat hepatocytes grown in cultures rapidly lose many specific activities including the cytochrome P-450 reductase–linked detoxification system in the endoplasmic reticulum, and the synthesis of tyrosine hydroxylase. The discovery and use of specific cofactors have improved the maintenance of differentiated properties in many types of cells (see Section 8) but there is no satisfactory general method of ensuring the continued expression of a specialised phenotype. Until such time as adequate techniques have been devised to maintain a selective pressure on cells to continue to express their differentiated characteristics, this must remain a severe limitation on the use of differentiated eukaryotic cells as a reliable source of technologically useful products.

6. TRANSFORMATION

Transformation is a term which is applied to alterations in the growth properties of cultured cells. Transformation is irreversible and appears to involve genetic change in that a specific piece of information that controls the neoplastic phenotype is added to the transformed host genome. In general terms, the alteration in the growth properties is one that enables cells to proliferate under conditions where untransformed cells are disadvantaged. Thus, transformed cells will grow to a higher density in conditions that limit the growth of normal cells, and this seems to be consequent on their reduced requirement for growth factors. There is evidence that the major alteration in transformation is one that affects the transport of nutrients across the cell membrane and this may make the cells less dependent on geometrical factors with regard to growth. Thus, whereas normal cells require to be adherent to a surface ("anchorage" dependence) to enable them to expose a large surface area to the medium in order to achieve the conditions necessary for cell growth and proliferation, transformed cells are able to grow under circumstances where their geometry is less favourable in terms of the surface-area-to-volume ratio. Hence transformed cells will grow in suspended culture, forming spherical clones. Thus, transformed cells can form tumours when injected in relatively small numbers into immunologically tolerant animals. For this reason, transformation is sometimes wrongly equated with malignant change. It must be stressed that the appearance and growth of cancer cells *in vivo* is a different process to *in vitro* transformation. Malignant tumours generally arise after a very prolonged period involving successive mutations that permit a clone to become established which is able to escape the *in vivo* regulatory processes.

The changes in nutrient transport through the transformed cell membrane seem to be accompanied by a number of other changes in membrane properties, notably alterations in lectin binding and other surface phenomena (Table IV). The properties listed in Table IV are often quoted as being those of malignant cells. However, the phenotypic characteristics of tumour cells can only totally be met by cells established from a tumour and not all are necessarily met by virus or other transformed cells. Some of these properties may be of value from the point of view of utilising the cells for biotechnological purposes, but the major advantages of transformed cells are their enhanced growth properties in terms of saturation density, method of growth and longevity of culture life. Unfortunately, as discussed in the previous section, the increased ability to proliferate is frequently accompanied by a reduction in the expression of the differentiated phenotypes. Capitalisation on this characteristic has only been made possible by hybridisation of these cells to non-growing cells with the required property.

6.1. Transforming Viruses and Tumours

Despite the sparsity of proof for the primary involvement of viruses in naturally occurring human cancers, experimental research has demonstrated the mechanisms whereby viruses interacting with normal cells are able to cause neoplastic changes. The tumour-causing (oncogenic) viruses are divided, according to their nucleic acid, into DNA and RNA oncogenic viruses. Among the DNA oncogenic viruses, the best known examples are the papova viruses, such as papilloma, polyoma and SV40; adenoviruses; herpes viruses, including herpes simplex type II; and the pox viruses. The RNA oncogenic viruses (the so-called oncorna viruses), include the sarcoma viruses, the avian, murine and feline leukaemia viruses, and the mammary tumour virus of the Bittner tumour. The molecular biology of these viruses is described in greater detail in Chapter 2 of Volume 2. The essential property of a tumour virus is its ability to cause a genetic alteration in the characteristics of the infected cell. This process is known as viral transformation. In the case of DNA viruses, the viral genome is inserted into the cell's genome and persists only in those cells which are non-permissive for the growth of the virus. In permissive cells, DNA viruses undergo replication and finally cause lysis of the cells. Clearly, the normal replicative life cycle of a DNA virus is incompatible with cell survival, although transformation by defective DNA viruses can occur in otherwise permissive cells. With RNA viruses, in order to incorporate the genetic information into the cellular genome, a DNA copy of the single-stranded viral RNA must first be made (reverse transcription process). This DNA copy of the viral genome then becomes inserted into the host cell genome and is known as a pro-virus. The pro-virus is present in cells transformed by RNA viruses, even in those cases where the virus does not produce an infectious particle. Like DNA viruses, the existence of the pro-virus can be demonstrated by nucleic acid hybridisation. It has also been shown that the DNA of normal cells may contain pro-virus material. This has been interpreted as indicating that normal cells carry the tumour virus genes without having been exposed to viral infection. In certain instances, these pro-virus–carrying normal cells can be induced to transform by treatment with chemical mutagens or ultra-violet light. These transforming segments of genetic material are known as oncogenes. It has been proposed that the process of integration of oncogenes into the host cell, being a form of recombination between the host cell and viral DNA, could be a source of mutations and lead to the transforming event. Viral integration may also lead to certain chromosomal aberrations.

Among the oncorna viruses it is possible to distinguish between those that cause rapid transformation (i.e. within hours or days) and those which are slowly transforming. Among the former are the sarcoma viruses, while the latter include several leukaemia viruses. The rapidly transforming sarcoma viruses possess a gene which seems to be essential for transformation. This is called the *src* (sarcoma) gene and is not present in slowly transforming viruses or leukaemia viruses. The *src* gene can be detected in the genome of both transformed cells and some untransformed cells. *Src* genes of normal cells are assumed to be suppressed for most of the life of the cell, but it may be that transformation by a "spontaneous" mechanism causes the expression of the gene product. The probable mechanism of action is that the *src* gene codes for a protein which transforms the cell. In some instances a phosphokinase has been identified as one of the major transformation-sensitive expressions of phenotype. At present it is not clear whether this is a protein not normally synthesised, or whether it is a normal protein which is present to a greater extent. One of the major actions of the phosphokinase is tyrosine phosphorylation of surface proteins, which has been shown to alter the characteristics of cells, particularly in relation to the uptake of certain nutrient materials from the exterior. Phosphorylation of surface proteins may alter the plasma membrane reactivity to lectin agglutination and other criteria that are used for transformation (Table IV). Other oncogenes besides *src* have been recognised and one of them, the *sis* gene, has been found to code for platelet-derived growth factor (PDGF). Thus some viruses carry genes that code for normal cell products. These oncogenes are also switched on by chemical mutagens, thus providing a link between viral and chemical-derived transformation; i.e. the oncogene may be the common cause or mechanism by which a cancer cell arises.

Transformation may be either viral, as described above, or "spontaneous". Most studies have been made on virally transformed cells and the properties of these cells, when transformed with well-known transforming viruses such as SV40 and polyoma, have been well described. Similar changes after spontaneous transformation have led to the proposal that spontaneous transformation is the result of activation of gene sequences (oncogenes), already present in the genome of the pre-transformed cell, similar to those existing in the viral genome. Many studies have indeed shown that spontaneously transformed cells express sequences similar to those present in transforming viruses, but recent work has indicated that similar (perhaps indistinguishable) changes are produced by point mutations in normal genes (32).

Since the properties of transformation that are described are those which are expressed by cells *in vitro*, which for reasons given elsewhere (see page 41) may not be equivalent to the spontaneous changes which give rise to cancer, it may be that in the future virally transformed lines will be regarded as acceptable material for the production of biotechnologically significant products for human use. However, at present, because of the possibility that *in vitro* transformation has connotations with regard to human cancer, material derived from such cultures would be regarded with suspicion.

7. AGEING

Ageing is a characteristic of cells which have a limited proliferative potential, i.e. have low saturation densities in ideal culture conditions, e.g. human diploid cell lines. Thus the increased growth ability of transformed cells, discussed in the previous section, means that transformation overcomes the ageing process. Theories of cellular ageing fall into two broad categories. One is the stochastic (clonal selection) theory, which postulates that a build-up of translational errors, or harmful effects, caused by mutation, radiation, infection etc. eventually causes death. The term "ageing", in this instance, is used as a description of a fundamental entropic time-related deterioration. The other theory is that death is genetically programmed and that once a cell no longer divides, but becomes specialised, it is dying; i.e. death is an extension of useful differentiation. Ageing is certainly genetically influenced as each species has a characteristic life-span, but variability within populations also suggests a phenotypic influence.

When the human diploid cell line WI-38 was established it was shown, for the first time, that cultured cells could exhibit ageing phenomena and a finite life-span (of 50 ± 10 population doublings for WI-38 cells) (17). Age-dependent changes that have been observed include lengthening of inter-mitotic intervals (from $19 \pm 25\%$ to $31 \pm 41\%$ hr) and changes in metabolism, enzyme levels and product expression. This work has led to extensive studies of ageing at the cellular level—a science now known as "cytogerontology". An initial interpretation of cellular ageing was that the growth rate of WI-38 cells was insufficient to offset the accumulating deleterious changes to the cells. That any population is inevitably destined to die out poses an interesting paradox since, in its most general form, this conclusion cannot be true. One possible solution to the paradox is to postulate a progenitor cell type that is immune to time-dependent deleterious events, the so-called germ cells, for example. A class of diploid cells that seem to have escaped ageing in culture is the haemopoietic lines derived from lymphocytes. An alternative approach (33) is to consider the effects of natural selection acting on individuals in a population that acquire transmissible defects. This approach is founded on the assumption that the proliferative efficiency of cells is reduced in proportion to their increased susceptibility to lethal events. A general model incorporating the significant features of a proliferating cell population has been discussed elsewhere (35). One of the important conclusions that can be reached is that continuation of the population is dependent upon the extent of the relative growth rates between individual cells within the population, and the saturation density. The higher the saturation density, then the more chance there is for the rapidly growing cells to increasingly outnumber the more slowly growing cells which are the result of deleterious genetic changes.

The conclusion that senescence in a cell population can be slowed by increasing the proliferation rate also applies to all theories of ageing based on the accumulation of errors, either by somatic mutation or protein variation [e.g., Orgel (24)] or a combination of these. It would seem to exclude, however, those theories of senescence which depend on programmed cell death, such as the theory advanced by Kirkwood and Holliday (19). Theories of programmed cell senescence are based on the notion of a cellular "clock", which regulates the actions of certain genes which are essential for cell reproduction and survival. The hybridisation data of Smith (39), in which somatic hybrids of cells possessing limited and unlimited life-spans gave rise to mortal (limited life-span) cell lines, indicate that if there are genes controlling such a mechanism of cellular senescence, they are dominant. These hybridisation studies showed that somatic hybrids of SV40-transformed cells mainly resulted in immortal lines, but hybrids of other immortal cells, including SV40-transformed cells as one fusion partner, resulted in mainly mortal cell lines. This suggests that at least two genes are involved in the control of the ageing mechanism. It is possible to interpret these complementation data in terms of a simplified model which employs two genes, a programming gene and a switching gene (see Fig. 9), each of which is inactivated either by SV40 or by spontaneous transformation. In this model, the action of the product of the switching gene is to prevent the expression of another gene which is responsible for producing an essential factor necessary for cell survival or fertility. Thus, in the absence of an intact programme and switching cascade, the life-extending gene continues to act and the cells are rendered immortal. By contrast, the mechanism envisaged for the stochastic senescence model is one of inactivation of the life-supporting gene. As the

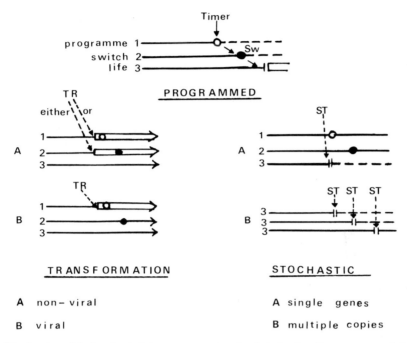

Fig. 9. A model of ageing to interpret programmed and stochastic cell senescence and the hybridisation data of Smith (39). It is suggested that SV40 transformation always inactivates the same gene (either 1 or 2) and other transformations either gene 1 or gene 2 randomly. This allows expression of gene 3 to continue rather than be switched off, or deleted by random events. ST, stochastic deletion; TR, transformation. (See text for full explanation.)

model shows, these two approaches are not necessarily mutually exclusive. However, the association between the life-span of the species and their genetic redundancy (i.e. the number of copies of genes in their nuclei) (22) may favour the stochastic model of ageing, since it would extend the period necessary for deleterious changes to affect all the copies and therefore extend the life-span. Multiple copies of the programming or switching genes would not alter the life expectancy of cells containing them. Nevertheless, because the life-span data apply to organismal ageing, this may not be an argument directly relevant to the senescence of cells. For example, there does not seem to be a clear relationship between the life-span of the species of origin (e.g. mouse or man) and the doubling potential of their fibroblasts in culture.

Limited-life-span cell lines, such as diploid fibroblast cells, are not ideal for production purposes, especially as they have to be used before any gross senescent changes take place. This, in practical terms, means the useful life of such cells is confined to between passages 12 (to allow for seed stock production) and 30. However, the ability to demonstrate that a cell has a limited lifetime does provide a quality control parameter for non-transformation, or non-malignancy. Transformed cells do not have this limited life-span and this confers another advantage to their use as a substrate for product generation, in addition to higher saturation densities, faster growth rates and ability to grow in suspension.

8. CONCLUSION

In this chapter, the composition and organisation of the animal cell have been described in relation to various aspects of cultural behaviour, such as motility, growth, adhesion, etc. The interrelationships of these functions have been discussed in order to emphasise the fact that any physiological change to one component has effects on all the others. This is especially noticeable when a cell is removed from its natural environment and transferred to the artificial conditions of a culture. The section on cell differentiation was included to explain the fact that cultured cells tend to become ubiquitous and capable of doing very little else but grow. This, until fairly recently, has been a limiting factor in the biotechnological exploitation of cells. That this problem has largely been overcome is reflected by the fact that cells are now an important and expanding area of biotechnology. One method used to achieve this is genetic recombination, either by chromosome exchange (somatic cell hybridisation, e.g. hybridomas) or by gene insertion. Another method of culturing cells, with expression of the desired characteristics, is to provide them with the requisite growth factors. For many years, cells have been grown in a medium with 30–50 defined ingredients plus calf serum, but the expectation that human cells would grow as well in calf, or horse, serum as in human serum, which is far more likely to provide the right factors, has been very naïve. In fact serum from one species is often inhibitory to cells of another species. The very detailed analyses of growth factors that are being carried out in many laboratories mean that many differentiated cells, which hitherto were considered impossible to cultivate, can now be used in culture. The list of factors is long (see Chapter 4 for examples) and each normal cell type probably has a different set of medium requirements. By including the correct ones, it is now possible to grow various types of epithelial cells, chrondrocytes, keratocytes, hepatocytes, mammary, bone and even some nerve cells, and also cells from many malignancies—not just an ubiquitous fibroblast-like or epithelial-like cell. Although public attention is focussed on the advances made by genetic engineering, the seemingly more mundane nutritional and physiological studies are equally important if animal cells are to be used to their full potential.

REFERENCES

1. Abell, C. W., and Monahan, T. M. (1973). The role of adenosine 3',5'-cyclic monophosphate in the regulation of mammalian cell division. *J. Cell Biol.* **59,** 549–558.
2. Abercrombie, M., and Heaysman, J. E. M. (1954). Observations on the social behaviour of cells in tissue culture. II. "Monolayering" of fibroblasts. *Exp. Cell Res.* **6,** 293–306.
3. Baserga, R. (1976). "Multiplication and Division in Mammalian Cells." Dekker, New York.
4. Baserga, R., Potten, C., and Ming, P. M. L. (1981). Cell fusion and the introduction of new information into temperature-sensitive mutants of mammalian cells. *In* "Cell Growth" (C. Nicolini, ed.), pp. 69–79. Plenum, New York.
5. Berridge, M. J. (1975). Control of cell division: a unifying hypothesis. *J. Cyclic Nucleotide Res.* **1,** 305–320.
6. Bradbury, E. M., Inglis, R. J., and Matthews, H. R. (1974). Control of cell division by very lysine rich histone (F1) phosphorylation. *Nature (London)* **247,** 257–261.
7. Bradshaw, R. A., and Rubin, J. S. (1980). Polypeptide growth factors: Some structural and mechanistic considerations. *J. Supramol. Struct.* **14,** 183–199.
8. Burger, M. M., and Noonan, K. D. (1970). Restoration of normal growth by covering of agglutination sites on tumor cell surfaces. *Nature (London)* **228,** 512–516.
9. Burger, M. M., Bombik, B. M., Breckenbridge, B. M., and Sheppard, J. R. (1972). Growth control and cyclic alterations of cyclic AMP in the cell cycle. *Nature (London)* **239,** 161–168.
10. Clarkson, B., and Baserga, R., eds. (1974). "Control of Proliferation in Animal Cells." Cold Spring Harbor Lab., Cold Spring Harbor, New York.
11. Cooper, S. (1981). Mini-review: The central dogma of cell biology. *Cell Biol. Int. Rep.* **5,** 539–549.
12. de Duve, C., Pressman, B. C., Gianetto, R., Wattiaux, R., and Applemans, F. (1955). Tissue fractionation studies. 6. Intracellular distribution patterns.
13. Griffiths, J. B. (1972). Role of serum, insulin and amino acid concentration in contact inhibition of growth of human cells in culture. *Exp. Cell Res.* **75,** 47–56.
14. Gurdon, J. B. (1963). Nuclear transplantation in Amphibia and the importance of stable nuclear changes in promoting cellular differentiation. *Q. Rev. Biol.* **438,** 54–78.
15. Gurney, T. (1969). Local stimulation of growth in primary cultures of chick embryo fibroblasts. *Proc. Natl. Acad. Sci. U.S.A.* **62,** 906–911.
16. Hakomori, S., Kijimoto, S., and Siddiqui, B. (1972). Glycolipids of normal and transformed cells. *In* "Membrane Research" (C. F. Fox, ed.), pp. 253–277. Academic Press, New York.
17. Hayflick, L. (1965). The limited in vitro lifetime of human diploid cell strains. *Exp. Cell Res.* **37,** 614–636.
18. Howard, A., and Pelc, S. R. (1953). Synthesis of DNA in normal and irradiated cells and its relation to chromosome breakage. *Heredity, Suppl.* **6,** 261–273.
19. Kirkwood, T. B. L., and Holliday, R. (1975). Commitment to senescence: A model for the finite and infinite growth of diploid and transformed human fibroblasts in culture. *J. Theor. Biol.* **53,** 481–496.
20. Laurence, E. B. (1981). The significance of chalones in epidermal growth. *In* "The Skin of Vertebrates" (R. I. C. Spearman and P. A. Riley, eds.), pp. 139–149. Academic Press, New York.

21. Leffert, H. L., and Koch, K. S. (1977). Control of animal cell proliferation. *In* "Growth, Nutrition and Metabolism of Cells in Culture" (G. H. Rothblat and V. J. Cristofalo, eds.), Vol. 3, pp. 225–294. Academic Press, New York.
22. Medvedev, Z. A. (1972). Possible role of repeated nucleotide sequences in DNA in the evolution of life spans of differentiated cells. *Nature (London)* **237**, 453–454.
23. Nicolini, C., ed. (1981). "Cell Growth." Plenum, New York.
24. Orgel, L. E. (1973). Ageing of clones of mammalian cells. *Nature (London)* **43**, 441–445.
25. Otto, A. M. (1982). Mini-review: Microtubules and the regulation of DNA synthesis in fibroblastic cells. *Cell Biol. Int. Rep.* **6**, 1–18.
26. Pardee, A. B. (1981). Molecular mechanisms of the control of cell growth in cancer. *In* "Cell Growth" (C. Nicolini, ed.), pp. 673–714. Plenum, New York.
27. Pinto da Silva, P., and Gilula, N. B. (1972). Gap junctions in normal and transformed fibroblasts in culture. *Exp. Cell Res.* **71**, 393–401.
28. Plattner, H. (1981). Mini-review: Membrane behaviour during exocytosis. *Cell Biol. Int. Rep.* **5**, 435–459.
29. Pollack, R. E., and Hough, P. V. C. (1974). The cell surface and malignant transformation. *Annu. Rev. Med.* **25**, 431–446.
30. Pollack, R. E., ed. (1981). "Readings in Mammalian Cell Culture." Cold Spring Harbor Lab., Cold Spring Harbor, New York.
31. Prescott, D. M. (1976). "Reproduction of Eukaryotic Cells." Academic Press, New York.
32. Reddy, E. P., Reynolds, R. K., Santos, E., and Barbacid, M. (1982). A point mutation is responsible for the acquisition of transforming properties by the T24 human bladder carcinoma oncogene. *Nature (London)* **300**, 149–152.
33. Riley, P. A. (1973). The principle of sequential dependence in cellular differentiation. *Differentiation* **1**, 183–189.
34. Riley, P. A. (1981). Control of proliferation of normal and neoplastic cells in culture. *In* "Regulation of Growth in Neoplasia" (G. V. Sherbet, ed.), pp. 131–198. Karger, Basel.
35. Riley, P. A. (1981). A theory of cellular senescence based on Darwinian principles in the light of Linnaeus. *In* "The Skin of Vertebrates" (R. I. C. Spearman and P. A. Riley, eds.), pp. 111–125. Academic Press, New York.
36. Ruoslahti, E. (1981). Fibronectin. *J. Oral Pathol.* **10**, 3–13.
37. Sheeler, P., and Bianchi, D. E. (1980). "Cell Biology." Wiley, New York.
38. Singer, S. J., and Nicolson, G. L. (1972). The fluid mosaic model of the structure of cell membranes. *Science* **175**, 720.
39. Smith, J. R., and Pereira-Smith, O. (1983). Dominance of finite vs. infinite in vitro lifespan in somatic cell hybrids. *In Vitro* **19**, 281–282.
40. Timpl, R., Engel, J., and Martin, G. R. (1983). Laminin—a multifunctional protein of basement membranes. *T.I.B.S.* **8**, 207–209.
41. Todaro, G. J., Lazar, G. K., and Green, H. (1965). The initiation of cell division in a contact-inhibited mammalian cell line. *J. Cell. Comp. Physiol.* **66**, 325–334.
42. Wittelsberger, S., and Folkman, J. (1981). Cell conformation and growth control. *In* "Cell Growth" (C. Nicolini, ed.), pp. 575–584. Plenum, New York.

3

Cell Biology: Experimental Aspects

J. B. GRIFFITHS

Vaccine Research and Production Laboratory
Public Health Laboratory Service
Centre for Applied Microbiology and Research
Salisbury, Wiltshire, United Kingdom

Animal Cell Biotechnology, Vol. 1

1. INTRODUCTION

This chapter is not intended to be a detailed description of the methods used in tissue culture. Instead, the aim is to discuss techniques in order to give the non-specialist an idea of what is involved. By giving a broad overview, it is hoped that it will stimulate the adaptation of old processes to new requirements and new technology to improve old methods. More detailed descriptions of methodology are referenced and there are many excellent books on the methods and applications of tissue culture (23, 41). This chapter is similar to the previous one in that the topics considered to be the most relevant to a basic understanding of animal cell biotechnology are reviewed. Topics described include how to initiate a cell culture *in vitro;* how to maintain it in culture; how to preserve it for long-term storage; and, finally, how to characterise a cell—in this section quality control factors are discussed with special reference to cell culture collections. In addition, the topics of measuring cell viability, quantifying cell growth, cloning and synchronisation are described as these are vital adjuncts to process initiation and control.

2. ESTABLISHMENT OF CELLS IN CULTURE

The first step is to dissociate organised tissues into single cell suspensions as efficiently and as gently as possible. This single cell suspension constitutes a primary culture which may be used as such, for instance in the batch production of viruses, or may be used as a starter culture to initiate a cell line, i.e. a culture of cells that increase in number and can be harvested to initiate new cultures. There are various categories of cell line depending upon their origin and subsequent culture, karyotype, longevity and culture characteristics (Fig. 1).

Eventual success is very dependent upon the initial dissociation of the starter tissue. There are many enzymic, chemical and mechanical treatments that can be used in addition to the classical explant outgrowth methods. The choice of method is largely influenced by the type of tissue and the extracellular matrix of the cells, as well as the logistical factors concerning quality and quantity of cells required. The most common sources of primary cells are kidneys from monkeys (250–750 million), dog (500 million) and rabbit (40 million), 14-day-old chick embryos (150 million) and 12- to 16-week-old human foetal lung (50 million) and kidney (15 million). Also to be taken into account are the availability of donor tissue, hence the reason for so many human cell lines derived from tonsils, the amnion and foreskin; the type of cell (differences between epithelial and fibroblast cells can be an important

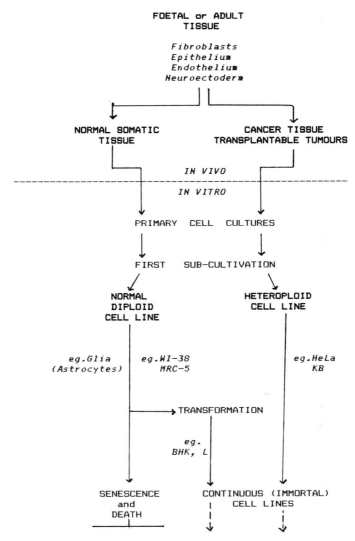

Fig. 1.   Derivation of cultured cells from somatic tissue.

factor in virus studies); and the age and size of the organ. Normally, embryonic cells are preferred since cell yield and quality decline with tissues from
older donors, probably due to increasing amounts of connective tissue present in ageing tissue. Also, the culture life of untransformed and non-malignant cells is approximately inversely proportional to the age of the donor
tissue.

2.1. Explant Techniques

Small fragments of tissue (1–3 mm^3) are placed on the surface of a culture vessel and sufficient growth medium is added so that the tissue is covered but will not float. In time cells will migrate out and form a monolayer, whilst the original tissue becomes necrotic. The cell monolayer can then be harvested with traditional techniques (see Section 4). A refinement is to embed the tissue fragments in a plasma clot on the substrate surface. This provides support, anchorage and a more suitable environment. Although these techniques allow only small numbers of cells to be produced, they are useful for initiating cultures of specialised fibroblast cells, e.g. genetic mutant lines from skin biopsies.

2.2. Dissociation of Tissues

The most commonly used agents for the dissociation of tissues are hydrolytic enzymes that break down the extracellular matrix. Although this method is efficient, it can damage the cell membrane by removing membrane components, thus causing cell lysis. Exposure time to these enzymes must be minimised and, because permeation through the tissue/organ is slow, the tissue must be minced or thinly sliced before the enzyme is added. This mechanical disruption causes not only cell loss, but also the problem that the contents of broken-up cells often clump the intact cells in gelatinous aggregates; these then may have to be treated with DNAase (35). However, there are specialised perfusion methods which allow these problems to be avoided or, alternatively, to avoid cell trauma altogether, an explant technique can be used. Finally, it should also be emphasised that freshly isolated cells are liable to carry a low level of microbial contamination, either from the donor animal or introduced during the involved handling procedures. All solutions used in the process should contain broad-spectrum antibiotics, such as gentamycin or neomycin, as well as the more specific ones (e.g. fungizone).

2.2.1. Enzymes Used for Dissociation

1. Trypsin (0.1–0.25%). A pancreatic enzyme which hydrolyses peptide bonds between carboxyl and amino groups of the basic amino acids. It is often used in association with chelating agents (e.g. EDTA) or other enzymes (especially collagenase). Trypsin supplied for cell culture purposes is, in fact, a mixture of proteases, nucleases, lipases and polysaccharidases, as purified trypsin is relatively ineffective against intercellular proteins. Optimum conditions for trypsin activity are the presence of Ca^{2+} and Mg^{2+} ions and a pH above 8, but for trypsinisation of cells, the pH should be in the range 7.2–7.6. Trypsin is rapidly neutralised by serum, thus allowing good control over exposure time.

2. Pronase (0.05%). A group of enzymes from *Streptomyces griseus* which act faster and give better separation into discrete cells than trypsin, which tends to leave cells in small clumps. For this reason, pronase is especially recommended for fibroblasts. Unlike trypsin, it cannot be inhibited and must either be washed out, using centrifugation, or diluted over 100-fold in growth medium.

3. Collagenase (0.01–0.15%). Hydrolyses the bonds between proline and other amino acids and works mainly on the extracellular matrix rather than the cell membrane. It is thus far less damaging than other enzymes and, also, it can be used in the presence of serum. This means that long exposure to the enzyme can be used in order to completely dissociate a tissue and allow epithelial cells to be collected after all the fibroblasts have been released. As Ca^{2+} ions are required, it should not be used with chelating agents.

4. Elastase (0.05%). A pancreatic enzyme that uniquely digests elastin, a constituent of connective tissue. This enzyme is also neutralised by serum.

5. Dispase (0.6–2.4 units/ml). A neutral protease from *Bacillus polymixa*, which is used in association with either EDTA or collagenase.

6. Hyaluronidase (0.1%). A polysaccharidase that specifically hydrolyses bonds common in hyaluronic acid and chondroitin sulphate A and C. These compounds are found in high concentration in all connective tissues.

7. Enzyme mixtures. A mixture is more efficient for many tissues, since one enzyme can weaken the stroma and allow the other to rapidly dissociate the cells. Examples are trypsin and collagenase (6) for liver and cartilage, or collagenase and hyaluronidase for epithelial cells.

2.2.2. Chemical Agents

The divalent cations of magnesium and calcium play a role in cell-to-cell adhesions. Chelating agents such as EDTA (ethylenediamine tetraacetic acid/versene), which chelates Ca^{2+} and Mg^{2+}, or EGTA (diaminoethoxyethane tetraacetic acid), which chelates Ca^{2+}, disrupt the cell-to-cell bonding. They can be used on their own (at 0.01–0.1 M) but, normally, are used in combination with trypsin at the reduced level of 0.005 M. The potassium complexing agent tetraphenylboron (5 mM) has also been used (43).

2.2.3. Mechanical Disruption

Mincing tissue with sharp instruments (scalpels, scissors etc.) is usually a prerequisite of chemical and enzyme treatments. Tissue can also be dissociated using a homogeniser (e.g. for chick embryos) or by squeezing the tissue through a nylon or stainless-steel mesh. This latter method is practical for embryonic tissues only, as older tissues have stronger intercellular bonds. It is also possible to get sufficient cells by perfusion of some organs (e.g. hepatocytes and pancreatic islet cells).

2.2.4. Dissociation Techniques

The crudest method is to chop the tissue into small cubes (2–3 mm), wash out the red blood cells with saline, add trypsin and stir in a baffled Ehrlenmeyer flask. Every 15–20 min the fluid is decanted and fresh trypsin added. The first collection, which usually contains a high proportion of damaged and red blood cells, should be discarded. This is a slow and laborious procedure which can be improved by adding trypsin slowly and continuously, and collecting the overflow into a storage receptacle at 4°C. An additional refinement is to circulate the cell/trypsin through a perfusion pump as this adds mechanical disruption to the procedure.

To improve both the yield and the quality of the cells, one of the following methods should be employed:

1. Cold trypsin diffusion. Trypsin is inactive at 4°C, but can diffuse through tissues at this temperature. Thus, if a kidney, for example, is immersed for 18–24 hr in trypsin at 4°C, it will diffuse throughout the organ. If it is then rapidly warmed to 37°C, trypsin will cause an even hydrolysis throughout the tissue, resulting in a very much reduced exposure time.

2. *In vivo* perfusion (22). The kidney of an anaesthetised animal can be perfused by means of a catheter in the aorta and another in the vena cava (Fig. 2). Blood cells can be washed out first and then warm trypsin circulated throughout the kidney. The kidney, when removed and decapsulated, falls apart with very little homogenisation and the yield, in the author's laboratory, is three to four times higher (in the order of 10/10 cells from a 2-year-old baboon kidney) than can be obtained with other methods (62).

2.3. Establishment of Haemopoietic Cells

Cells circulating in the blood and lymphatic system are very diverse, but are derived from a common precursor in the bone marrow—a stem cell (Fig. 3). The various haemopoietic cells vary in their growth potential. Leucocytes and erythrocytes have no potential for growth, lymphocytes need to be stimulated and macrophages show a limited degree of division and growth *in vitro*. The most common haemopoietic cell lines are derived from lymphocytes. *In vivo* lymphocytes do not divide unless stimulated by exposure to an antigen. The stimulated lymphocytes, known as lymphoblasts, then form a clone, the component cells of which are identical in that they secrete identical antibody. In culture, lymphocytes may be stimulated with mitogenic agents such as phytohaemagglutinin (PHA) or transformed with Epstein–Barr virus.

In order to establish a particular type of haemopoietic cell in culture, a selection technique is required. This can be by differential attachment to

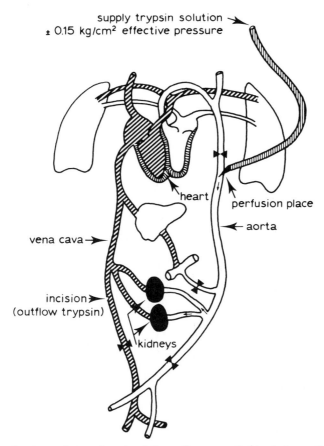

Fig. 2. *In vivo* perfusion of monkey kidneys (figure supplied by A. L. van Wezel).

glass bead columns, by density gradient centrifugation or both. When blood cells are passed through a glass bead column, the leucocytes, followed by the red blood cells and, finally, the platelets, are washed through. Monocytes (and granulocytes) will remain attached to the glass and can be removed by trypsin. Granulocytes cannot divide, and will die out, leaving the macrophages in culture. Red blood cells can be removed by haemolysis (e.g. with ammonium chloride) and lymphocytes will only divide and grow if stimulated with a mitogen (e.g. PHA). These selection techniques, therefore, will allow most cell types to be separated. Alternatively, an explant of bone marrow, brain or retina will yield a culture of monocytes/phagocytes. However, the most widely used method is density gradient centrifugation using non-ionic media for lymphocytes which aggregate leucocytes and

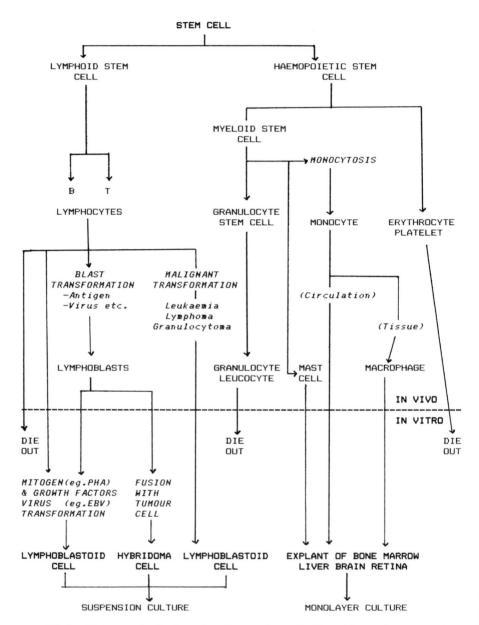

Fig. 3. Derivation of cell cultures from haemopoietic cells. The incidence of monocytes can be increased by inducing monocytosis (e.g. by injecting inactivated microorganisms into the donor animal). Macrophage production is controlled by lymphokines secreted by T lymphocytes. EBV, Epstein–Barr virus.

TABLE I Some Growth Factors Necessary for the Cultivation of Some Normal and Tumour Cells[a]

Insulin	Dexamethasone
Transferrin	Prostaglandin
Ethanolamine	Phosphoethanolamine
Hydrocortisone	Prolactin
Trace metals	Pituitary extract
Epidermal, fibroblast, nerve growth factors	

[a] These factors are available in MCDB series of media formulated by Ham (e.g. 56).

erythrocytes, thus allowing pure preparations of lymphocytes to be isolated (e.g. Lympho-prep and Ficoll-paque) (12).

2.4. Culture of Recently Established Cells

The fact that for many years all cells were expected to grow in a standard medium with a particular serum, usually calf, was discussed in the previous chapter (Chapter 2, Section 8). Many cells will, in fact, survive this treatment. However, the failure of establishing cells of particular types, or from particular organs or tumours, is now known not always to be an intrinsic factor of that cell but due to inadequate growth media. Many research groups have made painstaking studies of the factors and nutrients required for the successful cultivation of these cell types. Complicated media have evolved which are used for detoxification of the cells' environment and to supply the right type and concentrations of factors. This is a very complex subject and the reader is referred to the work of Ham and his colleagues (e.g. 56) and to a book entitled "Growth of Cells in Hormonally Defined Media" (45). Examples of some of the factors used are given in Table I.

3. CELL SEPARATION TECHNIQUES

Most of the dissociation techniques that have been described are either non-specific and give a mixed population of endothelial, fibroblast and epithelial cells, or allow fibroblasts to dominate and take over the culture. If cells other than fibroblasts are required, or a particular type of blood cell is wanted, then one of the following separation or selection methods has to be used:

1. Selective enzyme treatment. The collagenase digestion method described earlier increases the proportion of epithelial cells present. If brief

trypsinisation is then used to selectively remove fibroblasts (these detach more quickly than compact colonies of epithelial cells, which have stronger cellular bonds), eventually a pure epithelial culture will be obtained (36).

2. Specific inhibitors, e.g. sodium ethylmercurithiosalicylate (4) and iodoacetic acid (3 μg/ml), are far more highly toxic to fibroblast than to epithelial cells. Also, replacing L-valine with D-valine eliminates fibroblasts (15).

3. Transplantation. Alternate passage of cells *in vivo* and *in vitro* selects for tumour and ascites-type cells.

4. Cloning. The dilution of a cell suspension so that one cell grows to form a discrete colony of daughter cells (a clone) which can be recognised and then cultured in isolation from all other cells. Details of this technique are described in Section 6 of this chapter.

5. Centrifugation. Cells can be separated according to their density with isopycnic (density gradient) centrifugation, or to their size (velocity sedimentation) by rate-zonal and differential centrifugation. This latter method is usually preferred because the size range of cells (6–50 μm diameter) is greater than the density range (1.06–1.15) (49). Details of density gradient centrifugation, using commercially available support media that are non-ionic, are available from the manufacturing companies (12). These techniques enable the separation not only of cells of different types but also of sub-populations of cells at different stages of the cell's life-cycle.

6. Affinity chromatography. Cells that have differing surface properties can be separated using affinity chromatography with various immunoadsorbents. This technique is particularly useful for separating the various sub-populations of white blood cells.

7. Physical methods. These include electrophoresis and the widely used method of differential adherence to glass (e.g. endothelial cells attach more rapidly than fibroblasts, which attach before epithelial cells). Another method, which depends upon differences in membrane surface properties, particularly surface charge, is partition of cells in two-polymer aqueous phases (59). An example is dextran–polyethylene glycol, which gives polyethylene glycol in the upper and dextran in the lower phase.

8. Cell sorter. Cell sorters and flow cytometers (27) work by introducing suspension cells into a stream of liquid which passes through a focused beam of light, usually produced by a laser. As each cell passes through the light beam, signals are generated due to the scattering of light or the emission of fluorescence. These signals are received by photodetectors and amplified to generate the data required. Cell sorters work by imposing a surface charge on the droplet containing the cell, using a variety of selection criteria, e.g. cell size, fluorescence label. The signal generated by the cell is compared to criteria set by the operator and, if the criteria are met, a charge is applied to

the droplet so that it can be deflected, by charged plates, and collected. Fluorescent probes can be used to measure RNA and DNA content and the presence of particular cell surface antigens. The method has the advantage that in many cases the cells retain a high viability and can be subsequently cultured. Only if the fluorescent dye needs permeabilisation of the cells is viability affected. The resolution of cell sorters is such that cell organelles, for example metaphase chromosomes, can be separated.

4. SUBCULTURING AND HARVESTING

This technique is fundamental to cell biology—if a population of primary cells can be subcultured, then by definition it becomes a cell line. This constitutes a large source of cells capable of repeated build-up to seed large production equipment (Fig. 4). Even the human diploid cell lines, WI-38 and MRC-5, which have a limited life-span of approximately 50 doublings, give rise to enormous quantities of cells before dying out (for instance, one primary cell could give rise to 675 kg of cells in its lifetime).

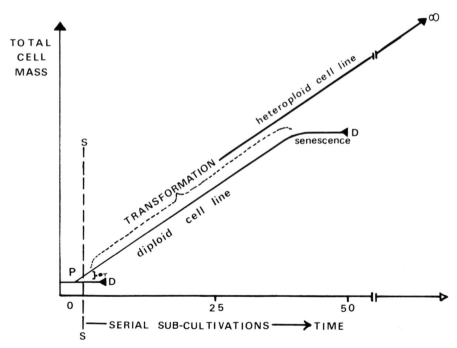

Fig. 4. Events influencing the culture lifetime of cells (based on human diploid cells). P, primary culture; D, death of whole cell preparation; S, first subcultivation.

The first time a cell population is subcultured it changes status from a primary culture to a cell line. Primary cells are often preferred to cell lines for diagnostic virology and vaccine production because after serial cultivation they tend to lose their susceptibility to viruses. However, for reasons of animal conservation, cost and to allow quality control to be carried out, cell lines should be used in preference to primary cells whenever possible. This is especially the case with monkey cells due to world shortages, import controls, high costs and potential microbiological hazards (e.g. herpes B virus). A compromise is to passage the cells two or three times, which increases the cell stock four- to eight-fold but enables them to be used before their virus susceptibility is too significantly reduced (57). A further advantage is that the cells are in a better physiological state and any excess can be freeze-preserved with a far higher recovery potential than is possible with primary cells.

4.1. Subculture Methods

Basically the same enzymes that are used for tissue dissociation are used for harvesting a monolayer of cells from the substrate. Trypsin is the most commonly used agent and the important factor is to keep the exposure time as short as possible. Good technique calls for:

1. Washing the cell sheet with phosphate-buffered saline (PBS) to remove all traces of serum before trypsin is added and
2. Keeping the cells and reagents at 37°C.

A useful refinement is to give the cell sheet a wash in trypsin (1 ml/20 cm^2) for 1 min and then to pour 90% of it away. After 10–15 min incubation at 37°C, the cells will be detached and the volume of trypsin small enough to allow immediate dilution into fresh culture medium. This avoids the stress caused by centrifugation.

With trypsin, fibroblast cells detach very irregularly and, usually, in clumps; so pronase is strongly recommended as a better option. Pronase gives a better single cell suspension and superior control over production procedures because the exposure time and cell quality are more uniform than with trypsin and EDTA preparations.

Many of the more specialised cell types die out after a passage or two in culture. One of the reasons for this is the inability to remove cells in a viable state from a monolayer. A non-destructive disaggregation method developed in Leibovitz's laboratory is to treat such monolayers with a hypo-osmolar medium [60% calcium- and magnesium-free minimum essential medium (MEM), plus 8% bovine serum albumin (BSA), 5% 360,000 MW polyvinylpyrrolidone (PVP) and 2% Methocel] and shake for 30 min at 100 rpm on an orbital shaker. This technique has been especially useful in the early passage subculturing of continuous cell lines from human tumours.

For certain work an enzyme cannot be used (e.g. harvesting cells for antigens) and, in such cases where viable cells are not required, cells can be harvested by:

1. Shaking with glass beads
2. Scraping off with a rubber-coated rod/spatula
3. Ultrasound (8)
4. Detergents, such as Nonidet
5. Increasing osmolality to 0.84 mM NaCl
6. Freezing and thawing

5. CELL QUANTIFICATION

A measure of total cell numbers is easily accomplished either by using haemocytometer counts of whole cells (stained and unstained) or stained nuclei (44), or with an electronic method. Likewise, determination of cell mass by measuring proteins, protein nitrogen or dry weight is relatively straightforward, provided that the cells have been washed free of serum and other medium components. It is far more difficult to get a reliable measure of cell viability because the methods employed either stress the cells, use a specific parameter of cell physiology or are carried out in artificial conditions. Cell growth may also be quantified indirectly by measuring a metabolic process. These methods are of particular application in culture systems where cell sampling, or visual observation, is impossible, e.g. immobilised cultures of glass beads or fibres. Finally, various radiochemical methods may be used.

5.1. Measurement of Cell Viability

The dye exclusion test is based on the premise that viable cells will prevent the uptake of certain dyes, whereas dead cells will be permeable to these dyes. Trypan blue (0.4%) is the most commonly used dye but has the disadvantage of staining protein. In the presence of serum, therefore, erythrosine B (0.4%) is to be preferred. Cells are enumerated in the standard manner using a haemocytometer, such as the Fuchs–Rosenthal, and both viable and stained cells can be counted to give a percentage viability.

Staining techniques interpret the ability of cells to resist dye uptake; i.e. they are a measure of cell permeability. Although there is a high degree of correlation between dye exclusion and ability to respire and divide, there are situations in which misleading results can be obtained. Two relevant examples are membrane leakiness caused by recent trypsinization, and freezing and thawing, especially in the presence of dimethyl sulphoxide. The uptake of dye is also pH- and concentration-dependent (39), and results can be influenced by serum concentration and the exposure time before counting. A disciplined and standardised procedure should be used.

The classic assay for viable cells is plating efficiency, which relies on the ability of a cell to divide and, in time, to form a colony. This technique requires a cell suspension to be diluted to a point where individual cells are well separated (50–200 cells/ml). Then, when these are added to a culture vessel, they will form discrete colonies and after 7–14 days can be stained and counted. The plating efficiency is the proportion of colonies to original cells. This method is probably the ultimate in viability assay in that it measures the cells capable of growth and division and, furthermore, requires only a very small quantity of cells. However, these low cell densities are extremely artificial and impose a great stress on the cell. In fact many cells, e.g. human diploid, will not tolerate dilutions below 10^3 cells/ml unless highly modified media are used. To combat this disadvantage, either very enriched media or conditioned media (i.e. media that have supported a limited amount of cell growth and are therefore rich in intermediate metabolites) should be used. Alternatively, the cells should be plated onto a feeder layer (a monolayer of cells which have been irradiated in order to destroy their mitotic capability, but which are still viable and metabolically active). Another disadvantage of the plating efficiency technique is the delay between setting up and obtaining a result. This can be reduced to a few days by using microscopical techniques to scan for, and detect, very small colonies (e.g. four- to eight-cell size). Cloning methods, it should be noted, select for cells best adapted for substrate attachment.

By no means such a good method as plating efficiency, but one which does not subject the cell to artificially low concentrations, is the recovery index method (19). Cells are plated out at normal densities (e.g. 10^4/ml), harvested and counted after a time interval that will permit one but no more division cycles. The recovery (viability) index is the increase in numbers as a proportion of the total number plated out.

Indirect measurements of viability are based on metabolic activity. The most commonly used parameter is glucose utilisation, but oxygen utilisation, lactic or pyruvic acid production, ATP concentration and carbon dioxide production can also be used, as can the expression of a required product (e.g. enzyme), or the rate of synthesis of proteins and nucleic acids. When cells are growing logarithmically, there is a very close correlation between nutrient utilisation and cell numbers. However, during other growth phases high utilisation rates, due to maintenance rather than growth, can give misleading results. The measurements obtained can be expressed as a growth yield (Y) or specific utilisation/respiration rate (QA). Knowledge of these values for a given cell line and culture conditions enables cultures to be set up which will not become nutrient-limited or, alternatively, which will produce a predicted cell number, e.g. in continuous-flow culture.

$$\text{Growth yield } (Y) = \frac{\text{biomass concentration (DX)}}{\text{substrate concentration (DS)}}$$

$$\text{Specific utilisation/respiration rate (QA)} = \frac{\text{substrate concentration (DS)}}{\text{time (DT)} \times \text{cell mass/numbers (X)}}$$

Typical values of growth yields for glucose (10^6 cells/g) are 385 (MRC-5), 620 (Vero) and 500 (BHK).

Continuous measurement of the redox potential with a special electrode can be used to detect changes in the growth rate of cells [see review (17)]. Redox is a reflection of the charge of the medium and is a net result of pH, dissolved oxygen and the balance of oxidative and reducing chemicals in the medium. A value of +75 mV has been found to be near the optimum for many cell lines and, if this value is maintained throughout the culture, enhanced cell yields are obtained. However, if redox is not controlled, it can be used to indicate the phase of growth that the cell population is in (e.g. logarithmic, stationary etc.). As the redox value falls to a minimum value about 24 hr before the end of logarithmic growth and then begins to increase, it can be used to predict the end of cell growth and the beginning of the stationary phase. This has particular value in making a decision when to infect or induce cells, or when to change media (17).

5.2. Quantitative Measurements

There are a whole range of chemical and physical measurements that can be used to quantify cell mass (Table II), but the most commonly used method is the measurement of total cell protein by the modified Lowry technique (37). The disadvantage of the chemical methods is that large samples are required to get accurate and meaningful measurements. The advantages are that they are very simple techniques, some are suitable for automation and others are capable of being adapted to micro-techniques (e.g. micro-Kjeldahl).

The physical methods are more sensitive. Radioactive chromium (51) exhibits a differential binding pattern depending upon whether chromium is reduced (chromic), in which case it binds to non-viable protein, or oxidised (chromate), in which case it binds to viable protein (19, 46). Chromium binding can thus be used as a quantitative measure of cell protein, either of

TABLE II Chemical and Physical Measurements for Determining Cell Mass

Methods	Comments	References
Packed cell volume	Haematocrit centrifugation	60
Wet, dry weight	Insensitive and tedious	38
DNA	Indole or diphenylamine	38
	Microfluorometric	55
Protein	Modified Lowry technique	37
Protein N	Nessler's reagent	38
	Micro-Kjeldahl	
Microabsorptiometry		2
Microinterferometry		50
Chromium-51		46
Electronic counters	e.g. Coulter counter	
Intracellular volume	Radiochemical method	58
Microcalorimetry		3
(bioactivity monitor)		
Turbidity	Photometry at 590–650 nm	42

living cells or, in the presence of ascorbic acid, of dead cells. This procedure has to be carried out under controlled conditions as reduced chromium can be measured only in serum-free media and, also, there is a continuous turnover of the chromium label from the cells.

Electronic cell counters give rapid and accurate results of cell numbers and are especially useful when there are many samples to be counted. However, they do not distinguish between individual and small clumps of cells, or between viable and dead cells. Cell suspensions, therefore, should be examined for the incidence of clumping to see if they are suitable for this procedure.

Many of the techniques listed in Table II are included for specialist requirements, rather than for routine measurement.

5.3. Summary of Cell Quantification

There are many methods of measuring cell growth, each with its own limitations. Ideally, two methods should be employed to get a balanced picture of cell growth. The methods selected will depend upon the criterion of cell physiology important to that particular study, i.e. whether it is total cell mass, in which case numbers would be misleading due to variation in cell size, or cells capable of dividing or cells able to function biochemically. Other factors which dictate the method are whether cells are in suspension, and thus easily sampled, or in monolayer, in which case the whole cell sheet will have to be sacrificed unless an indirect parameter is chosen; whether cell sampling is possible at all; whether the measurement is of an increase or decrease in cell numbers (e.g. in toxicity studies); and the time delay which can be allowed before the result is required.

6. CLONING

Cloning is the technique of establishing a population of cells that are all derived from one original cell. This subject is also described in Chapter 3 of Volume 2 with reference to establishing clones of cells that all secrete the identical antibody.

The purpose of cloning is to derive a homogeneous population of cells. This allows selection of the most efficient subpopulation, and stabilisation of the cell population, with regard to the required cellular function. A good example of this is the 10-fold range in productivity for foot-and-mouth disease virus (FMDV) found within the same population of BHK cells (51). This heterogeneity is very typical of heteroploid cells. Diploid cells are far more homogeneous, which is fortunate because it is unrealistic to clone them in view of their limited life-span.

To set up clones from a cell population, the individual cells have to be separated so that the colonies which they form are discrete and can be physically isolated, or removed, from all the others. The simplest way of achieving this is by serial dilution until, theoretically, there is one cell per unit volume of medium. This unit volume is then added to a micro-titre plate or a similar vessel. The wells in this plate are observed microscopically and those in which there is only one cell are marked, so that if a colony develops it will be known to be derived from a single cell. As previously discussed with regard to cell viability (Section 5.1), many cells will not survive when over-diluted and many modifications have been introduced to overcome this problem, as well as the use of conditioned media and feeder layers. These include the use of glass capillary tubes (to reduce the volume of medium), the use of small carriers which can be individually removed from a culture containing large numbers of them (e.g. small fragments of glass coverslips, glass beads or microcarrier beads) and isolating cells on semi-solid agar (0.4%) and, when sufficient growth has occurred, removing the individual colonies.

Fibroblast cells migrate across a culture surface and do not produce discrete colonies like epithelial cells. Therefore serial dilution methods which do not isolate single cells are unsuitable for fibroblasts.

7. SYNCHRONISATION

Normally, within a population, cells are distributed randomly between the various cell cycle phases (mitosis, G1, S, G2). For many studies it is beneficial to have all the cells, or at least a high proportion, in the same phase, e.g. when carrying out investigations into cell growth and division and the molecular biology of cell regulation. The value of synchronised populations in biotechnology is both for understanding and defining cellular mechanisms involved in product generation and also, possibly, for increasing productivity.

The two basic methods for obtaining synchronised cells are blocking (induction) and selection (separation).

1. Induction involves blocking the cells at a fixed point in the life-cycle and allowing the cells to accumulate at this point. The cells are then released from this block (usually by washing) so that they can proceed in synchrony. The compounds most commonly used to block cells are colcemid or vinblastine (inhibitors of mitosis), or thymidine, amethopterine, 5-aminouracil or hydroxyurea (inhibitors of DNA synthesis). However, best results are achieved by arresting cells either at an interface between life-cycle phases or in the shortest phase. The S phase can be from 1 to 8 hr in duration; thus, cells after treatment could still be this degree of time apart. Techniques such as cold shock, starvation (general or specific with a single nutrient such as leucine) only partially synchronise cells (parasynchronisation). Synchronisation is never complete and usually refers to 20–50% of the cells. This incidence can be increased by repeated blocks.

Induction methods affect cell metabolism and can result in unbalanced growth and even chromosome aberrations, especially as many of the agents used are cytotoxic drugs.

2. Selection involves physically removing part of the cell population. This can be done by mitotic shake-off or by velocity centrifugation. The basis for mitotic shake-off is that monolayer cells round up and become loosely attached to the substrate, and each other, during mitosis (54). The degree of attachment varies between cells, so the amount of force required to detach these cells has to be found empirically. Detachment can be facilitated by using Ca^{2+}- and Mg^{2+}-free medium, and a hypotonic medium (61). A high degree of synchrony is obtained (80–90%), especially if the force used for detachment is minimal so that only metaphase cells are released. However, the yield is always low as, theoretically, only 1–4% of the cells are in mitosis at any given time. Repeated collections from the same population can be carried out and the cells stored at 4°C, or even at −196°C. Alternatively, a colcemid block can be used to increase the proportion of mitotic cells.

Mitotic shake-off is, of course, only suitable for monolayer cells. Suspension cells can be selected by velocity sedimentation on the basis that cell volume increases twofold as the cell progresses from mitosis to G-2 (48). Centrifugation in BSA, sucrose, Ficoll or Nycodenz gradients (12) will select a band of cells of the same volume, and a high proportion will be of similar life-cycle age. However, there are inherent size distributions within any population so overlap will occur but, even so, a 60–75% synchronisation can be expected.

Cells also have an inherent variation in the time taken to traverse the cell cycle and, therefore, even the most highly synchronised population of cells will become asynchronous after only a generation or two.

In order to test for the degree of synchronisation within a population, a mitotic index measurement is made. This involves fixing the cells, smearing and drying on a coverslip, staining (e.g. acetoorcein), washing and then examining them microscopically for those in mitosis (showing condensed chromatin).

8. IMMOBILISATION, ENTRAPMENT OR MICROENCAPSULATION OF CELLS

The entrapment of cells in semi-solid matrices, or spheres, has many applications but the basic function is to stabilise the cell and thus protect it from sub-optimal conditions. Cells can be immobilised by adsorption, covalent bonding, cross-linking or entrapment in a polymeric matrix. Materials that can be used are gelatin, polylysine, alginate and agarose; the choice largely depends upon the problem being addressed. These techniques have been used to protect cells that are being transported or posted between laboratories (63), to store cells at 4°C for an extended period of time (e.g. 5 weeks) (40), to avoid immune rejection of transplanted cells (25) and to protect fragile cells (e.g. hybridomas) from mechanical stress in large-scale culture equipment (34, 47). The latter application not only makes it possible to use such equipment but also allows production of hormones, antibodies, immunochemicals and enzymes over much longer periods than are possible in homogeneous suspension culture. The matrix allows free diffusion, between the enclosed microenvironment and the medium, of nutrients and generated product. To protect cells from trauma and variable temperature encountered in postal transportation, and to minimise risks of leakage, it is sufficient to add them to a nutrient gelatin mixture (1.5% gelatin in complete growth medium) at 37°C at a density of 2×10^6/ml. Rapid solidification is achieved by transferring the mixture to 4°C. Cells can be released by warming to 37°C and diluting in fresh medium. This technique allows human diploid and primary monkey cells to be distributed worldwide from the author's laboratory (63).

Alginate is a polysaccharide and is cross-linked with Ca^{2+} ions. The rate of cross-linking is dependent on the concentration of Ca^{2+} (e.g. about 30 min with 10 mM $CaCl_2$). A recommended technique (40) is to suspend the cells in isotonic NaCl buffered with Tris (1 mM) and 4% Na-alginate, and to add this mixture dropwise into a stirred solution of isotonic NaCl, 1 mM Tris, 10 mM $CaCl_2$ at pH 7.4. The resulting spheres are 2–3 mm in diameter. The entrapped cells can be harvested by dissolving the polymer in 0.1 M EDTA or 35 mM sodium citrate. Disadvantages of alginate are that calcium must be present and phosphate absent and that large molecules, such as monoclonal antibodies, cannot diffuse out. For these reasons, agarose in a suspension of paraffin oil provides a more suitable alternative. Five percent agarose in Ca^{2+}- and Mg^{2+}-free PBS is melted at 70°C, cooled to 40°C, and mixed with cells suspended in their normal growth medium. This mixture is added to an equal volume of paraffin oil and emulsified with a vibromixer. The emulsion is cooled in an ice-bath, growth medium added and, after centrifugation, the oil is removed. The spheres (80–200 μm) are washed in medium, centrifuged and, after removing the remaining oil, transferred to the culture vessel (34, 47).

The use of entrapped cells in fermenter culture is an attractive technique for product generation. It has many of the advantages of microcarriers (spheres with cell growth on the outer surface) in that medium changes, and alterations in the cell-to-medium volume ratio, can be performed easily. In addition, entrapment can be used to facilitate perfusion, or medium changes, and products can be harvested cell-free.

9. PRESERVATION OF ANIMAL CELLS

Cells do not survive freeze-drying, nor do they withstand storage at temperatures above −25°C (see Chapter 11, Volume 2). Some survival may occur during short-term storage at −140°C or even at −70°C. However, as physico-chemical changes do not become negligible until the temperature falls below −140°C, the most convenient storage system for cells is in liquid nitrogen at −196°C. Cells are completely stable at this temperature and, therefore, any loss of viability during preservation must occur during the freezing and thawing process.

9.1. The Need for Cell Preservation

Cells have a tendency to lose, or change, their properties over a period of time in culture and differences occur between cells from the same source handled in different laboratories. The preservation of reference stocks of cells is vital for research programmes where data collected over a period of time need to be consistent, for genome conservation in biotechnology, for keeping seed stock which has been quality-controlled for product manufacture, for prolonging the availability of low-passage stock of finite-life cells (e.g. human diploid lines), for genetic studies to preserve parental stock and as a reserve against loss by contamination. In addition, there is the advantage that cells need not be subcultivated when not immediately required, thus saving labour, media costs and risk of contamination, and allowing the clones of fusion experiments to be investigated at leisure. If cells are to be preserved, it is important to do this as optimally as possible because a low recovery means that selection has taken place. Also, with finite-life cells frozen at passage 8, a 10% recovery would mean the loss of 3 to 4 passage equivalents—or in terms of a production run at passage 30, the difference of between 10^{12} and 10^{11} cells.

9.2. Physico-chemical Considerations

During the freezing process there are three events that can occur, either sequentially or together, depending upon the cooling rate. These are the formation of ice crystals, the removal of water and the increase in solute concentration. The two factors which cause cell damage and loss of viability are intracellular ice formation (which occurs at fast cooling rates, e.g. $>100°C/min$); and high salt concentration (which occurs at slower cooling rates, e.g. $<1°C/min$). The optimum cooling rate is thus a compromise between salt damage on the one side and intracellular ice damage on the other (26).

When a cell suspension is cooled, the temperature falls to the freezing point of water but, because of the salts present, the freezing point is depressed and the system becomes super-cooled (see Chapter 11, Volume 2). For ice crystallisation to occur a nucleus is needed, and the higher the temperature, the larger the nucleus that is required. In practice this occurs between $-3°$ and $-11°C$. When ice crystals form, latent heat of fusion is released due to the reduction in the energy of water molecules when their movement is inhibited. This brings about a rise in temperature to above the melting point (i.e. ice crystallisation temperature). This means that no more nuclei are available for initiating crystallisation and those that have already formed will continue to grow until all the free water is frozen. Only then will the temperature continue to fall. This phenomenon gives rise to a long phase-transition period during which time the cell is subjected to fluctuations of temperature above and below the eutectic point (see Fig. 5). Whilst this is occurring, the salt concentration is continually increasing in the remaining water, thus allowing osmotically induced loss of water from cells and shrinkage (dehydration) of cells (Fig. 6). This process is dependent upon the

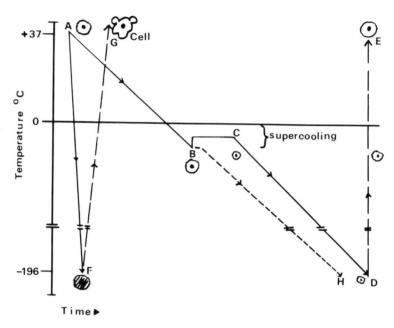

Fig. 5. Cooling and warming profiles of cells. (A–D) Slow cooling rate. Ice crystallisation begins at B; water is lost from the cells, which consequently shrink. After phase transition (B–C) the cell is dehydrated and reaches −196°C with no intracellular ice. On warming (D–E) the cell resumes its original size. (A–F) Fast cooling rate. There is insufficient time for dehydration to occur; thus intracellular ice is formed. On warming (F–G) the cell is damaged, showing blebbing, lysis etc. (A–H) Cooling profile produced when a programmable biological freezer is used.

cell membrane remaining intact and acting as a barrier to ice crystals. If this process proceeds at the correct rate, the cell is dehydrated with minimum damage and a high recovery will result. If the cooling rate is too slow, then excessive exposure to these hypertonic conditions will cause damage. If the cooling rate is too fast, then the latent heat of fusion is absorbed and the temperature will continue falling, thus exposing many more nuclei for crystallisation. A large number of ice crystals will rapidly form and convert all the free water, and there will not be enough time for controlled dehydration of the cell to occur. At the critical temperature, the water still within the cell will freeze and intracellular ice crystals will form (Fig. 7). As intracellular ice is far more damaging than solute concentration, it is desirable to achieve the fastest cooling rate that avoids any risk of this happening. This rate varies between cell types but, as a generalisation, the following are recommended when cryoprotectants are present: 0.5°–2°C/min for lymphocytic cells, 1°–3°C/min for fibroblastic cells and 2°–10°C/min for epithelial cells.

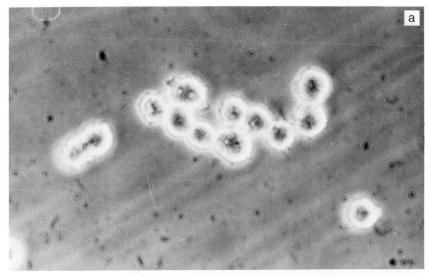

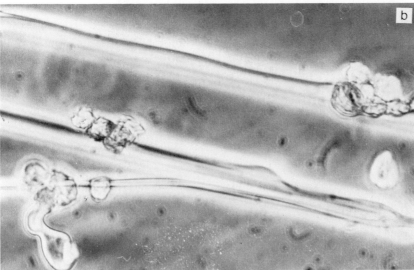

Fig. 6. CHO cells cooled at 2°C/min (*18*). (a) Cells at 20°C before cooling; (b) cells at −15°C; (c) cells at 1°C after thawing. Reproduced with permission of Academic Press Inc. (*continued*).

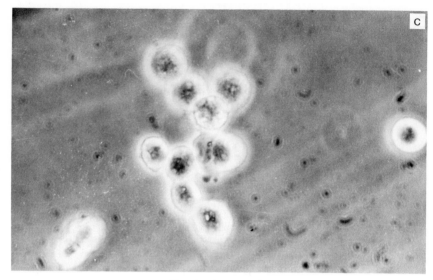

Fig. 6.   (*Continued.*)

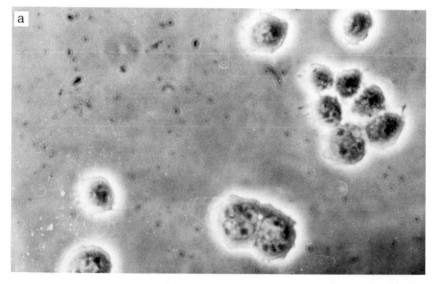

Fig. 7.   CHO cells cooled at 100°C/min (*18*). (a) Cells at 20°C before cooling; (b) cells at −95°C; (c) cells at 25°C after thawing. Reproduced with permission of Academic Press Inc.

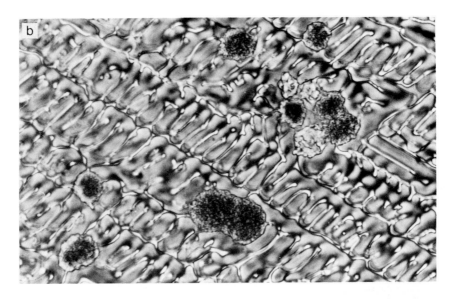

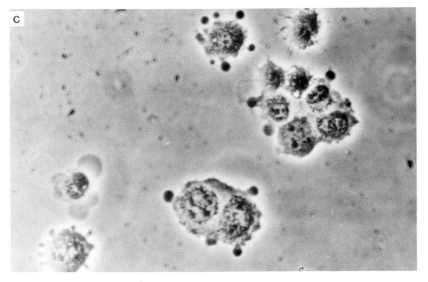

Fig. 7. *(Continued.)*

Whatever damage is inflicted upon the cell during cooling, it cannot be manifested until the cell is thawed. During rewarming the cells are subjected to hypertonic stress, followed by rehydration as the cells take in water to return to their normal size. The amount of damage sustained during cooling, added to that occurring during rewarming, determines whether a cell will survive or whether the injury is sufficient to lead to dilution shock damage (11) caused by excessive entry of water. These effects can be mini-mised by using as fast a warming rate as possible in order to minimise the time spent in hypertonic conditions and to prevent growth of small ice crystals into larger ones.

9.3. Practical Aspects of Freeze Preservation

9.3.1. Container and Volume

To avoid a significant temperature gradient through the freezing mixture, which would cause variable freezing and thawing rates, as small a volume as possible should be used. Heat transfer is more efficient through glass than plastic and, for practical purposes, 1-ml glass ampoules are suitable. This volume is not so large that the cooling/thawing rate will be significantly slower in the centre compared to the perimeter. Screw-cap plastic vials are not recommended because leakage can occur, which can lead to a loss of material, a risk of contamination and could be a potential biological safety hazard. If the volume to be frozen is too large to allow 1-ml ampoules to be conveniently used, then the plastic bags used for storage of blood are conve-nient as the contents will be spread thinly with a large surface area for efficient heat exchange.

9.3.2. Formulation

Generally cells are suspended in a mixture of growth medium and cryo-protectant at a concentration between 3×10^6 and 1×10^7/ml. Cryoprotec-tive agents, of which there are many, all have a high solubility in water, which means that they remain in solution at temperatures well below the freezing point of water. There are two types of cryoprotectant: penetrating [e.g. dimethyl sulphoxide (DMSO) and glycerol] and non-penetrating (e.g. PVP and hydroxyethyl starch). Those that penetrate the cell have the ability to reduce the temperature at which ice is formed, postponing cation leakage to higher osmolalities so that ice, when formed, is glass rather than crystalline, and preventing over-shrinkage of the cells. These effects are a consequence of cryoprotectants reducing the proportion of the system that converts to ice. They thus protect against solution, rather than ice, damage and allow cooling rates to be used which are too slow to allow ice formation. The action of the non-penetrating cryoprotectants is not so clear but they are believed to alter the permeability characteristics of membranes. One conse-quence of this could be a reversible solute leak which would counteract osmotic stress. For a review of cryptoprotective agents see Meryman (29).

It is essential to include a cryoprotectant and, despite its toxicity, DMSO (10%) is the most efficient because it diffuses in and out of cells very rapidly.

Polyvinylpyrrolidone has been recommended for lymphocytic cell lines but, in the author's experience, it has no advantage over DMSO as far as cryoprotection is concerned—unless a non-toxic substance is required. If DMSO is used, the freezing mixture must be diluted at least 40-fold immediately on thawing to reduce DMSO toxicity, and a medium change given as soon as the cells are attached to the substrate.

9.3.3. Cooling Rates

To obtain the best results, a commercial biological freezer should be used. These allow the setting up of a cooling rate which is maintained throughout phase transition without permitting prolonged eutectic point fluctuations to occur (Fig. 5). The cooling rate should be controlled to at least −50°C, at which point it is permissible to gradually increase the cooling rate. At −120°/−140°C the cells can be taken from the cooling chamber and put into liquid nitrogen.

Some biological freezers work on the principle that the temperature differential between the freezing chamber and the inside of the ampoule can be programmed. The larger the differential, then the greater the input of nitrogen vapour and the faster the fall in temperature. The cells can be driven straight through the phase-transition point by increasing the temperature differential, thus absorbing the latent heat of fusion. The alternative to biological freezers is to lower ampoules slowly through nitrogen vapour into the liquid nitrogen. Tables are available (e.g. from Union Carbide for the O-ring technique) for calculating the cooling rate at a range of distances above the surface of liquid nitrogen.

9.3.4. Resuscitation

As stated previously, cultures should be rewarmed as fast as possible (20). Transferring a 1-ml ampoule from liquid nitrogen into water at 37°C gives a warming rate of approximately 400°C/min and this is the minimum rate to aim for. Should the sealing be faulty, then this is the time when the ampoule would explode and precautions should always be taken during thawing; e.g. put the ampoule in a container and wear the proper safety clothing.

When cells have thawed, the ampoule is opened and they are placed in normal growth medium pre-warmed to 37°C. Even the most optimally freeze-thawed cells are extremely vulnerable at this time due to membrane damage which has to be repaired. Thus the use of supplemented (rather than minimal) medium, and even conditioned medium, results in higher recovery and faster establishment in culture.

9.3.5. Special Methods

The above description of freeze-preservation is suitable for routine purposes. For cells of special importance, or known fragility, alternative freezing methods can be used to increase the chances of a high recovery rate. An example is the two-step freezing method of Farrant (*10*). In this procedure the cell suspension is frozen as rapidly as possible to a temperature between $-15°$ and $-40°C$ (the exact temperature has to be found empirically for a given cell and system), held at this temperature for 10 min and then frozen rapidly to $-196°C$. The basis of this technique is that the cell is pushed rapidly through the hypertonic phase, stopped before intracellular ice is formed, dehydrated, then frozen rapidly again in the knowledge that no intracellular ice can be formed in a dehydrated cell.

10. CELL CHARACTERISATION

The purpose of preserving cell lines for long periods, in order to provide a reference stock and to conserve the genome for future applications, has been described. Equally important is the need to have reliable data on the cell line, so that one is confident that the cell line is what it is claimed to be and will perform according to the information pertaining to it. Incidences of laboratory workers finding that for many years they had been using the ubiquitous HeLa cell, as opposed to the cell line they thought it to be or the one with which they started, are commonplace (*14*). Cell culture work is labour-intensive and, therefore, gives ample opportunity for errors due to the human factor, e.g. mislabelling and cross-infection of cultures. It is thus important to check a cell's authenticity, either by performing characterisation checks in the laboratory or by using fully characterised cells obtained from culture collections.

Cell culture collections provide the facilities to fulfil the following requirements:

—To store contaminant-free and well-characterised reference cell lines so that investigators can refer back to the same passage material over a long time-span.
—To provide a stock of cell types of which investigators would not otherwise have knowledge or access. This particularly applies to hybridomas, where a panel is often needed to fully characterise the antigen source; this may be possible only by collecting clones derived in a range of laboratories.
—To provide insurance against loss of irreplaceable material.
—To constantly develop methodology for characterisation, quality control and storage of cells.

Cell culture collections are important for future developments in biotechnology and they also provide much of the needed characterisation techniques and data. For this reason, the following description of cell characterisation criteria is in the format used and published by culture collections. In general, the information given for each cell line comes into one of the following four categories: (1) history and genealogy, (2) identification, (3) growth characteristics and properties and (4) quality control.

Basically, the same quality control procedures are used for cells, tissue culture media and end product evaluation. Thus many of the tests for bacterial, fungal, mycoplasmal and viral contamination, and for tumourigenicity, can be found in Chapter 4 (Media) in this volume and Chapters 12 and 13 (End Product Quality Control) in Volume 2, and will not be repeated in this section. However, there are additional techniques for cells, such as testing for virus contamination by reverse transcriptase activity (1) and electron microscopy.

10.1. Cell History and Genealogy

In this category, information is given on the organ of derivation and the age, sex and species of the donor organism, together with any abnormalities or disease states of the donor. Also included is the culture history of the cell line since its derivation, e.g. number of passages, transformation information and any other relevant events. This is primarily background information but does allow a selection to be made, e.g. when it is known that a particular organ or species is especially susceptible to virus infection or transformation, or when screening for a particular biochemical property or when ageing or disease would have an influence.

10.2. Identification

One of the first criteria used to authenticate a cell line is to identify the species of origin followed, if possible, by finger-printing tests for individual cell lines, or different clones of the same cell line. These tests can be immunological, biochemical or genetical.

10.2.1. Immunological Tests

The most convenient is the fluorescent-labelled antibody test (52). A fluorescent dye (e.g. fluorescein isothiocyanate) is conjugated with antibody purified from antisera prepared by inoculating well-washed (i.e. serum-free) cells into laboratory animals. Other suitable tests are the mixed agglutination, cytotoxic-antibody dye exclusion (16) and indirect haemagglutination tests and the histocompatability antigen (HLA) assay. Histocompatibility antigens are membrane-associated transplantation antigens which are present on the surface of all nucleated cells. This assay is very sensitive and can be used to distinguish cells with the same karyotype. Methods available include the fluorochromatic microcytotoxicity test (5) and the mixed hae-madsorption technique (9). Immunological tests are an effective means of identifying the species of origin of cultured cells but they are very time-consuming and antisera are always expensive to produce.

10.2.2. Biochemical Tests

Probably the most powerful method of identifying a cell line is by iso-enzyme electrophoresis (13, 32). This can discriminate between the same tissue of different species, different tissues of the same species, different cell lines and different clones, or monolayer and suspension cells of the same cell line.

Gel electrophoresis of enzymes is a finger-print method as characteristic banding patterns (zymograms) are obtained for each cell line. The first choice of enzymes is glucose-6-phosphate dehydrogenase (G6PD), lactate dehydrogenase (28) and nucleoside phosphorylase. However, if differences cannot be detected with these, many others can be used (28), e.g. esterase, isocitric and malic dehydrogenases, hexokinase, acid and alkaline phos-phatase and lipase. Human cells have two variants of G6PD, one of which is typical of Caucasian and the other of some Negroid races. The recognition of this distinction was responsible for showing that many cell lines were in fact HeLa (of Negroid origin), and not what they purported to be.

Zyograms can be readily photographed to provide a permanent record for reference purposes. When cell lines, rather than just species of origin, need to be identified it is important to produce zymograms very early in the cell's life history.

10.2.3. Genetic Tests

Karyology (i.e. the counting and identification of the chromosome comple-ment) is a means of species identification and also gives an indication of whether the cell is normal or malignant. Normal (untransformed) cells are diploid (or, in the case of genetic mutant lines, near diploid) and thus a straightforward count will indicate the donor species (e.g. 46 chromosomes for man). However, to be absolutely sure that a cell line is normal, despite its diploidy, chromosome aberrations (e.g. breaks, translocations) have to be shown to be absent (33). As transformed cells exhibit heteroploidy, the chromosome count will be very variable and cover the range between dip-loid and tetraploid numbers. To characterise a heteroploid cell line, a fre-quency histogram is constructed from chromosome counts of 100–200 indi-vidual cells. This enables a modal chromosome number and range to be quoted for that cell line.

To allow visualisation and counting of chromosomes, the cells have to be arrested in metaphase using an anti-mitotic agent, such as colchicine or colcemid [at 0.05 μg/ml for 2–4 hr or Velban (0.01 μg/ml)]. The treated cells are harvested, concentrated by centrifugation and then swollen in hypotonic salt solution for 5–15 min. This causes the chromosomes to spread out within the nucleus and thus avoids overlapping. The time factor is critical because the nucleus must remain intact so that the chromosome complement of each cell is delineated. The swollen cells are then fixed, smeared and dried onto a coverslip and stained, usually with Giemsa (10%) or acetic orcein (2% in 45% acetic acid). Species identification can then be carried out by counting or by recognition of the chromosome morphology.

To gain the maximum information from a chromosome preparation, the morphology of each chromosome is studied and identified. To aid recognition of individual chromosomes, and chromosome aberrations, a chromosome banding technique should be used. The staining of metaphase chromosomes with the fluorochrome quinacrine mustard results in distinct banding patterns (Q bands) when viewed under a fluorescence microscope (31). This method is expensive and the fluorescence has a limited life so no permanent preparations can be achieved. Alternative banding methods have been developed which are less expensive, easier to use and give a permanent record. These include trypsin–Giemsa (G) banding (24), in which nucleoproteins are removed by treatment with trypsin (0.025%) followed by staining with Giemsa, or barium–Giemsa (C) banding (53).

Karyotype analysis is a time-consuming and, therefore, expensive test but it is essential for any cells used for the production of human vaccines (21).

10.3. Growth Characteristics and Cell Performance Data

This category of information is of vital importance to the biotechnologist (Table III) and is an area which will have to be expanded to meet present and

TABLE III Characterisation Data on Cell Growth and Performance

Parameter	Comments
Morphology	Fibroblast or epithelial
Growth pattern	Monolayer/suspension/orientation
Growth potential	Cell yield per time interval
Life-span	Immortal or limited
Plating efficiency	
Preferred medium	Special requirements
Serum	Type and concentration
Freeze–thaw data	Recovery, freezing-cocktail
Virus spectrum	Susceptibility to viruses
Special functions	Drug susceptibility
	Hormone/enzyme secretion

future demands. Characterisation tests for conditions and levels of product generation, stability of gene expression, identification of the product (especially from hybridoma cell lines), fragility, adhesion to various substrates and respiratory quotients are examples of the types of information on the cell line which the biotechnologist requires to know. In addition, suitable characterisation tests for genetically engineered cells will be needed. Much of this is beyond the resources of a culture collection and reliance will have to be placed upon data feedback from the users into a data-bank. Cell culture collections provide an authenticated cell line of guaranteed standard. Databank information will not carry such guarantees but the wealth of new information on each cell line, and the access to the contents of the many private collections in laboratories throughout the world, will more than offset this.

The importance of this concept is recognised—cell culture collections are beginning to fulfil this requirement and so, also, are independent data-banks such as Mirdab (30) and the ones specifically for hybridomas (7).

REFERENCES

1. Aldrich, C. D., and Macik, P. (1979). Reverse transcriptase assays for detection of RNA oncogenic viruses on cultured cells. *Tissue Cult. Assoc. Man.* 5, 1147–1150.
2. Barer, R., and Smith, F. (1972). Microscope for weighing bits of cells. *New Sci.* 24, 380–384.
3. Bioactivity Monitor-LBK Instruments, S-161 26 Bromma, Sweden.
4. Braaten, J. T., Lee, M. S., Schenk, A., and Mintz, D. H. (1974). Removal of fibroblastoid cells from primary monolayer cultures of rat neonatal endocrine pancreas by sodium ethylmercurithio salicylate. *Biochem. Biophys. Res. Commun.* 61, 476–482.
5. Brautbar, C. (1973). Serological identification of cells in culture. *In* "Tissue Culture: Methods and Applications" (P. F. Kruse, Jr. and M. K. Patterson, eds.), Chapter 14. Academic Press, New York.
6. Cahn, R. D., Coon, H. G., and Cahn, M. B. (1967). *In* "Methods of Developmental Biology" (F. H. Wilt and N. K. Wessells, eds.), pp. 493–530. Crowell-Collier, New York.
7. Codata, International Council of Scientific Unions, Paris, France. The Hybridoma Databank.
8. Conger, A. D., and Ziskin, M. C. (1981). Detachment of tissue-culture cells by diagnostic ultrasound. *Radiology* 139, 233.
9. Espmark, J. A., Ahlqvist-Roth, L., Sarne, L., and Persson, A. (1978). Tissue typing of cells in culture. III. HLA antigens of established human cell lines. *Tissue Antigens* 11, 279–286.
10. Farrant, J., Walter, C. A., Lee, H., and McGann, L. E. (1977). Use of two-step cooling procedure to examine factors affecting cell survival following freezing and thawing. *Cryobiology* 14, 273–286.
11. Farrant, J., and Morris, G. J. (1973). Thermal shock and dilution shock as the cause of freezing injury. *Cryobiology* 10, 134–140.
12. Ficoll-paque, Percoll-Pharmacia Fine Chemicals, Sweden: Lympho-Prep, Nycodenz-Nyegaard, Norway.
13. Fish, D. C., Dobbs, J. P., and Carter, R. C. (1971). Identification of clones of mammalian cells by isoenzyme distribution patterns. *Experientia* 27, 37–39.

14. Gartler, S. M. (1968). Apparent HeLa cell contamination of human heteroploid cell lines. *Nature (London)* **217**, 750–751.
15. Gilbert, S. F., and Migeon, B. R. (1975). D-Valine as a selective agent for normal human and rodent epithelial cells in culture. *Cell* **5**, 11–17.
16. Greene, A. E., Coriell, L. L., and Charney, J. (1964). A rapid cytotoxic antibody test to determine species of cell cultures. *J. Natl. Cancer Inst. (U.S.)* **32**, 779–786.
17. Griffiths, J. B. (1983). Oxidation–reduction potential. *Dev. Biol. Stand.* **55**, 113–116.
18. Griffiths, J. B., Cox, C. S., Beadle, D. J., Hunt, J., and Reid, D. S. (1979). Changes in cell size during the cooling, warming and post-thawing periods of the freeze–thaw cycle. *Cryobiology* **16**, 141–151.
19. Harris, L. W., and Griffiths, J. B. (1974). An assessment of methods used for measuring recovery of mammalian cells from freezing and thawing. *Cryobiology* **11**, 80–84.
20. Harris, L. W., and Griffiths, J. B. (1977). Relative effects of cooling and warming rates on mammalian cells during the freeze–thaw cycle. *Cryobiology* **14**, 662–669.
21. Jacobs, J. P., Magrath, D. I., Garrett, A. J., and Schild, G. C. (1981). Guidelines for the acceptability, management and testing of serially propagated human diploid cells for the production of live virus vaccines for use in man. *J. Biol. Stand.* **9**, 331–342.
22. Kammer, H. (1969). Cell dispersal methods for increasing yield from animal tissues. *Appl. Microbiol.* **17**, 524–527.
23. Kruse, P. F., Jr., and Patterson, M. K., eds. (1973). "Tissue Culture: Methods and Applications." Academic Press, New York.
24. Lavappa, K. S. (1978). Trypsin–Giemsa banding procedure for chromosome preparations from cultured mammalian cells. *Tissue Cult. Assoc. Man.* **4**, 761–764.
25. Lim, F., and Sun, A. M. (1980). Microencapsulated islets as bioartificial endocrine pancreas. *Science* **210**, 908–910.
26. Mazur, P. (1970). Cryobiology: The freezing of biological systems. *Science* **168**, 939–949.
27. Melamed, M. R., Mullaney, P. F., and Mendelsohn, M. L. (1979). *In* "Flow Cytometry and Sorting," p. 157. Wiley, New York.
28. Melnick, P. J. (1973). Introduction to cytoenzymological methods and isoenzymes. *In* "Tissue Culture: Methods and Applications" (P. F. Kruse, Jr. and M. K. Patterson, eds.), Chapter 17. Academic Press, New York.
29. Meryman, H. T. (1971). Cryoprotective agents. *Cryobiology* **8**, 173–183.
30. Microbiological Resource Databank (MIRDAB). Excerpta Medica, Amsterdam.
31. Miller, O. J., Miller, D. A., Allderdice, P. W., Der, V. G., and Grewal, M. S. (1971). Quinacrine fluorescent karyotypes of human diploid and heteroploid cell lines. *Cytogenetics* **10**, 337–346.
32. Montes de Oca, F., Macy, M. L., and Shannon, J. E. (1969). Isoenzyme characterisation of animal cell cultures. *Proc. Soc. Exp. Biol. Med.* **132**, 462–469.
33. Moorhead, P. S., Nichols, W. W., Perkins, F. T., and Hayflick, L. (1974). Standards of karyology for human diploid cells. *J. Biol. Stand.* **2**, 95–101.
34. Nilsson, K., Scheirer, W., Merten, O. W., Ostberg, L., Liehl, E., Katinger, H. W. D., and Mosbach, K. (1983). Entrapment of animal cells for production of monoclonal antibodies and other biomolecules. *Nature (London)* **302**, 629–630.
35. Norrby, K., Knutson, F., and Lundin, P. M. (1966). On the single cell state in enzymatically produced tumour cell suspensions. *Exp. Cell Res.* **44**, 421–428.
36. Owens, R. B. (1976). Selective cultivation of mammalian epithelial cells. *In* "Methods in Cell Biology" (D. M. Prescott, ed.), Vol. 14, pp. 341–354. Academic Press, New York.
37. Oyama, V. I., and Eagle, H. (1956). Measurement of cell growth in tissue culture with a phenol reagent. *Proc. Soc. Exp. Biol. Med.* **91**, 305–307.
38. Paul, J. (1958). Determination of the major constituents of small amounts of tissue. *Analyst* **83**, 37–42.

39. Phillips, H. F. (1973). Dye exclusion tests for cell viability. *In* "Tissue Culture: Methods and Applications" (P. F. Kruse, Jr. and M. K. Patterson, eds.), pp. 406–408. Academic Press, New York.

40. Pilwat, G., Washausen, P., Klein, J., and Zimmermann, U. (1980). Immobilization of human red blood cells. *Z. Naturforsch.* **35**, 352–356.

41. Prescott, D. M., ed. (1967–1976). "Methods in Cell Biology," Vols. I–XIV. Academic Press, New York.

42. Rappaport, C. (1957). Colorimetric method for estimating number of cells in monolayer cultures without physiological damage. *Proc. Soc. Exp. Biol. Med.* **96**, 309–316.

43. Rappaport, C., and Howze, G. B. (1966). Dissociation of adult mouse liver by sodium tetraphenylboron, a potassium complexing agent. *Proc. Soc. Exp. Biol. Med.* **121**, 1010–1016.

44. Sandford, K. K., Earle, W. R., Evans, V. J., Waltz, H. K., and Shannon, J. E. (1951). The measurement of proliferation in tissue cultures by enumeration of cell nuclei. *J. Natl. Cancer Inst. (U.S.)* **11**, 773–795.

45. Sato, G. H., Pardee, A. B., and Sirbasku, D. A., eds. (1982). "Growth of Cells in Hormonally Defined Media," Cold Spring Harbor Conf. Cell Proliferation, Vol. 9. Cold Spring Harbor Lab., Cold Spring Harbor, New York.

46. Scaife, J. F., and Brohee, H. A. (1967). A quantitative measurement of cell damage by means of CR'51 binding. *Exp. Cell Res.* **46**, 612–615.

47. Scheirer, W., Nilsson, K., Merten, O. W., Katinger, H. W. D., and Mosbach, K. (1983). Entrapment of animal cells for the production of biomolecules such as monoclonal antibodies. *Dev. Biol. Stand.* **55**, 155–162.

48. Shall, S. (1973). Selective synchronisation by velocity sedimentation separation of mouse fibroblast cells grown in suspension culture. *In* "Methods in Cell Biology" (D. M. Prescott, ed.), Vol. 7, pp. 269–285. Academic Press, New York.

49. Sheeter, P., and Doolittle, M. H. (1980). Separation of mammalian cells by velocity sedimentation. *Int. Lab.* March, pp. 71–84.

50. Smith, F. H. (1972). A laser-illuminated scanning microinterferometer for determining the dry mass of living cells. *Microscope* **20**, 153–160.

51. Spier, R. E., and Clarke, J. B. (1980). Variation in the susceptibility of BHK populations and cloned cell lines to three strains of foot-and-mouth disease virus. *Arch. Virol.* **63**, 1–9.

52. Stulberg, C. S. (1973). Extrinsic cell contamination of tissue culture. *In* "Contamination in Tissue Culture" (J. Fogh, ed.), pp. 2–23. Academic Press, New York.

53. Sumner, A. T., Evans, H. J., and Buckland, R. A. (1971). New technique for distinguishing between human chromosomes. *Nature (London), New Biol.* **232**, 31–32.

54. Tobey, R. A., Anderson, E. C., and Petersen, D. F. (1967). Properties of mitotic cells prepared by mechanically shaking monolayer cultures of Chinese hamster cells. *J. Cell. Physiol.* **70**, 63–68.

55. Tobey, R. A., Crissman, H. A., and Kraemer, P. M. (1972). Microfluorometric analysis of DNA content. *J. Cell Biol.* **54**, 638–643.

56. Tsao, M., Walthall, B., and Ham, R. (1982). Clonal growth of normal human epidermal keratocytes in a defined medium. *J. Cell. Physiol.* **110**, 219–229.

57. van Wezel, A. L., Van der Velden-de Groot, C. A. M., and Van Herwaarden, J. A. M. (1980). The production of inactivated polio vaccine on serially cultivated kidney cells from captive-bred monkeys. *Dev. Biol. Stand.* **46**, 151–158.

58. Vasquez, B., Ishibashi, F., and Howard, B. V. (1982). Measurement of intracellular volume in monolayers of cultured cells. *In Vitro* **18**, 643–649.

59. Walter, H. (1975)..Partition of cells in two-polymer aqueous phases. *In* "Methods in Cell Biology" (D. M. Prescott, ed.), Vol. 9, pp. 25–49. Academic Press, New York.

60. Waymouth, C. (1956). A rapid quantitative hematocrit method for measuring increase in cell population of strain L (Earle) cells cultivated in serum-free nutrient solutions. *J. Natl. Cancer. Inst. (U.S.)* **17**, 305–311.
61. Wheatley, D. N. (1976). Hypertonicity and synchronisation of mammalian cells in mitosis. *In* "Methods in Cell Biology" (D. M. Prescott, ed.), Vol. 14, Chapter 27. Academic Press, New York.
62. Wilton-Smith, P. (1983). Unpublished data.
63. Wilton-Smith, P. (1983). Sending cells safely through the mail. *Trends Biotechnol.* **1**, 101.

4

Cell Growth Media

K. J. LAMBERT
J. R. BIRCH
Celltech Ltd.
Slough, Berkshire
United Kingdom

Animal Cell Biotechnology, Vol. 1

1. INTRODUCTION

Media and nutritional requirements have been reviewed by Ham and McKeehan (48), Litwin (74), Waymouth *et al.* (132) and Sato *et al.* (114).

The functions of the cell growth medium may be defined as (1) maintenance of suitable physiochemical conditions (e.g. pH and osmolarity) and (2) provision of nutrients for synthesis of cell biomass and products. A nutrient is defined as a substance that enters cells and is utilised as a metabolic substrate or cofactor. The known essential nutrients (those which cultured cells cannot synthesise) include amino acids, vitamins, dissolved gases and inorganic ions as well as an energy source (Section 3).

A culture medium may have to provide an environment for clonal (single cell) growth, mass growth and product formation or maintenance of non-dividing cells (Section 6).

Early media were based on the analysis of plasma and other biological fluids. They led to the design of *complex* media, such as 199 (89), containing some components which have still not been shown to be essential for *in vitro* cell growth. An alternative approach to medium design is the reduction of the number of components to the minimum shown to be essential for growth, i.e. a *minimal* medium. This technique is exemplified by the systematic studies of Eagle, which gave rise to the minimal essential medium (MEM) (33) illustrated in Table I.

Both complex and minimal approaches to medium design relied on *qualitative* definition of nutrient requirements, i.e. which (rather than how much of) basic medium components were required. Nutrients defined in this way are discussed in Section 3. Since 1967 *quantitative* data, i.e. measurements of the utilisation of individual components of cell culture media, have been available. These results (Section 4) have shown that cell types differ in their quantitative requirements.

TABLE I Eagle's Minimum Essential Medium (MEM)

Component	Amount[a]	Component	Amount[a]
L-Amino acids		Vitamins	
Arginine	0.6	Choline	8.3
Cystine	0.1	Folic acid	2.3
Glutamine	2.0	Inositol	11.0
Histidine	0.2	Nicotinamide	8.2
Isoleucine	0.4	Pantothenate	4.6
Leucine	0.4	Pyridoxal	6.0
Lysine	0.4	Riboflavin	0.27
Methionine	0.1	Thiamine	3.0
Phenylalanine	0.2	Inorganic ions	
Threonine	0.4	NaCl	116
Tryptophan	0.05	KCl	5.4
Tyrosine	0.2	$CaCl_2$	1.8
Valine	0.4	$MgCl_2 \cdot 6H_2O$	1.0
Carbon source		$NaH_2PO_4 \cdot 2H_2O$	1.1
Glucose	5.5	$NaHCO_3$	23.8
Supplement			
Whole or dialysed serum	5–10% v/v		

[a] All values are millimolar except for vitamin amounts, which are micromolar.

MEM, 199, DMEM (32), RPMI 1640 (88) and F12 (46), although based on the qualitative rather than the quantitative requirements of cells, are traditional cell culture media of broad current application. Their widespread use relies on supplementation with animal serum or other non-defined medium supplements (Section 5). Problems associated with the use of serum have led to recent extensive development of serum-free media (Section 8). Only a few established cell lines can grow in protein-free, "chemically defined" media (Section 9). Since 1970 many specialised serum-free media for both primary and established lines have been reported. Serum is replaced by complex mixtures of purified proteins, lipids, trace elements, peptides and other nutrients (Section 10). These reports have shown that there is no single universally applicable serum-free culture medium. Furthermore, some serum substitutes are more expensive than serum. For large-scale industrial applications it has been common to use inexpensive autoclavable undefined serum substitutes such as peptones (Section 9).

Cell growth media require particular care in their preparation, sterilisation, storage and quality control (Sections 11 and 12).

2. CHOICE OF MEDIUM

The following factors should be considered:

1. Precedent. A reliable and rapid approach is to use a medium previously shown to be suitable for that cell type.

2. Cell type. Transformed and established cell lines usually grow in simpler media or require less serum than non-transformed or primary lines.

3. Commercial availability. Specialist suppliers provide a wide range of traditional media and a few serum-free formulations (e.g. Iscove's medium). Growth factors and pre-mixed supplements for serum-free media are available commercially (e.g. from Collaborative Research, Inc.).

4. Culture system. Nutrient requirements for a given cell type are influenced by the culture system (*100*) (Section 6).

5. Cost. The expense of serum has to be balanced against the high cost of supplements for most serum-free media.

6. Degree of chemical definition required. Serum-free medium may be a prerequisite for certain research projects.

3. BASIC COMPOSITION OF MEDIA

3.1. Energy Sources

Carbohydrate metabolism was reviewed for vertebrate cells by Morgan and Faik (*90*), Eagle *et al.* (*35*), and Gregg (*42*) and for insect cells by Vaughn (1972).

The traditional carbon source is glucose at 5–20 mM. Glutamine (0.7–5 mM) is the other major source of energy in most media (*109, 141*): up to 30% of the energy requirements of human diploid fibroblasts can be provided by glutamine in conventional medium with serum (*142*). Glucose and glutamine metabolism in cultured cells was reviewed by McKeehan (*78*).

A complication in the use of glucose, particularly at high levels, is that glycolytic conversion to lactate may rapidly lead to a reduction in pH to inhibitory levels. In agitated culture systems pH can be automatically controlled by addition of acid (for example CO_2) or alkali. A low initial glucose concentration with daily (or more frequent) feeding is an additional means of controlling the production of lactate. The degree of conversion to lactate is less when glutamine is used as the principal energy source (*109, 140*).

The use of fructose or galactose (2–20 mM) instead of glucose is another means of controlling lactate production (*28, 109*). Galactose is the carbon source in Leibovitz L-15 medium (*73*), and supported growth of WI-38 human diploid fibroblasts (6). Fructose maintained high-density microcarrier culture of MDCK cells at constant pH (54).

Wice *et al.* (135) have described continuous growth of several cell lines on uridine or cytidine as the carbon source.

3.2. Amino Acids

Amino acid metabolism has been reviewed by Patterson (101) for animal cells and by Vaughn (127) and Millam–Stanley (86) for insect cells.

Thirteen amino acids (listed in Table I) are considered essential (33) for cultured cells. The requirement for the essential amino acids is reduced when "non-essential" amino acids are included (45). The non-essential amino acids commonly added to media are alanine, asparagine, aspartic acid, proline, glycine and serine, normally used at 0.1–2.0 mM. In some situations there is a specific requirement for an amino acid normally considered non-essential, for example serine for lymphoblastoid cell lines (12).

Glutamine plays a vital role in metabolism, both as an energy source (Section 3.1) and as a precursor for nucleic acid synthesis. Cells have been adapted to use glutamic acid, which is more stable in culture media, instead of glutamine (45); for human diploid cells such adaption was associated with a change in morphology (44).

The balance between amino acids is important (81).

3.3. Vitamins

3.3.1. Water-soluble Vitamins

Requirements for water-soluble vitamins were reviewed by Blaker (16) and Higuchi (50). The B group vitamins, most of which are cofactors for enzymes, have been the most closely studied. Biotin, folic acid, nicotinamide, pantothenic acid, pyridoxine, riboflavin and thiamine are required by most cultured cells and included in most media. Biotin is not included in some media (e.g. MEM, Table I) normally used with serum but a biotin requirement for human diploid fibroblasts has been demonstrated (22). Coenzymes can generally be substituted for the parent vitamins, but with the possible exception of folinic acid (47, 50) they offer no special advantage.

Choline and *meso*-inositol are strictly speaking not vitamins, since they function as substrates for lipid synthesis rather than as cofactors. Growth limitation by choline or inositol causes an alteration in cellular membrane composition and decreased cellular adhesion (76).

Requirements for other water-soluble vitamins (ascorbic acid, *p*-aminobenzoic acid, cobalamin/B_{12}) are less clear. Ascorbic acid has been reported to increase the yield of interferon from microcarrier cultures of human cells (26). Vitamin B_{12} (included in F12 medium) has been reported as a specific requirement for some cells (50, 56).

3.3.2. Fat-soluble Vitamins

Information on the effects of the fat-soluble vitamins (A, D, E, K and ubiquinone) on cultured cells is limited [for review, see Smith (117)]. Vitamin A has been the most intensively studied (75) and both stimulatory and inhibitory effects of vitamin A have been demonstrated (116). Vitamin A and vitamin E (as α-tocopherol) are included in medium 199.

3.4. Major Inorganic Ions

Animal cells require sodium, potassium, calcium, magnesium, phosphate, bicarbonate and chloride. The concentrations given in Table I are typical. Sulphate at 3 μM to 0.9 mM is often included, although sulphur is provided by the methionine and cystine. The inorganic ions have essential physiological functions, such as maintaining osmotic pressure and membrane potentials and buffering, as well as promoting cell attachment or acting as cofactors for enzyme reactions. Bicarbonate is generally included in media as both a nutrient and pH buffer. Concentrations of inorganic ions have traditionally been based upon those in general-purpose balanced salt solutions [for details, see Paul (102)]. However, cell yield and growth rate may be modified by the ratios of different ions, particularly sodium/potassium ratios (131).

3.5. Trace Elements

Trace element requirements for cultured cells were reviewed by Nielsen (98) and Hutner (53). Most conventional media do not contain trace elements in the recipe. Fifteen trace elements (Co, Cu, I, Fe, Mn, Mo, Zn, Se, Cr, Ni, V, As, Si, F, Sn) have been established as essential or beneficial for animals. A requirement for iron has been clearly defined for many cells in culture. Iron is an essential component of, for instance, cytochromes and iron deficiency can be demonstrated through its effect on growth rate as well as cell yield (14). When serum is used in media, iron is present in the iron-transport protein transferrin (siderophilin). For some cell types, iron is added as a transferrin–ferric iron complex to synthetic media (55). In some media, inorganic salts of iron are used (128).

Serum functions both as a source of other trace elements and as a means of binding them so that they are non-toxic but available to cells. Metal chelating agents, such as ethylenediamine tetraacetic acid and nitrilo triacetic acid, can complex metals and act as metal buffers in media (106). Chelation of iron at 1 mol EDTA/mol Fe has been used in culture media (71). Chelation, as pointed out by Hutner (53), is a principle often overlooked in medium design. Many common medium constituents, such as amino acids and hydroxy acids, act as chelating agents.

Trace elements are derived from chemicals or glass used in medium preparation and have been shown to be present as contaminants of commercially prepared media (38). Trace element requirements which have been defined during the development of serum-free media are discussed in Section 9.

3.6. Lipids

Lipid metabolism in cultured cells was reviewed by King and Spector (63); cholesterol requirements for growth were reviewed by Chen and Kandutsch (24).

Serum contains two classes of lipid transport protein: albumin, which carries free fatty acids, and the lipoproteins, which carry phospholipids, triglycerides and cholesterol. Serum is therefore an excellent source of lipids for cell culture and minimal traditional culture media do not contain additional lipids. The complex media F12 and 199 contain linoleic acid and cholesterol, respectively.

Traditional techniques in solubilising lipids in a non-toxic and stable form were reviewed by Morton (91) and include ethanol as a solvent with derivatives of Tween (polyoxyethylenesorbitan) as sources of fatty acids or as non-toxic stabilising detergents. Newer lipid delivery techniques and requirements of cultured cells for lipids in the absence of serum are discussed in Section 10.

3.7. pH and Buffers

Techniques for measurement and control of pH are discussed elsewhere (Part III, Chapter 11, this volume). Methods for controlling the accumulation of lactate were described in Section 3.1.

Different cell types have different optima for maximal growth within the range 6.9 to 7.8. Optimising the pH can allow sparing of nutrients such as serum (21) and glucose (11).

Traditionally, cell culture media have been buffered by a CO_2/HCO_3 system [for nomograms relating pH to bicarbonate concentration at 5% CO_2 or 10% CO_2, see Waymouth (131)]. The pKa of 6.1 for $NaHCO_3$ results in suboptimal buffering throughout the physiological pH range. Another disadvantage is that constant equilibration with CO_2 at 5% or more in air is usually required to maintain pH and this may add considerable expense and inconvenience to small-scale culture. Particularly for such systems, where automatic pH control is not available, use of organic buffers such as HEPES or α-glycerophosphate should be considered. HEPES gives near optimal buffering in the range 6.8–7.2 and can be used without carbon dioxide. HEPES has most frequently been used at 10–25 mM, although 50 mM was not inhibitory for human diploid fibroblasts (82) when appropriate osmotic correction was made. In using HEPES without carbon dioxide, it should be remembered that there is a requirement for bicarbonate independent of its buffering action: if cultures are sparse, there may be insufficient CO_2 generated by the cells to supply this requirement.

A range of buffer solutions (including organic buffers) for maintaining specific pH values in culture media was given by Eagle (34). Phenol red is widely used (at 3–55 μM) as an indicator of pH for small-scale culture systems. Phenol red may be significantly toxic to some cells (74), particularly in serum-free systems.

3.8. Gases

Requirements for gases in mammalian cell culture were reviewed by McLimans (83). Methods for control of oxygen tension are discussed in this volume (Part III, Chapter 11).

3.8.1. Oxygen

Oxygen is an essential element for growth of mammalian cells and is typically provided in small-scale cultures by overlaying the medium with air or a mixture of 95% air/5% CO_2. It should be noted that oxygen is very poorly soluble in culture media (approximately 7 ppm at saturation with air). Mammalian cells at populations usually encountered in culture will rapidly deplete this oxygen [see Fleischaker and Sinskey (37) for oxygen consumption rates of various mammalian cell types]. Oxygen probably limits maximum population density in cultures more frequently than is generally realised, even on the small scale. One should therefore pay some attention to the oxygen mass transfer characteristics of a given culture system. McLimans et al. (84) have emphasised this in relation to the depth of the culture fluid overlay on cells cultured under static conditions. The problem of oxygen supply is of particular significance in designing large-scale culture systems and various techniques have been devised for increasing the efficiency of oxygen transfer in fermenters [see Boraston et al. (19) for discussion]. Whilst all animal cells probably require oxygen, the actual dissolved oxygen tension (DOT) which is the optimum for growth seems to vary from one cell type to another. Hence, whilst the growth kinetics of a mouse hybridoma cell line were relatively unaffected over a range of DOTs from 10 to 100% of saturation with air (19), other cell types show a distinct optimum within this range (56, 122, 123).

3.8.2. *Carbon Dioxide*

Definition of the optimal CO_2 tension for growth is complicated by the interrelation between CO_2, bicarbonate and pH (Section 3.7). Carbon dioxide can form an important source of bicarbonate through solution in the medium. McLimans (83) suggests that a gas mixture of 0.5–2.0% CO_2 in air may be optimal for some cultured cells.

3.9. Nucleic Acid Precursors

Cells *in vitro* are ordinarily capable of synthesising purines and pyrimidines (61). However, cell yields may be improved and nutrients such as folic acid spared by inclusion of purine and pyrimidine sources. The nucleic acid precursors most commonly used have been adenine, hypoxanthine, cytidine and thymidine at 10^{-5} to 10^{-7} M.

3.10. Antibiotics

The use of antibiotics in cell culture media was reviewed by Perlman (103). Ideally, antibiotics should be avoided since resistant microorganisms may develop, and antibiotics often have adverse effects on cell growth and function. Antibiotics can reduce cell yield, growth rate and longevity (e.g. 39) and the cytotoxicity of antibiotics may be increased under serum-free conditions. However, in large-scale systems the economic consequences of contamination may outweigh the adverse affects of antibiotics.

Penicillin (100 IU/ml), streptomycin (50 μg/ml) or gentamycin (50 μg/ml) as antibacterial agents and amphotericin B (25 μg/ml) or nystatin (25 μg/ml) as antifungal agents are the most frequently used antibiotics.

The antibiotic amphotericin B is toxic for some insect cell lines (127).

4. QUANTITATIVE STUDIES

Quantitative studies, i.e. measurements of the utilisation and breakdown of individual medium components, are essential for full optimisation of the growth medium. The basic quantitative information required is (1) the effect of nutrient concentration on cell growth rate and (2) the yield of cells per unit weight of nutrient utilised (Fig. 1). Sources of growth yield data for mammalian cells are listed in Table II.

Growth yields from glucose reported for mammalian cells are listed in Table III. The 10-fold range in the efficiency of glucose utilisation reflects differences in cell type, culture system and conditions used. The rate of

(a)

The specific growth rate μ of a cell population is defined as

$$\mu = \frac{1}{n} \cdot \frac{dn}{dt}$$

where n = cell number/ml, t = time.

Thus $\mu = 1n\ 2/t_d$ where t_d is population doubling time.

(b)

The growth yield (Y) is defined by the quotient

$$\frac{\Delta x}{\Delta s} = Y$$

where Δx = increase in biomass (as cell numbers or weight) resulting from utilisation of an amount Δs of substrate.

Fig. 1.   Parameters of growth and nutrient utilisation. (a) Cell growth rate or doubling time. (b) Growth yield. For a growth-limiting substrate, a plot of maximum biomass obtained against initial substrate concentration is typically a straight line with slope Y.

glucose utilisation has been shown to vary with pH and the stage of the growth cycle (11).

Growth yields from essential amino acids for three cell types growing in MEM with serum are listed in Table IV. Cystine and glutamine are the most highly utilised amino acids for a variety of cultured cells (e.g. 20). The high requirement for glutamine is partly due to its instability in culture media: decomposition to pyrrolidone, carboxylic acid and ammonia can occur (125). Table IV lists data for three systems: stirred large-scale suspension (BHK 21), unstirred batch suspension (LS) and unstirred monolayer (MRC-5). For a given system, growth yields from amino acids have been shown to vary with factors such as growth rate (45), cell density (20) and serum concentration (70).

Growth yields from vitamins when they are present in excess are listed in Table V. The growth yield from a vitamin is often higher when it is growth-limiting than when it is supplied in excess (18). Folic acid, although apparently not essential for LS cells, was utilised by MRC-5 to a greater extent than biotin, pantothenic acid and thiamine. Serum forms an important source of the most highly utilised vitamins, choline and inositol (13, 70); supplementation with these essential components is necessary when serum is reduced.

TABLE II Explicit Growth Yield Data for Cultured Mammalian Cells

Nutrient	A. Y determined in the absence of serum Cell type	Culture system	Reference
Choline	LS mouse fibroblast	Unstirred suspension	13
Glucose	HeLa human cervical carcinoma	Stirred suspension	17
Vitamins	LS mouse fibroblast	Stirred suspension	18
Glucose, phosphorus, potassium	LS mouse fibroblast	Unstirred suspension	15

Nutrient	B. Y determined in the presence of serum Cell type	Culture system	Reference
Amino acids	LS mouse fibroblast	Unstirred suspension, chemostat	45
Amino acids, glucose	WI-38 human diploid fibroblast	Unstirred monolayer	43
Amino acids, vitamins, glucose	MRC-5 human diploid fibroblast	Unstirred monolayer	>70
Amino acids	MOPC-31C mouse myeloma	Stirred suspension	110
Amino acids	MDCP canine kidney	Microcarrier	20
Amino acids, glucose	BHK 21 clone 13 hamster kidney	Stirred suspension	1
Oxygen	Mouse hybridoma	Stirred suspension, chemostat	19

TABLE III Growth Yields (Y) from Glucose

Cell type	Y^a	Reference
HeLa	0.295	17
LS	0.76	15
WI-38	0.075	43
MRC-5	0.09	70
BHK 21	0.28	1
Bri-8	0.16	11

[a] Grams dry weight of biomass per gram of glucose utilised.

TABLE IV Growth Yields (*Y*) from Amino Acids

Amino acid	BHK 21 clone 13		LS		MRC-5	
	Y^a	Percentage utilisation	Y	Percentage utilisation	Y	Percentage utilisation
L-Arginine	17.1	38	8.7	42	7.5	21
L-Cystine	81.2	100	20.0	80	5.6	74.5
L-Glutamine	5.4	97	0.72	98	0.9	73.0
L-Histidine	—	—	52.6	25	102.5	4.0
L-Isoleucine	18.1	62	10.2	65	4.0	33.0
L-Leucine	7.6	33	10.1	67	4.6	36.0
L-Lysine	21.7	17	13.0	60	6.2	26.0
L-Methionine	32.5	33	52.6	57	19.6	33.0
L-Phenylalanine	130.0	5	30.4	39	155.0	6.0
L-Threonine	38.2	22	26.4	33	6.3	35.0
L-Tryptophan	108.3	17	91.0	48	58.6	23.5
L-Tyrosine	20.3	45	71.5	17	17.3	13.0
L-Valine	20.3	24	17.0	50	5.9	30.0
	Arathoon and Telling (1)		Griffiths and Pirt (45)		Lambert and Pirt (70)	

[a] Grams dry weight of biomass per gram of amino acid utilised.

Quantitative data on inorganic ions and trace elements are limited. Birch and Pirt (15) reported that a threshold value of potassium of 0.4 mM had to be exceeded before growth of LS cells could occur; for maximum growth rate, 0.53 mM potassium was required. For magnesium, the threshold was 0.05 mM, with maximum growth rate at 0.2 mM. Birch and Pirt (15) also reported growth yields of 75 μg dry weight biomass/μg K and 69 μg dry weight biomass/μg P for potassium and phosphorus, respectively.

Optimum concentrations for growth of mammalian cells have been reported for iron (0.6–1.4 μM), zinc (0.6 μM) and selenium (30 nM) (80, 124).

Oxygen utilisation rates for a range of cell types in batch culture were listed by Fleischaker and Sinskey (37). Boraston et al. (19) reported quantitative requirements of mouse hybridoma cells in batch culture and in oxygen-limited chemostat culture. Under oxygen-limited conditions the values for oxygen utilisation rate (ca. 6.5 μg/10^6 cells/hr) at growth rates approaching μ_{max} (the theoretical maximum growth rate; for definition of growth rate, see Fig. 1) in the chemostat were close to values obtained for cells growing in batch culture at dissolved oxygen tensions of 8–60%. The growth yield for oxygen was 0.0072 × 10^6 cells/μg O_2 for growth rates between 0.03 hr^{-1} and μ_{max}; growth yields were reduced at lower dilution rates.

This section has emphasised the difference between cell types in their

TABLE V Growth Yields (Y) from Vitamins

Vitamin	Cell type[a]	Y[b]
Biotin	Mouse LS	2.0×10^5
	MRC-5	2.5×10^3
Folic acid	Mouse LS	2.4×10^5
	MRC-5	7.7×10^2
Nicotinamide	Mouse LS	4.8×10^2
	MRC-5	7.2×10^2
Pantothenic acid	Mouse LS	6.0×10^3
	MRC-5	1.4×10^3
Pyridoxine	Mouse LS	4.36×10^2
	MRC-5	3.7×10^2
Riboflavin	Mouse LS	2.0×10^4
Thiamine	Mouse LS	3.2×10^4
	MRC-5	1.3×10^3
Inositol	Mouse LS	1.56×10^2
	MRC-5	1.3×10^2
Choline	Mouse LS	0.7×10^2
	MRC-5	1.2×10^2

[a] From Blaker and Pirt (18), Blaker (16) for LS cells, Lambert and Pirt (70) for MRC-5 cells.
[b] Grams dry weight of biomass per gram of vitamin utilised.

quantitative requirements. Detailed quantitative studies will become increasingly important as mammalian cells become more widely used in industrial processes. The potential of systematic quantitative studies in optimising media was demonstrated by Pirt *et al.* in their development of a defined medium designed to support high yields of LS cells (13, 14, 45).

Practical application of growth yield data is of course limited by the interrelation of nutrient utilisation with other factors in the culture system (summarised in Fig. 2) which may limit cell yield and growth rate. Continuous culture systems (Part III, Chapter 10, this volume; Pirt, 1975) are particularly suitable for analysis of these interrelating factors (106).

5. NON-DEFINED MEDIUM SUPPLEMENTS

5.1. Serum

Commercially prepared foetal calf, newborn calf and horse serum, in concentrations from 0.5 to 30% v/v, are the most commonly used serum supplements. Foetal calf serum (FCS), in spite of high price, has particular advantages. Its generally low concentration of immunoglobulin is important

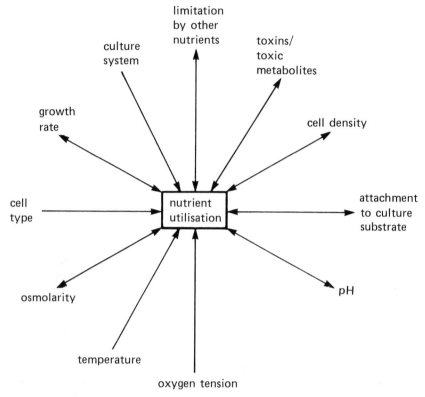

Fig. 2. Factors affecting utilisation of an individual nutrient in cell culture.

for *in vitro* monoclonal antibody production. Foetal calf serum often has the ability to promote the growth of more fastidious cell types at low densities. The reason for this is not known, but FCS has high levels of biotin (92), which is not present in some media (e.g. Eagle's MEM).

Sparing of serum may be achieved through optimisation of other environmental factors (Fig. 2), such as pH or gas tension. Serum may also be partially replaced by peptones (1) or chemically defined factors such as hormones (67). Cheaper sera, such as newborn calf or swine sera, may entirely substitute for FCS (10) or partially replace FCS after an initial (low-density) stage of culture (25).

The known functions of serum in culture media are summarised in Table VI and discussed in detail in Section 10.1. Since serum contains more than 1000 proteins and other substances, it is not surprising that the relationship between *in vitro* growth promotion and biochemical or biophysical properties of serum has not yet been completely clarified. Current progress in serum-free media is discussed in Sections 8, 9 and 10.

TABLE VI Functions of Serum

1. Carrier/buffer/chelator for labile or water-insoluble nutrients
2. Binds and neutralises toxins
3. Provides protease inhibitors which inactivate trypsin
4. Assists attachment of cells to substrate
5. Provides essential low-molecular-weight nutrients
6. Provides hormones and peptide growth factors
7. Protective effect (e.g. in agitated suspension culture)

Problems associated with the use of serum in culture media are listed in Table VII. Serum can be highly variable in composition (108) and growth-promoting ability (70). Potential contamination with infectious agents is a major objection to the use of serum for preparation of cell products for human use. Bacteriophages have been detected in commercial sera (85) and bovine viruses have been detected in FCS (66). Serum is the major source of mycoplasma contamination in cell culture (4). Preparation of serum is discussed in Section 11.1.

5.2. Others

Non-defined medium supplements (such as peptones, egg yolk emulsion, lactalbumin hydrolysate, tryptose phosphate and milk) have been used in serum-free media and are discussed in Sections 8 and 9.

6. EFFECT OF CULTURE SYSTEM ON MEDIUM DESIGN

6.1. Suspension Culture

Suspension culture systems are discussed in Part III, Chapter 10, this volume. The main factors affecting medium design for suspension systems are (1) the relatively high cell yields to be supported, (2) cell aggregation, (3)

TABLE VII Problems with Serum

1. Variability in composition
2. High cost and periodic scarcity
3. Source of contamination (mycoplasmas, bacteriophages, viruses)
4. Source of toxins
5. Modification of cell surface by adsorption/incorporation of serum proteins
6. Prevents definition of nutritional environment
7. Interference in product purification (e.g. by immunoglobulins in immunological products)
8. Encourages overgrowth by fibroblasts in primary culture of other cell types

mechanical damage to cells and (4) precipitation of serum protein and foam formation.

Cells that can be grown either as monolayers or suspended tend to require higher nutrient concentrations or more serum protein when grown in suspension. In one report, nutritional demands were increased by agitation (9). When cells are grown up to high density, medium design should aim to minimise exhaustion of key nutrients and accumulation of toxic metabolites. Enhanced buffering and control of pH and gas supply will be required. Reduced levels of calcium (e.g. from 1.8 to 2.0 mM $CaCl_2$) have been used to reduce cell aggregation or attachment to the culture vessel. Polymers such as methylcellulose and Pluronic F68 (a copolymer of polyoxyethylene and polyoxypropylene) have been used at 1 g/litre in suspension systems since the early observations that they could protect cells in agitated serum-free suspension culture (69) or prevent precipitation of serum proteins (121).

Silicone antifoam agents (e.g. aqueous emulsion of polydimethylsiloxane at a final dilution of 1/10,000) are effective non-toxic antifoam agents for mammalian cells.

6.2. Monolayer Culture

Monolayer culture techniques have been discussed in detail elsewhere (Part III, Chapters 7 and 8, this volume). Pharmacia Fine Chemicals (104) includes details of media for microcarrier systems. Medium considerations for substrate-attached growth include (1) requirements for minimising damage during subculture, especially for serum-free media (discussed in Section 10.2), and (2) specific requirements for adhesion, including divalent cations (particularly calcium) and attachment proteins.

Mechanisms of cellular adhesion and the composition of the extracellular matrix have been extensively studied [for review, see Kleinman et al. (65)]. Fibronectin is a major protein involved in cell surface interactions and is also active in promoting in vitro cell attachment at 1–5 µg/ml. Culture media supplemented with 10% foetal calf serum contain 2–3 µg fibronectin/ml (49). Although many non-transformed cell types, including normal diploid fibroblasts, can synthesise fibronectin, attachment is greatly stimulated by exogenous fibronectin. Fibronectin has been included in serum-free media for fibroblasts (56, 133).

Coating of the culture surface with a positively charged polymer such as poly-D-lysine (79) has been used to improve cell attachment, particularly in the absence of serum (8, 105, 128).

Attachment to the culture surface and medium nutrients have a reciprocal relationship in supporting cell growth (Fig. 2). For example, the extracellular matrix has been shown to affect requirements for growth factors (*40*); conversely, ascorbic acid is necessary for synthesis of collagen, an important component of the extracellular matrix.

6.3. Clonal Growth

Feeder layers and conditioned media are frequently used in cloning (deriving a cell population from a single cell) with medium containing serum.

"Conditioned medium" is medium which has been removed from actively growing or confluent cultures and clarified by centrifugation. It may be used diluted typically 3 : 1 or 1 : 1 with fresh medium for cells which are very difficult to grow in culture or where very low cell densities are involved. "Conditioning" essentially compensates for the inadequacies of the recipient medium. Possible mechanisms by which high-density populations condition medium are: leakage of low-molecular-weight "population-dependent" nutrients (e.g. pyruvate, CO_2), release of growth-stimulating macromolecules and neutralisation of inhibitors.

A feeder layer is a high-density (non-proliferating) cell population which provides conditioning functions by direct co-culture with the recipient low-density cells. Feeder layers may be prepared from any cell population. Proliferating monolayers (e.g. fibroblasts) should be irradiated with 4000–5000 rads (cobalt-60 or cesium-135) so that division is halted before the cells to be supported are added. Alternatively, non-dividing feeder cells such as peritoneal macrophages, thymocytes or lymphocytes may be used. The effects of different feeder layers on cloning efficiency of transformed human cells were reported by Namba *et al.* (97). Astaldi (2) has described a human endothelial cell supernatant (HECS) which is claimed to support superior yields and stability of hybridoma clones compared to those obtained with conventional feeder layers.

Further research on cellular nutrition should in the future add to the list of cells which can be cloned without feeders (e.g., *46*) and render conditioned medium and feeder layers, with their disadvantages of lack of definition and time-consuming preparation, less necessary.

6.4. Maintenance Media

For virus cultures or maintaining cells in the non-growing state for biochemical studies, it is common to use media with low levels of serum. For specific product (e.g. interferon) production, particular conditions may be required which are beyond the scope of this review (see Chapter 1, Volume 2).

7. MEDIA FOR INSECT CELLS

Media for insect cells were reviewed by Vaughn (127), Millam-Stanley (86) and Hink (52). Although a few insect lines grow in media designed for vertebrates (e.g. Leibovitz L-15 medium), 30 or more formulations designed for insects are available. Grace's basal medium and Mitsuhashi and Maramorosch's medium for mosquito cells are examples [details in Hink (52)]. The insect media differ generally from vertebrate ones in having higher osmolarity, lower pH, higher amino acid concentrations and phosphates as buffers rather than carbonates. Supplementation of basal media with 5–20% serum and other undefined ingredients such as lactalbumin hydrolysate has superseded the use of insect hemolymph. Large-scale production of insect cells has been supported by medium containing serum (87, 134). Recent serum-free media for insect cells are listed in Table VIII and discussed in Section 9.

TABLE VIII Examples of Serum-free Media 1980-1983

Polymer	Type I (chemically defined) Application	Reference
—	Established mouse hybridoma attached monolayer	27
PVP	Established rat hepatoma monolayer	99
—	Established insect monolayer	7
Methylcellulose	Established insect monolayer	136

Protein/peptide	Type II (defined with animal proteins) Application	Reference
Insulin EGF[a]	Human diploid fibroblast monolayer	8
Insulin Transferrin Albumin Fibronectin EGF	Human diploid fibroblast monolayer	25
Insulin Transferrin Fibronectin EGF PDGF[a]	Human diploid fibroblast monolayer	105
Insulin EGF	Human diploid fibroblast clones	128
Insulin Transferrin Albumin	Human diploid fibroblast monolayer	133

TABLE VIII (*Continued*)

Protein/peptide	Application	Reference
EGF Thrombin Insulin Transferrin Albumin Fibronectin	Human diploid fibroblast monolayer	56, 138
EGF Thrombin Insulin Transferrin α-Cyclodextrin Fibronectin	Human diploid fibroblast monolayer	139
EGF Insulin Transferrin	Established mouse hybridoma suspension	23
Insulin Casein Transferrin	Established mouse and human hybridoma suspension	30
Transferrin Albumin	Established mouse and rat-mouse hybridoma agitated suspension	36
Insulin Transferrin	Established mouse hybridoma suspension	94
Insulin Transferrin Albumin	Established mouse hybridoma agitated suspension	95
	Established mouse lymphoblastoid suspension	9
Insulin Casein Transferrin	Established mouse T-lymphoma suspension	29, 30
Insulin Transferrin	Established mouse T-lymphoma suspension	31
Transferrin Albumin	Established human lymphoblastoid suspension	96
Insulin Transferrin Albumin	Established human lymphoblastoid suspension	115
Transferrin Albumin	Human T- and B-lymphoblastoid suspension	126
Insulin Transferrin	Established human lymphoblastoid suspension	139

	Type III (non-defined with low-cost supplement)	
Low-cost supplement	Application	Reference
Colostrum	Established canine kidney monolayer; primary bovine kidney monolayer	64

(*continued*)

TABLE VIII (*Continued*)

Low-cost supplement	Type III (non-defined with low-cost supplement)	Reference
	Application	
Albumin	Established human lymphoblastoid suspension	72
Peptone		
Pluronic F68		
Milk	Established rat fibroblast suspension	120
Insulin	Established insect line	93
Lactalbumin hydrolysate		
yeast extract		
Egg yolk emulsion	Established insect line	112

a Epidermal growth factor and platelet-derived growth factor are discussed in Section 10.1.

8. SERUM-FREE MEDIA: GENERAL FEATURES

Serum-free media have been reviewed by Higuchi (50), Katsuta and Ta-kaoka (58), Keay (60), Waymouth (130) and Ham and McKeehan (48).

These media may be arbitrarily divided into three groups according to the way in which the basal composition (as discussed in Section 3) is supplemented (Table IX). Only one group (type I) is regarded as completely chemically defined (within the limits imposed by the purities of the substances used). Low-molecular-weight components and "defined" polymers (Section 6.1) which can carry trace elements are included in type I media. Such media are relatively cheap but limited in application to a few established cell lines (Section 9). Type II media are distinguished from type I by the inclusion of purified proteins, hormones and lipids. The definition of the medium is decreased (since proteins can carry lipids and vitamins as well as trace elements) and costs are increased. The majority of serum-free media come into the type II category and are discussed in Section 10. Type III media differ from the others in that low cost and simplicity, rather than definition, are the objectives. Autoclavable serum substitutes have the particular advantage, for large-scale industrial applications, of eliminating potential infections (viral and mycoplasmal) associated with serum.

At the time of writing, serum-free media are not generally used for routine cell culture because of high cost, lack of commercial availability and differences between cell lines in nutritional requirements.

TABLE IX Classification of Serum-Free Media

Type	Degree of definition	Advantages	Criticisms
I	No protein; all components of known molecular structure and configuration	"Completely" chemically defined	Application limited to a few established cell lines
II	Purified proteins and hormones (i.e. insulin, albumin, transferrin) included	Wide application: diploid, primary and established lines	Proteins may contain trace elements, lipids, vitamins; relatively expensive
III	Partial—serum replaced by non-defined biological supplements (such as peptone)	Low cost; many are autoclavable	Lack of definition

9. SIMPLE SERUM-FREE MEDIA FOR ESTABLISHED
CELL LINES

It has been possible for many years to grow a limited number of established cell lines in simple protein-free chemically defined (type I) media. Examples were developed by Higuchi and Robinson (51) and Birch and Pirt (14). Recent type I media are listed in Table VIII. The simplest formulations are for established lines which have no lipid requirements in the absence of serum. The medium of Niwa *et al.* (99) consists of MEM plus biotin supplemented at 0.1% v/v with the polymer polyvinylpyrrolidone. Cleveland *et al.* (27) have reported a protein-free medium supporting attached growth and antibody production for eight mouse or rat–mouse hybridomas. Their medium, which contains 22 trace elements, would not, however, support low-density cultures or stationary suspension cultures.

The type I media listed in Table VIII for insects are distinguished by the presence of the vitamin carnitine (a growth factor for some insects; carnitine can be synthesised by vertebrates). The medium of Becker and Landureau (7), which contains nucleotide precursors and an optimised combination of B vitamins, supported 10–12 subcultures of *Periplaneta americana*. The complex medium of Wilkie *et al.* contains lipids (stabilised by ethanol and Tween) and supports continued passage of *Spodoptera frugiperda* and five passages of *Aedes aegypti* and *Anopheles gambiae*.

Some simple media have proved to be useful when supplemented with cheap non-defined components other than serum. These are listed in Table VIII under type III media and in Section 5.2. The use of peptones as serum substitutes was reviewed by Keay (60) and Rutzky (113). Use of peptones led to completely autoclavable media for cultivation of cells used for vaccine production (59).

Bovine colostrum and milk were reported to support long-term culture of epithelial cell monolayers (64) and transformed fibroblast lines in suspension (120) in serum-free media.

A type III medium for *Drosophila melanogaster*, containing lactalbumin hydrolysate and yeast extract was reported by Mosna (93). Roder (112) described continuous passage of lepidopteran cells (and replication of viruses) in a low-cost serum-free medium containing 1% egg yolk emulsion.

10. SPECIALISED SERUM-FREE MEDIA

10.1. Introduction

Specialised type II (protein-containing) serum-free media have been reviewed by Barnes and Sato (5) and Sato *et al.* (114). We only have space to

comment on human diploid fibroblasts, hybridomas and lymphoblastoid
lines; recent type II media for these cell types are listed in Table VIII.
Serum-free media for many other established and primary lines have been
published in such journals as *In Vitro, Experimental Cell Research, Cell
Biology International Reports, Journal of Cellular Physiology, Journal of
Immunological Methods* and *Developments in Biological Standardization*.

The purified proteins (e.g. albumin, transferrin, insulin, casein, fibronectin) or peptides included in these media provide a range of functions associated with serum in media (nos. 1, 2 and 4–7 in Table VI). Albumin (included
in many type II media at concentrations up to 5 g/litre) is known to provide
many of these functions. It can bind toxins such as excess trace elements
(56), fatty acids (139) or hydrogen peroxide (31). Albumin has a protective
(mechanical) effect and carries essential nutrients such as fatty acids, trace
elements and hormones. In some applications casein can substitute for albumin (30).

Transferrin is used in type II media at 1–100 μg/ml as an iron transport
protein, but may have other functions *in vitro* (such as chelation of toxic
trace metals).

Insulin is included in most type II media [but not that of Iscove and
Melchers (55)] at about 1 μM. Insulin deficiency in HeLa cells was demonstrated by Blaker *et al.* (17) to occur after 13 generations in serum-free
medium.

Many type II media have included polypeptide growth factors such as
epidermal growth factor (EGF) and platelet-derived growth factor (PDGF).
Polypeptide growth factors have been isolated from various non-serum
sources (41) and shown to stimulate division of a number of cell types.

Fibronectin is discussed in Section 6.2.

Attempts to replace proteins and especially albumin in type II defined
media have centred on defining lipid requirements, trace element requirements and toxic effects for each individual cell type and system.

Lipids, in addition to utilisation as an energy source, are likely to be
specifically required for incorporation into cell membranes and as precursors
for prostaglandin synthesis. Candidate essential lipids are the fatty acids
oleic (18:1n-9), linoleic (18:2n-6), linolenic (18:3n-3) and arachidonic
(20:4n-6), also cholesterol, cholesterol precursors or phospholipids. Specific
requirements for oleic acid, cholesterol or linoleic acid have been reported
for some cell types (58a, 63, 81).

Traditional techniques for solubilising lipids in culture medium were discussed in Section 3.6. Liposomes are vesicles formed by sonication of phos-
pholipid-containing mixtures and have been used to incorporate lipids such
as soybean lecithin (phosphatidylcholine) and cholesterol into cell media (36,
55). Liposomes have been used to combine a mixture of lipids with a stabilising antioxidant, vitamin E (8). High-affinity carriers for fatty acids in culture

media include albumin (56, 57, 138), α-cyclodextrin (139) and β-lactoglobulin (118). Use of a non-toxic derivative of a fatty acid [as used by Darfler and Insel (31)] can avoid the use of a carrier. Mammalian cells utilise ethanolamine to synthesise phosphatidylethanolamine, a major constituent of membrane phospholipids. Ethanolamine has been shown to be required for growth of hybridoma cells (94).

It is possible that the 15 trace elements listed in Section 3.6 may be beneficial for cultured animal cells, although specific requirements for most of them will be difficult to demonstrate. For nutritional studies, actual measurement of the trace element content of complete filtered type II media (e.g. 56) is essential. Selenium, since its identification as an essential nutrient for human diploid fibroblasts (80), has been widely used in media. Requirements of cultured cells for copper, manganese, molybdenum and vanadium have been reported (82). Cobalt is often added to media as vitamin B_{12}.

Serum may have a detoxifying function in culture media and it is also possible that some serum substitutes have a similar role. Selenium is an integral part of the enzyme glutathione peroxidase, which destroys lipid peroxides. Selenium may thus function in serum-free media as a component of a detoxifying system. Darfler and Insel (31) have used catalase, which destroys H_2O_2, as a way of eliminating albumin or casein from type II media and suggest that Na_2SeO_3 can, under certain conditions, substitute for catalase.

10.2. Human Diploid Fibroblasts

Minimising fibroblast damage during subculture by trypsinisation is a problem since inhibitors are absent in serum-free media. Ham and co-workers (81) have developed subculture procedures, using minimal treatment with crystalline trypsin at low temperature, which have been used for long-term serial transfer of human diploid fibroblasts in serum-free medium (56). Alternatively, soybean trypsin inhibitor (0.5 mg/ml) may be used when harvesting cells in serum-free medium.

Fibronectin or poly-D-lysine coating of the culture surface (Section 6.2) is important for serum-free growth of human diploid fibroblasts.

Putrescine was reported as a growth factor for human diploid fibroblasts by Pohjanpelto and Raina (107) and is a component of Ham's F12 medium.

Kan and Yamane (56) obtained long-term mass culture (76 population doublings) of human diploid fibroblasts in serum-free medium containing 5 g/litre bovine serum albumin and 10 mg/litre fibronectin. Cell growth was nearly as good (58 population doublings) when albumin was replaced by α-cyclodextrin (139).

Growth of MRC-5 cells on microcarriers in type II serum-free medium has been described (25).

Ham and co-workers have demonstrated how systematic qualitative and quantitative modification of the medium and culture conditions can lead to development of type II serum-free medium for clonal growth of human diploid fibroblasts (8, 47, 81).

10.3. Hybridomas

Serum-free type II media for hybridomas (HB101 and HB102) are commercially available from Hana Biologics Inc.

A simple type II medium (23) with insulin and transferrin, but no lipids, supported long-term growth of a hybridoma line derived from P3X63Ag8. Darfler and Insel's CITTL medium supported static suspension culture of mouse hybridomas derived from NS1 (31).

Sato's group have developed type II media for hybridomas derived from the MPC11 plasmacytoma. Their ITES medium (containing insulin, transferrin, ethanolamine and selenite) supported long-term hybridoma growth in static suspension (94), but for agitated suspension culture, growth was very slow unless soybean or egg yolk phospholipids were added (95). This result was attributed in part to a requirement for linoleic acid as well as ethanolamine in agitated culture. KSLM serum-free medium supporting growth of NS-1 myeloma cells and NS-1 derived hybridomas has been reported (58a).

Fazekas de St. Groth (36) reported serum-free continuous growth and antibody production by 10 mouse and rat–mouse hybridomas in agitated suspension (cytostat). His medium is a modified version of that of Iscove and Melchers (55) and includes albumin and transferrin as well as sonically dispersed soybean lipid and cholesterol. Differences in nutritional requirements between hybridoma lines (even those with the same parent cell line) were demonstrated.

10.4. Lymphoblastoid Lines

Iscove's serum-free medium (55), consisting of DMEM plus albumin, transferrin and soybean lipid, is available commercially. Although originally developed for short-term cultures, it also supports long-term culture of human B-lymphoblastoid lines of normal and malignant origin (96). Uittenbogaart et al. (126) have shown that some human T-lymphoblastoid and B-lymphoblastoid lines can grow in long-term culture in Iscove's medium without the lipids.

Darfler and Insel (30) developed CITTL serum-free medium (DME/F12 plus casein, insulin, transferrin, testosterone and linoleic acid) to support long-term mass growth of a variety of lymphoid cells. For clonal growth (*31*) Na$_2$SeO$_3$ was included, linoleic acid was replaced by dilinoleoylphosphatidylcholine and casein was replaced by catalase (see Section 10.1).

Yamane's group (*115*) report that RITC 55-9 medium (enriched DMEM with transferrin, insulin and bovine serum albumin) supports long-term growth and α-interferon production by human lymphoblastoid cells. This medium contains reducing compounds such as L-cysteine, glutathione and ascorbic acid. Their medium RITC 56-1 (*139*) successfully replaces albumin with α-cyclodextrin for long-term culture of UMCL human lymphoblastoid lines.

11. MEDIUM PREPARATION

11.1. General Principles

Sera and many media are available commercially (e.g. from Flow Laboratories Ltd. and Gibco Laboratories). Serum can be obtained as a filter-sterilised product (see also Section 12), tested for absence of certain viruses and mycoplasma. Heat inactivation (56°C for 30 min) can be used as an added safeguard against mycoplasma and some viruses but it should be noted that this treatment also destroys some growth-enhancing components (*108*). Serum has been sterilised by chemical treatment (*3*).

Commercially supplied liquid medium is convenient for small-scale work. However, powdered medium is less expensive and allows convenient storage of large quantities of stable pre-tested media.

Commercial suppliers can provide data on preparation of standard media. For complete in-house assembly of a medium from individual components, exact details of composition (e.g. salt forms and degree of hydration) are required. Original papers, the *Journal of Tissue Culture Methods* (formerly the TCA manual) or step-by-step guides (e.g., *52, 102*) should be consulted.

The main considerations in medium preparation are solubility of components, freedom from toxicity and stability.

11.1.1. Solubility

A common problem is the formation of a precipitate of calcium phosphate, which is sparingly soluble, particularly in alkaline solutions. Thus during medium preparation, calcium and phosphate should be kept separate until the solution is quite dilute. Tyrosine and cystine are less soluble than the other amino acids. For preparation of concentrated (×100) stock solutions, they may be dissolved in 0.1 N HCl.

11.1.2. Freedom from Toxicity

Water is the largest single constituent of the medium. A proven method of preparation is double distillation from glass or quartz vessels. Alternatively, water which has been purified by passage through an ion exchange resin and activated charcoal connected in series with a glass distillation or reverse osmosis system can be used (e.g. systems from Elga Ltd. and Millipore Ltd.). Best quality (e.g. Analar grade) chemicals should be used. A certificate of analysis for bulk lots of commercially supplied amino acids and vitamin mixtures should be obtained. Hospital preparations of antibiotics, which may contain toxic preservatives or fillers, should be avoided. Where seals or tubing come into contact with medium, silicone rubber or non-toxic medical quality PVC tubing should be used. Gases for tissue culture should be of medical or biological grade from a certified source and be free from contaminants such as nitrous oxide, carbon monoxide and hydrocarbons.

11.1.3. Stability

Antibiotics, once dissolved (except nystatin), and glutamine are the most labile non-protein constituents and are routinely prepared as sterile $\times 100$ concentrates for addition to complete medium. Bicarbonate loses CO_2 on autoclaving, thus is usually sterilised by filtration. Serum should be stored at $-20°C$ or lower and is stable for at least 1 year. Protein supplements such as insulin and other hormones are best stored at $-20°C$ and added to the medium just before use.

For preparation from individual stock solutions, stability should be considered when grouping ingredients; for example ascorbic acid is best stored in solution with cysteine (which is nutritionally interchangeable with cystine), glutathione or other reducing agents. Many lipids, such as linoleic and linolenic acid, are subject to oxidation and should be stored under nitrogen (see also Sections 3.6 and 10).

Stability during use should also be considered. Glutamine and many antibiotics (103) are subject to decomposition at 37°C. Insulin rapidly loses activity in serum-free medium (5). Of the vitamins, thiamine, riboflavin and pyridoxine are the least stable. Thiamine was reported to be 50% oxidised to a biologically inert form in 7 days at 37°C (18). Riboflavin, folic acid and pyridoxine are light-sensitive. Waymouth (131) reported significant loss of growth-supporting ability by a medium exposed for three periods of 4 hr to room light, as compared to a control medium protected from light by wrapping its container in aluminium foil.

11.2. Sterilisation

Techniques for sterilisation are discussed in detail elsewhere (Chapter 5, this volume; 102). There are few completely autoclavable media in common use, although versions of Eagle's media are available commercially which can be autoclaved prior to the addition of one or two labile components. Certain components of standard media may be autoclaved. However, serum or proteins, which are used in many media, cannot be autoclaved. It is therefore often convenient to sterilise the complete medium by filtration.

11.2.1. Autoclaving

Inorganic salts (except bicarbonate) may be prepared as ×20 stock solutions sterilised at 121°C for 15 min. Calcium and magnesium salts should be autoclaved separately from phosphates to avoid precipitation. Methylcellulose (×20), carboxymethylcellulose (×10), Pluronic F68 (×20), silicone antifoams (×1000) and phenol red (water-soluble, ×100) can also be autoclaved. Methylcellulose possesses the unusual property of being less soluble at high temperature than at room (or refrigerator) temperature. Sugars are subject to decomposition when autoclaved in the presence of inorganic salts and organic compounds; the sugars are generally stable when autoclaved in distilled water. Attention should be paid to the quality of steam supplies to autoclaves; in particular, the presence of volatile boiler preservatives (e.g. stripping amines) should be avoided.

11.2.2. Filtration

Pressure filtration (using a syringe, peristaltic pump or pressurised vessel to force medium through a 0.10- or 0.22-μM pore size sterilising membrane) has virtually replaced earlier methods [discussed by Paul (102)]. Disposable filter units, pre-sterilised or autoclavable and specifically designed for sterilising tissue culture media, are available from companies such as Millipore Ltd., Gelman Sciences Ltd., Schleicher and Schull GmbH, Pall Filtration Ltd. and Sartorius GmbH. Disposable units are convenient for small- to medium-scale applications and can be supplied pre-tested for sterility, toxicity and pyrogenicity and complying with standards for medical use. The cost of disposables should be balanced against the costs involved in cleaning, assembling and testing a stainless steel filter housing with replaceable membranes; the latter becomes more cost-effective as scale increases. Cartridge units, with large surface area, are available for large-scale applications.

12. QUALITY CONTROL

A detailed account of quality control measures relevant to media is given as section XIV in Kruse and Patterson (68). The basic physiochemical quality control parameters for media are pH and osmolarity. Tests for sterility of the medium (i.e. lack of contamination by bacteria, fungi, mycoplasmas or viruses) and for ability of the medium to permit cell growth, production and purification procedures required are also advisable.

12.1. pH and Osmolarity

Buffers and pH were discussed in Section 3.7. It is most important that the final pH measurement on the medium (or appropriate sample of the medium) be made at 37°C after sterilisation and under the conditions to be used in practice (including equilibration with the correct gas mixture). Sodium hydroxide and HCl (or CO_2) are normally used for large adjustments in pH; since NaCl is the largest component in the balanced salt solution, the smallest relative change in inorganic ion composition is made.

Osmolarity is a measure of the total osmotic activity contributed by ions and non-ionised molecules in the medium, expressed as milliosmoles per volume of solvent. Osmolality is the same parameter related to weight of solvent. Further details and osmolarities for different media are given by Waymouth (129). The osmolarity may be calculated by using tables of osmotic coefficients (e.g., 111). For quality control purposes, it is preferable to directly measure the osmolarity of a complete medium sample using a suitable osmometer (with an accuracy $>\pm5$ mOsm/litre). If changes in medium recipe require an adjustment to the osmolarity, this may conveniently be achieved by using a solution of sodium chloride. For example, a stock NaCl solution of 1 mg/ml has osmolarity 32 mOsm/ml. The amount required in millilitres is given by:

$$\frac{\text{required osmolarity} - \text{measured osmolarity (mOsm/litre)}}{32}$$

Osmolarity measurements are useful for checking errors in medium preparation, since a particular formulation should be reproducible batch to batch.

The NaCl content of the medium should be adjusted when major changes in medium formulation, which will affect osmolarity, are made.

12.2. Sterility Testing

Sterility testing details are available as part of the quality control specification for commercially supplied media. Sera can be tested before purchase for sterility in parallel with the growth promotion tests described in Section 12.3. Contamination of media prepared in-house may occur through faulty filtration apparatus or technique. For large-scale filtration, routine sterility testing to check for filter failure should include 1% and for exacting work 10% of the total throughout.

12.2.1. Mycoplasmas

Mycoplasmas have a wide range of deleterious effects on cell cultures. Methods of testing for mycoplasmas are given by McGarrity (77), Stanbridge and Schneider (119), and Kihara et al. (62). Mycoplasmas, if present, are likely to be in low concentration in serum. Direct culture, which is the most sensitive detection method, is therefore appropriate. Most species of mycoplasmas can be cultured in a suitable medium. Mycoplasma cultures are somewhat difficult to perform and really require the services of a microbiologist. A large sample of serum or medium (25 ml per 100 ml of test medium) should be used for initial inoculation into mycoplasma enrichment broth. Mycoplasmas present in the broth can be identified microscopically after subculture of the broth onto agar or by a fluorochrome technique after broth co-culture with a test monolayer cell culture. Mycoplasma test and stain kits are available commercially.

12.2.2. Bacteria and Fungi

Contamination of serum and medium by bacteria or fungi can be detected by simple culture methods. A 10.0-ml test sample is sufficient to inoculate several different broth media (e.g. thioglycollate, trypticase soy, Sabouraud dextrose) and spread several types of plate (e.g. blood agar, deoxycholate). Duplicate sets can be inoculated at 37° and 30°C for 14 days. Precipitation or crystals can be distinguished from growth by carrying out a Gram stain and/or attempting to make subcultures.

12.2.3. Viruses

Screening by individual laboratories for viruses in serum is not easy unless specialised facilities are available. Many serum suppliers test their product for contamination by certain viruses and it is possible to request specific test methods and results.

12.3. Testing for Growth Promotion

Pre-testing of serum before purchase for its ability to promote cell growth and product formation and non-interference with purification or assay procedures is necessary. Most commercial suppliers will reserve large amounts of several serum batches for a limited time while aliquots from each batch are tested by the customer. The candidate serum should be tested in parallel with a standard known to promote the function(s) required. Details of individual tests obviously depend on individual requirements; however, general points on testing of sera for growth promotion are:

1. Sufficient length of test (i.e. multiple passages).

2. Quantitative measurement [ideally growth rate and cell yield per unit amount of serum. The growth yield for serum can vary between batches, as shown by Lambert and Pirt (70)].

3. Cells, growth substrate and media used for testing must be identical to those to be used in practice.

Simple growth promotion tests are described by McGarrity (77) and Wolf (137). For definition of growth yield, see Section 4.

REFERENCES

1. Arathoon, W. R., and Telling, R. C. (1982). Uptake of amino acids and glucose by BHK 21 clone 13 suspension cells during cell growth. *Dev. Biol. Stand.* **50**, 145–154.

2. Astaldi, G. C. B. (1983). Use of human endothelial culture supernatant (HECS) as a growth factor for hybridomas. *In* "Methods in Enzymology" (J. J. Langone and H. Van Vunakis, eds.), Vol. 92, pp. 39–46. Academic Press, New York.

3. Bahnemann, H. G. (1976). Inactivation of viruses in serum with binary ethyleneimine. *J. Clin. Microbiol.* **3**, 209–210.

4. Barile, N. F., and Kern, J. (1971). Isolation of *Mycoplasma argini* from commercial bovine sera and its implication in contaminated cell cultures. *Proc. Soc. Exp. Biol. Med.* **138**, 432–437.

5. Barnes, D., and Sato, G. (1980). Serum-free culture: a unifying approach. *Cell* **22**, 649–655.

6. Baugh, C. L., and Tytell, A. A. (1967). Propagation of the human diploid cell WI-38 in galactose medium. *Life Sci.* **6**, 371–380.

7. Becker, J., and Landureau, J. C. (1981). Specific vitamin requirements of insect cell lines (*P. americana*) according to their tissue origin and in vitro conditions. *In Vitro* **17**, 471–479.

8. Bettger, W. J., and Ham, R. G. (1982). The crucial role of lipids in supporting clonal growth of human diploid fibroblasts in a defined medium. *In* "Growth of Cells in Hormonally Defined Media" (G. H. Sato, A. B. Pardee, and D. A. Sirbasku, eds.), pp. 61–64. Cold Spring Harbor Lab., Cold Spring Harbor, New York.

9. Birch, J. R. (1980). The role of serum in the culture of a human lymphoblastoid cell line. *Dev. Biol. Stand.* **46**, 21–27.

10. Birch, J. R., Cramer, F., Edwards, D. J., Cartwright, T., and Gould, H. (1978). Serum requirement for growth and immunoglobulin synthesis by cultured human lymphocyte cell lines. *Dev. Biol. Stand.* **42**, 165–169.

11. Birch, J. R., and Edwards, D. J. (1980). The effect of pH on the growth and carbohydrate metabolism of a lymphoblastoid cell line. *Dev. Biol. Stand.* **46**, 59–63.

12. Birch, J. R., and Hopkins, D. W. (1977). The serine and glycine requirements of cultured human lymphocyte lines. *In* "Cell Culture and its Application" (R. T. Acton and J. D. Lynn, eds.), pp. 503–511. Academic Press, New York.

13. Birch, J. R., and Pirt, S. J. (1969). The choline and serum protein requirements of mouse fibroblast cells (strain LS) in culture. *J. Cell Sci.* **5**, 135–142.

14. Birch, J. R., and Pirt, S. J. (1970). Improvements in a chemically defined medium for the growth of mouse cells (strain LS) in suspension. *J. Cell Sci.* **7**, 661–670.

15. Birch, J. R., and Pirt, S. J. (1971). The quantitative glucose and mineral nutrient requirements of mouse LS (suspension) cells in chemically defined medium. *J. Cell Sci.* **8,** 693–700.

16. Blaker, G. J. (1971). The vitamin nutrition of mammalian cells in culture. Ph.D. Thesis, University of London.

17. Blaker, G. J., Birch, J. R., and Pirt, S. J. (1971). The glucose insulin and glutamine requirements of suspension cultures of HeLa cells in a defined culture medium. *J. Cell Sci.* **9,** 529–537.

18. Blaker, G. J., and Pirt, S. J. (1971). The uptake of vitamins by mouse fibroblast cells (strain LS) during growth in a chemically defined medium. *J. Cell Sci.* **8,** 701–708.

19. Boraston, R., Thompson, P. W., Garland, S., and Birch, J. R. (1984). Growth and oxygen requirements of antibody producing mouse hybridoma cells in suspension culture. *Dev. Biol. Stand.* **55,** 103–111.

20. Butler, M., and Thilly, W. G. (1982). MDCK microcarrier cultures: Seeding density effects and amino acid utilisation. *In Vitro* **18,** 213–219.

21. Ceccarini, C. (1975). Effect of pH on plating efficiency, serum requirement and incorporation of radioactive precursors into human cells. *In Vitro* **11,** 78–86.

22. Chalifour, L. E., and Dakshinamurti, K. (1982). The biotin requirement of human fibroblasts in culture. *Biochem. Biophys. Res. Commun.* **104,** 1047–1053.

23. Chang, T. H., Steplewski, Z., and Koprowski, H. (1980). Production of monoclonal antibodies in serum-free medium. *J. Immunol. Methods* **39,** 369–375.

24. Chen, H. W., and Kandutsch, A. A. (1981). Cholesterol requirements for cell growth: Endogenous synthesis versus exogenous sources. *In* "The Growth Requirements of Vertebrate Cells in Vitro" (C. Waymouth, R. G. Ham, and P. J. Chapple, eds.), pp. 327–342. Cambridge Univ. Press, London and New York.

25. Clark, J. M., Gebb, C., and Hirtenstein, M. D. (1982). Serum-supplements and serum-free media: Applicability for microcarrier culture of animal cells. *Dev. Biol. Stand.* **50,** 81–91.

26. Clark, J. M., and Hirtenstein, M. D. (1981). High yield culture of human fibroblasts on microcarriers: A first step in production of fibroblast-derived interferon (human beta interferon). *J. Interferon Res.* **1,** 391–400.

27. Cleveland, W. L., Wood, I., and Erlanger, B. L. (1983). Routine large-scale production of monoclonal antibodies in a protein-free culture medium. *J. Immunol. Methods* **56,** 221–234.

28. Cristofalo, V. J., and Kritchevsky, D. (1965). Growth and glycolysis in the human diploid cell strain WI-38. *Proc. Soc. Exp. Biol. Med.* **118,** 1109–1113.

29. Darfler, F. J., Murakami, H., and Insel, P. A. (1980). Growth of T-lymphoma cells in serum-free medium: Lack of involvement of the cyclic AMP pathway in long-term cultures. *Proc. Natl. Acad. Sci. U.S.A.* **77,** 5993–5997.

30. Darfler, F. J., and Insel, P. A. (1982). Growth of normal and transformed lymphoid cells and hybridomas in serum-free medium. *In* "Growth of Cells in Hormonally Defined Media" (G. H. Sato, A. B. Pardee, and D. A. Sirbasku, eds.), pp. 717–726. Cold Spring Harbor Lab., Cold Spring Harbor, New York.

31. Darfler, F. J., and Insel, P. A. (1983). Clonal growth of lymphoid cells in serum-free media requires elimination of H_2O_2 toxicity. *J. Cell. Physiol.* **115,** 31–36.

32. Dulbecco, R., and Freeman, G. (1959). Plaque production by the polyoma virus. *Virology* **8,** 396–397.

33. Eagle, H. (1959). Amino acid metabolism in mammalian cell cultures. *Science* **130,** 432–437.

34. Eagle, H. (1971). Buffer combinations for mammalian cell culture. *Science* **174,** 500–503.

35. Eagle, H., Barban, S., Levy, M., and Schulze, H. O. (1958). The utilization of carbohydrates by human cell cultures. *J. Biol. Chem.* **235**, 551–557.
36. Fazekas de St. Groth, S. (1983). Automated production of monoclonal antibodies in a cytostat. *J. Immunol. Methods* **57**, 121–136.
37. Fleischaker, R. J., Jr., and Sinskey, A. J. (1981). Oxygen demand and supply in cell culture. *Appl. Microbiol. Biotechnol.* **12**, 193–197.
38. Glassman, A. B., Rydzewski, R. S., and Bennett, C. E. (1980). Trace metal levels in commercially prepared tissue culture media. *Tissue Cell* **12**, 613–617.
39. Goetz, I. E., Moklebust, R., and Warren, C. J. (1979). Effect of some antibiotics on the growth of human diploid skin fibroblasts in cell culture. *In Vitro* **15**, 114–119.
40. Gospodarowicz, D., and Ill, C. R. (1980). Do plasma and serum have different abilities to promote cell growth? *Proc. Natl. Acad. Sci. U.S.A.* **77**, 2726–2730.
41. Gospodarowicz, D., and Moran, J. S. (1976). Growth factors in mammalian cell culture. *Annu. Rev. Biochem.* **45**, 531–558.
42. Gregg, C. (1972). Some aspects of the energy metabolism of mammalian cells. *In* "Growth, Nutrition and Metabolism of Cells in Culture" (G. H. Rothblat and V. J. Cristofalo, eds.), Vol. 1, pp. 83–135. Academic Press, New York.
43. Griffiths, J. B. (1970). The quantitative utilisation of amino acids and glucose and contact inhibition of growth in cultures of the human diploid cell WI-38. *J. Cell Sci.* **6**, 739–749.
44. Griffiths, J. B. (1973). The effects of adapting human diploid cells to grow in glutamic acid media on cell morphology, growth and metabolism. *J. Cell Sci.* **12**, 612–629.
45. Griffiths, J. B., and Pirt, S. J. (1967). The uptake of amino acids by mouse cells (strain LS) during growth in batch culture and chemostat culture: The influence of cell growth rate. *Proc. R. Soc. London, Ser. B* **168**, 421–438.
46. Ham, R. G. (1965). Clonal growth of mammalian cells in a chemically defined synthetic medium. *Proc. Natl. Acad. Sci. U.S.A.* **53**, 288–293.
47. Ham, R. G. (1974). Nutritional requirements of primary culture. A neglected problem of modern biology. *In Vitro* **10**, 119–129.
48. Ham, R. G., and McKeehan, W. L. (1979). Media and growth requirements. *In* "Methods in Enzymology" (W. B. Jakoby and I. H. Pastan, eds.), Vol. 58, pp. 44–93. Academic Press, New York.
49. Hayman, E. G., and Ruoslahti, E. (1979). Distribution of fetal bovine serum fibronectin and endogenous rat cell fibronectin in extracellular matrix. *J. Cell Biol.* **83**, 255–259.
50. Higuchi, K. (1973). Cultivation of animal cells in chemically defined media, a review. *Adv. Appl. Microbiol.* **16**, 111–136.
51. Higuchi, K., and Robinson, R. C. (1973). Studies on cultivation of mammalian cell lines in a serum-free, chemically defined medium. *In Vitro* **9**, 114–121.
52. Hink, W. F. (1979). Cell lines from invertebrates. *In* "Methods in Enzymology" (W. B. Jakoby and I. H. Pastan, eds.), Vol. 58, pp. 450–465. Academic Press, New York.
53. Hutner, S. H. (1972). Inorganic nutrition. *Annu. Rev. Microbiol.* **26**, 313–346.
54. Imamura, T., Crespi, C. L., and Brunengraber, H. (1980). Utilization of carbohydrate by Madin–Darby canine kidney (MDCK) cells in high density suspension of dextran microcarriers. *Fed. Proc., Fed. Am. Soc. Exp. Biol.* **39**, 2145p.
55. Iscove, N. N., and Melchers, F. (1978). Complete replacement of serum by albumin, transferrin and soybean lipid in cultures of lipopolysaccharide reactive B-lymphocytes. *J. Exp. Med.* **147**, 923–933.
56. Kan, M., and Yamane, I. (1982). In vitro proliferation and lifespan of human diploid fibroblasts in serum-free BSA-containing medium. *J. Cell Physiol.* **111**, 155–162.
57. Kan, M., and Yamane, I. (1982). Effects of sulphydryl groups and oxygen tension on the cell proliferating activity of bovine serum albumin in culture. *Cell Struct. Func.* **7**, 133–143.

58. Katsuta, H., and Takaoka, T. (1973). Cultivation of cells in protein and lipid-free synthetic media. *Methods Cell Biol.* **6**, 1–42.
58a. Kawamoto, T., Sato, J. D., Le, A., McClure, D. B., and Sato, G. H. (1983). Development of a serum-free medium for growth of NS-1 mouse myeloma cells and its application to the isolation of NS-1 hybridomas. *Anal. Biochem.* **130**, 445–453.
59. Keay, L. (1977). The growth of L-cells and Vero cells on an autoclavable MEM–peptone medium. *Biotechnol. Bioeng.* **19**, 399–411.
60. Keay, L. (1978). The cultivation of animal cells and production of viruses in serum-free systems. *Methods Cell Biol.* **20**, 169–209.
61. Kelley, W. N. (1972). Purine and pyrimidine metabolism of cells in culture. In "Growth, Nutrition and Metabolism of Cells in Culture" (G. H. Rothblat and V. J. Cristofalo, eds.), Vol. 1, pp. 211–255. Academic Press, New York.
62. Kihara, K., Ishida, A., and Okumura, W. (1981). Detection of mycoplasmal contaminants in sera. *J. Biol. Stand.* **9**, 243–251.
63. King, M. E., and Spector, A. A. (1981). Lipid metabolism in cultured cells. In "The Growth Requirements of Vertebrate Cells In Vitro" (C. Waymouth, R. G. Ham, and P. J. Chapple, eds.), pp. 293–312. Cambridge Univ. Press, London and New York.
64. Klagsbrun, M. (1980). Bovine colostrum supports the serum-free proliferation of epithelial cells but not of fibroblasts in long-term culture. *J. Cell Biol.* **84**, 808–814.
65. Kleinman, H. K., Klebe, R. J., and Martin, G. R. (1981). Role of collagenous matrices in the adhesion and growth of cells. *J. Cell Biol.* **88**, 473–485.
66. Kniazeff, A. J., Wopschall, L. J., Hopps, H. E., and Morris, C. S. (1975). Detection of bovine viruses in fetal bovine serum used in cell culture. *In Vitro* **11**, 400–403.
67. Krömer, E., Scheirer, W., and Katinger, H.W. D. (1982). Hormone supplemented cell growth media requiring only 2% foetal calf serum for reduced cost of routine cell culture. *Dev. Biol. Stand.* **50**, 355–359.
68. Kruse, P. F., Jr., and Patterson, M. K., eds. (1973). "Tissue Culture: Methods and Applications." Academic Press, New York.
69. Kuchler, R. J., Marlow, M. L., and Merchant, D. J. (1960). The mechanism of cell binding and cell-sheet formation in L-strain fibroblasts. *Exp. Cell Res.* **20**, 428–437.
70. Lambert, K., and Pirt, S. J. (1975). The quantitative requirements of human diploid cells (strain MRC-5) for amino acids, vitamins and serum. *J. Cell Sci.* **17**, 397–411.
71. Lambert, K., and Pirt, S. J. (1979). Growth of human diploid cells (strain MRC-5) in defined medium: Replacement of serum by a fraction of serum ultrafiltrate. *J. Cell Sci.* **35**, 381–392.
72. Lazar, A., Reuveny, S., Traub, A., Reuveny, S., Traub, A., Minai, M., Grosfeld, H., Feinstein, S., Gez, M., and Mizrahi, A. (1982). Factors affecting the large scale production of human lymphoblastoid interferon. *Dev. Biol. Stand.* **50**, 167–171.
73. Leibovitz, A. (1963). The growth and maintenance of tissue cell cultures in free gas exchange with the atmosphere. *Am. J. Hyg.* **78**, 173–180.
74. Litwin, J. (1979). A survey of various media and growth factors used in cell cultivation. *Dev. Biol. Stand.* **42**, 37–45.
75. Lotan, R. (1980). Effect of vitamin A and its analogs (retinoids) on normal and neoplastic cells. *Biochim. Biophys. Acta* **605**, 33–91.
76. Lynch, R. D., Schneeberger, E. E., and Geyer, R. P. (1979). Alterations in L-fibroblast lipid metabolism and morphology during choline deprivation. *Exp. Cell Res.* **122**, 103–113.
77. McGarrity, G. J. (1979). Detection of contamination. In "Methods in Enzymology" (W. B. Jakoby and I. H. Pastan, eds.), Vol. 58, pp. 18–28. Academic Press, New York.
78. McKeehan, W. L. (1982). Glycolysis, glutaminolysis and cell proliferation. *Cell Biol. Int. Rep.* **6**, 635–650.

79. McKeehan, W. L., and Ham, R. G. (1976). Stimulation of clonal growth of normal fibroblasts with substrata coated with basic polymers. *J. Cell Biol.* **71**, 727–734.
80. McKeehan, W. L., Hamilton, W. G., and Ham, R. G. (1976). Selenium is an essential trace nutrient for growth of WI-38 diploid human fibroblasts. *Proc. Natl. Acad. Sci. U.S.A.* **73**, 2023–2027.
81. McKeehan, W. L., Hamilton, W. G., and Ham, R. G. (1981). The relationship between defined low molecular weight substances and undefined serum derived factors in the multiplication of untransformed fibroblasts. *In* "The Growth Requirements of Vertebrate Cells In Vitro" (C. Waymouth, R. G. Ham, and P. J. Chapple, eds.), pp. 223–243. Cambridge Univ. Press, London and New York.
82. McKeehan, W. L., McKeehan, K. A., Hammond, S. L., and Ham, R. G. (1977). Improved medium for clonal growth of human diploid fibroblasts at low concentrations of serum protein. *In Vitro* **13**, 399–416.
83. McLimans, W. F. (1972). The gaseous environment of the mammalian cell in culture. *In* "Growth, Nutrition and Metabolism of Cells in Culture" (G. H. Rothblat and V. J. Cristofalo, eds.), Vol. 1, pp. 137–170. Academic Press, New York.
84. McLimans, W. F., Crouse, E. J., Tunnah, K. V., and Moore, G. E. (1968). Kinetics of gas diffusion in mammalian cell culture systems. I. *Exp. Biotechnol. Bioeng.* **10**, 725–740.
85. Merrill, C. R., Friedman, T. B., Attallah, A. F., Geier, M. R., Kreil, K., and Yarkin, R. (1972). Isolation of bacteriophages from commercial sera. *In Vitro* **8**, 91–93.
86. Millam-Stanley, M. S. (1972). Cultivation of arthropod cells. *In* "Growth, Nutrition and Metabolism of Cells in Culture" (G. H. Rothblat and V. J. Cristofalo, eds.), Vol. 2, pp. 327–371, Academic Press, New York.
87. Miltenberger, H. G., and David, P. (1980). Mass production of insect cells in suspension. *Dev. Biol. Stand.* **46**, 183–186.
88. Moore, G. E., Gerner, R. E., and Franklin, H. A. (1967). Culture of normal human leucocytes. *JAMA, J. Am. Med. Assoc.* **199**, 519–524.
89. Morgan, J. F., Morton, H. J., and Parker, R. C. (1950). Nutrition of animal cells in tissue culture. I. Initial studies on a synthetic medium. *Proc. Soc. Exp. Biol. Med.* **73**, 1–8.
90. Morgan, M. J., and Faik, P. (1981). Carbohydrate metabolism in cultured animal cells. *Biosci. Rep.* **1**, 669–682.
91. Morton, H. J. (1981). How to prepare aqueous media containing fat-soluble substances. *In* "The Growth Requirements of Vertebrate Cells In Vitro" (C. Waymouth, R. G. Ham, and P. J. Chapple, eds.), pp. 353–357. Cambridge Univ. Press, London and New York.
92. Moskowitz, M., Cheng, D. K. S., Moscatello, D. K., and Otsuka, H. (1980). Growth stimulation of BHK cells in culture by free biotin in serum. *JNCI, J. Natl. Cancer Inst.* **64**, 639–644.
93. Mosna, G. (1981). Insulin can completely replace serum in *Drosophila melanogaster* cell cultures in vitro. *Experientia* **37**, 466–467.
94. Murakami, H., Masui, H., Sato, G. H., Sneko, N., Chow, T. P., and Kano-Sueoka, T. (1982). Growth of hybridoma cells in serum-free medium: Ethanolamine is an essential component. *Proc. Natl. Acad. Sci. U.S.A.* **79**, 1158–1162.
95. Murakami, H., Masui, H., and Sato, G. H. (1982). Suspension culture of hybridoma cells in serum-free medium: Soybean phospholipids as the essential components. *In* "Growth of Cells in Hormonally Defined Media" (G. H. Sato, A. B. Pardee, and D. A. Sirbasku, eds.), pp. 711–715. Cold Spring Harbor Lab., Cold Spring Harbor, New York.
96. Muzik, H., Shea, M. E., Lin, C. C., Jamro, H., Cassol, S., Jerry, L. M., and Bryant, L. (1982). Adaption of human long-term B lymphoblastoid cell lines to chemically defined serum-free media. *In Vitro* **18**, 515–524.
97. Namba, M., Fukushima, P., and Kimoto, T. (1982). Effects of feeder layers made of

human, mouse, hamster and rat cells on the cloning efficiency of transformed human cells. *In Vitro* **18**, 469–475.

98. Nielsen, F. H. (1981). Consideration of trace element requirements for preparation of chemically defined media. *In* "The Growth Requirements of Vertebrate Cells In Vitro" (C. Waymouth, R. G. Ham, and P. J. Chapple, eds.), pp. 68–81. Cambridge Univ. Press, London and New York.

99. Niwa, A., Yamamoto, K., Sorimachi, K., and Yasumura, Y. (1980). Continuous culture of Reuber hepatoma cells in serum-free, arginine-, glutamine- and tyrosine-deprived chemically defined medium. *In Vitro* **16**, 987–993.

100. Okuda, A., Kajiwara, Y., and Kimura, G. (1983). Difference in growth factor requirements of rat 3Y1 cells among growth in mass culture, clonal growth in low density culture and stimulation to enter S phase in resting culture. *In Vitro* **19**, 376–384.

101. Patterson, M. K. (1972). Uptake and utilization of amino acids by cell cultures. *In* "Growth, Nutrition and Metabolism of Cells in Culture" (G. H. Rothblat and V. J. Cristofalo, eds.), Vol. 1, pp. 171–209. Academic Press, New York.

102. Paul, J. (1975). "Cell and Tissue Culture," 5th ed. Livingstone, Edinburgh and London.

103. Perlman, D. (1979). Use of antibiotics in cell culture media. *In* "Methods in Enzymology" (W. B. Jakoby and I. H. Pastan, eds.), Vol. 58, pp. 110–116. Academic Press, New York.

104. Pharmacia Fine Chemicals (1981). "Microcarrier Cell Culture: Principles and Methods." Pharmacia Fine Chemicals AB, Uppsala, Sweden.

105. Phillips, P. D., and Cristofalo, V. J. (1981). Growth regulation of WI-38 cells in a serum-free medium. *Exp. Cell Res.* **134**, 292–302.

106. Pirt, S. J. (1975). "Principles of Microbe and Cell Cultivation." Blackwell, Oxford.

107. Pohjanpelto, P., and Raina, A. (1972). Identification of a growth factor produced by human fibroblasts in vitro as putrescine. *Nature (London) New Biol.* **235**, 247–249.

108. Price, P. J., and Gregory, E. A. (1982). Relationship between in vitro growth promotion and biophysical and biochemical properties of the serum supplement. *In Vitro* **18**, 576–584.

109. Reitzer, L. J., Wice, M. B., and Kennell, D. (1979). Evidence that glutamine, not sugar, is the major energy source for cultured HeLa cells. *J. Biol. Chem.* **256**, 2669–2676.

110. Roberts, R. S., Hsu, H. W., Lin, K. D., and Young, T. D. (1976). Amino acid metabolism of myeloma cells in culture. *J. Cell Sci.* **21**, 609–615.

111. Robinson, R. A., and Stokes, R. H. (1959). "Electrolyte Solutions." Butterworth, London.

112. Roder, A. (1982). Development of a serum free medium for cultivation of insect cells. *Naturwissenschaften* **69**, 92–93.

113. Rutzky, L. P. (1981). Peptone growth factors for serial cell proliferation in the absence of serum. *In* "The Growth Requirements of Vertebrate Cells In Vitro" (C. Waymouth, R. G. Ham, and P. J. Chapple, eds.), pp. 277–292. Cambridge Univ. Press, London and New York.

114. Sato, G. H., Pardee, A.B., and Sirbasku, D. A., eds. (1982). "Growth of Cells in Hormonally Defined Media," Cold Spring Harbor Conf. Cell Proliferation, Vol. 9. Cold Spring Harbor Lab., Cold Spring Harbor, New York.

115. Sato, T., Minamoto, Y., Yamane, I., Kudo, T., and Tachibana, T. (1982). Spontaneous interferon production and growth of lymphoblastoid cells in serum-free medium. *Exp. Cell Res.* **138**, 127–134.

116. Schröder, E. W., Rapaport, E., Kabcenell, A. K., and Black, P. H. (1982). Growth inhibitory and stimulatory effects of retinoic acid on murine 3T3 cells. *Proc. Natl. Acad. Sci. U.S.A.* **79**, 1549–1552.

117. Smith, J. R. (1981). The fat-soluble vitamins. *In* "The Growth Requirements of Vertebrate Cells In Vitro" (C. Waymouth, R. G. Ham, and P. J. Chapple, eds.), pp. 343–352. Cambridge Univ. Press, London and New York.

118. Spieker-Polet, H., and Polet, H. (1981). Requirement of a combination of a saturated and an unsaturated free fatty acid and a fatty acid carrier protein for in vitro growth of lymphocytes. *J. Immunol.* **126**, 949–954.

119. Stanbridge, E. J., and Schneider, E. L. (1977). The need for non-cultural methods for the detection of mycoplasma contaminants. *Dev. Biol. Stand.* **37**, 191–200.

120. Steimer, K. S., and Klagsbrun, M. (1981). Serum-free growth of normal and transformed fibroblasts in milk: Differential requirements for fibronectin. *J. Cell Biol.* **88**, 294–300.

121. Swim, H. E., and Parker, R. F. (1968). Effect of pluronic F68 on growth of fibroblasts in suspension on a rotary shaker. *Proc. Soc. Exp. Biol. Med.* **103**, 252–254.

122. Taylor, W. G., Camalier, R. F., and Sanford, K. K. (1978). Density dependent effects of oxygen on the growth of mammalian fibroblasts in culture. *J. Cell. Physiol.* **95**, 33–40.

123. Taylor, W. G., Richter, A., Evans, V. J., and Sanford, K. K. (1974). Influence of oxygen and pH on plating efficiency and colony development of WI-38 and Vero cells. *Exp. Cell Res.* **86**, 152–156.

124. Thomas, J. A., and Johnson, M. J. (1967). Trace metal requirements of NCTC clone 929 strain L cells. *J. Natl. Cancer Inst. (U.S.)* **39**, 337–345.

125. Tritsch, G. L., and Moore, G. E. (1962). Spontaneous decomposition of glutamine in cell culture media. *Exp. Cell Res.* **28**, 360–364.

126. Uittenbogaart, C. H., Cantor, Y., and Fahey, J. L. (1983). Growth of human malignant lymphoid cell lines in serum-free medium. *In Vitro* **19**, 67–72.

127. Vaughn, J. L. (1971). Cell culture media and methods. *In* "Invertebrate Tissue Culture" (G. Vago, ed.), Vol. 1, pp. 4–40. Academic Press, New York.

128. Walthall, B., and Ham, R. G. (1981). Multiplication of human diploid fibroblasts in a synthetic medium supplemented with EGF, insulin and dexamethasone. *Exp. Cell Res.* **134**, 303–311.

129. Waymouth, C. (1973). Determination and survey of osmolality in culture media. *In* "Tissue Culture: Methods and Applications" (P. F. Kruse, Jr., and M. K. Patterson, eds.), pp. 703–708. Academic Press, New York.

130. Waymouth, C. (1981). Requirements for serum-free growth of cells: comparison of currently available defined media. *In* "The Growth Requirements of Vertebrate Cells in Vitro" (C. Waymouth, R. G. Ham, and P. J. Chapple, eds.), pp. 33–47. Cambridge Univ. Press, London and New York.

131. Waymouth, C. (1981). Major ions, buffer systems, pH, osmolarity and water quality. *In* "The Growth Requirements of Vertebrate Cells in Vitro" (C. Waymouth, R. G. Ham, and P. J. Chapple, eds.), pp. 105–117. Cambridge Univ. Press, London and New York.

132. Waymouth, C., Ham, R. G., and Chapple, P. J., eds. (1981). "The Growth Requirements of Vertebrate Cells in Vitro." Cambridge Univ. Press, London and New York.

133. Weinstein, R., Hoover, G. A., Majure, J. *et al.* (1982). Growth of human foreskin fibroblasts in a serum-free defined medium without platelet-derived growth factor. *J. Cell Physiol.* **110**, 23–28.

134. Weiss, S. A., Smith, G. C., Kalter, S. S., and Vaughn, J. L. (1981). Improved method for the production of insect cell cultures in large volume. *In Vitro* **17**, 495–502.

135. Wice, B. M., Reitzer, L. J., and Kennell, D. (1981). The continuous growth of vertebrate cells in the absence of sugar. *J. Biol. Chem.* **256**, 7812–7819.

136. Wilkie, G. E. I., Stockdale, H., and Pirt, S. J. (1980). Chemically defined media for production of insect cells and viruses in vitro. *Dev. Biol. Stand.* **46**, 29–37.

137. Wolf, K. (1979). Laboratory management of cell cultures. *In* "Methods in Enzymology" (W. B. Jakoby and I. H. Pastan, eds.), Vol. 58, pp. 116–118. Academic Press, New York.

138. Yamane, I., Kan, M., Hoshi, H., and Minamoto, Y. (1981). Primary culture of human diploid cells and its long term transfer in a serum free medium. *Exp. Cell Res.* **134**, 470–474.

139. Yamane, I., Kan, M., Minamoto, Y., and Amatsuji, Y. (1982). Alpha-cyclodextrin: A partial substitute for bovine serum albumin in serum-free culture of mammalian cells. *In* "Growth of Cells in Hormonally Defined Media" (G. H. Sato, A. B. Pardee, and D. A. Sirbasku, eds.), pp. 87–92. Cold Spring Harbor Lab., Cold Spring Harbor, New York.

140. Zielke, H. R., Sumbilla, C. M., Sevdalian, D. A., Hawkins, R. L., and Ozand, P. T. (1980). Lactate: A major product of glutamine metabolism by human diploid fibroblasts. *J. Cell. Physiol.* **104**, 433–441.

141. Zielke, H. R., Ozand, P. T., Tildon, J. T., Sevdalian, D. A., and Cornblath, M. (1976). Growth of human diploid fibroblasts in the absence of glucose utilization. *Proc. Natl. Acad. Sci. U.S.A.* **73**, 4110–4114.

142. Zielke, H. R., Ozand, P. T., Tildon, J. T., Sevdalian, D. A., and Cornblath, M. (1978). Reciprocal regulation of glucose and glutamine utilization by cultured human diploid fibroblasts. *J. Cell. Physiol.* **95**, 41–48.

5

Equipment Sterilisation

G. THRELFALL
S. G. GARLAND
Celltech Ltd.
Slough, Berkshire
United Kingdom

1. INTRODUCTION

1.1. Importance of Sterilisation

Sterilisation of equipment and media is a central requirement for all work involving microbiological cell cultures. It is especially important in animal cell cultures because of their great susceptibility to contamination and over-

Animal Cell Biotechnology, Vol. 1
Copyright © 1985 by Academic Press, Inc.

growth by bacterial or fungal cells. The fast growth rate of these cells relative to animal cells can result in extremely rapid overgrowth originating from very small numbers of contaminating cells. Great care must therefore be taken to exclude all possible means of contamination during initiation and maintenance of animal cell cultures at all scales of operation, and consequently much effort has been directed towards developing efficient and reliable methods of sterilising equipment and materials to be used for this purpose.

1.2. Definition of Sterilisation

It is relevant at this point to consider what is implied by the term "sterilisation". This is an absolute term meaning the complete destruction of all living organisms, and a widely-accepted definition is that of Sykes (9), who defined sterilisation as "the total inactivation of all forms of microbial life in terms of their ability to reproduce". He distinguished it from "disinfection", which implies only the reduction of contaminating microroganisms to a "safe", i.e. non-infective, level, but not necessarily their total destruction.

1.3. Sterilising Agents

They are often classified as either chemical or physical agents, but these distinctions are not precise and it is preferable to discuss each on its individual merits with regard to the particular environmental conditions under which it can be effectively used. Incidentally, ultra-violet radiation is commonly referred to as a sterilising agent though many data indicate its limitations for this purpose and its use will not be included in the subsequent discussion.

The number of sterilising agents that can be usefully and practically employed for animal cell culture equipment is relatively small. They can be summarised as follows:

1. Heat, including steam and hot air
2. Ionising radiations, principally gamma rays
3. Toxic chemicals, principally ethylene oxide

By far the most important of these agents is steam, for reasons which will become apparent during the subsequent discussion. However, there are some situations in which it cannot be used, e.g. where a piece of equipment cannot withstand exposure to high temperature or humidity, and a different agent must be used. If high temperature is the problem, for instance with certain plastics such as polyethylene, which cannot be safely steam-sterilised at 121°C, it may be possible to sterilise in an autoclave, using low-temperature steam combined with the biocidal effects of ethylene oxide. In the example quoted, however, it is probably more practical to use similar equipment, vessels etc. made of polypropylene, which will withstand autoclaving at 121°C.

2. MECHANISM OF STERILISATION

2.1. Death of Microorganisms

Most of the studies on this subject have examined the lethal effects of heat on a microbial population. The studies have indicated that the kinetics of thermal destruction of many microorganisms follows a logarithmic course, approximating to a first-order (unimolecular) reaction mechanism (1). The practical implication of this mode of killing is that the larger the number of microorganisms in the population to be destroyed, the greater the severity of the sterilising activity required to be effective. Thus where given sterilisation conditions are known to destroy say 99.99% of a cell population, to effect a further 10-fold reduction, i.e. to 99.999%, would require approximately double the efficiency of the sterilising conditions. This could be achieved, for instance by an appropriate increase in steam temperature or dose of radiation, or by doubling the exposure time. It is worth noting that treatment with steam under readily-attainable sterilising conditions ensures a reduction of the cell population of $>10^{15}$, whereas with most other sterilisation processes this is often of the order 10^8–10^9, e.g. with sterilisation by ethylene oxide (9).

2.2. Resistance of Bacterial Spores

The greater heat resistance of bacterial spores compared with vegetative cells of the same species is well known, though the reasons for it are not completely understood. The spore cells are surrounded by a thicker membrane, which could make access of steam more difficult, but it seems that an important factor is likely to be their lower water content so that the cell protein is in a relatively dehydrated state. This will make the protein molecules inherently less susceptible to inactivation by heat, a point which is discussed later (see Section 2.3).

Since most vegetative bacterial cells can be killed by holding at 80°C for a few minutes, it is evident that the main concern in equipment sterilisation must be to ensure that conditions lethal to spores are achieved. The results of several authors (8,9) indicate that practically all of a wide range of bacterial spores was killed by heating at 121°C for 15 min, though certain rarer species, e.g. *Clostridium botulinum*, might take somewhat longer. Incidentally, to inactivate spores of this species by dry heat in a time of 15 min would require a temperature of 170°C (9).

2.3. Effects at the Molecular Level

All of the agents used depend on the inactivation of essential intracellular constituents required for growth and reproduction of the cell. It is well known that many cell proteins, especially the enzymes, which are intricately involved in all aspects of cellular development and proliferation, are sensitive to the denaturing effects of heat. They are also susceptible to molecular changes induced by ionising radiations or toxic chemicals, which are furthermore known to have adverse effects of the cell's genetic material, i.e. the chromosomes.

2.3.1. Heat

The most universally applicable agent for sterilisation is heat, either as moist heat in the form of saturated steam or dry heat, as hot air. It is well known that the efficiency of sterilisation by steam is considerably greater than by dry heat in terms of the temperature and time of exposure required, and it has been postulated that this is due to a difference in mechanism of action of these agents on the cellular proteins. Whereas the action of steam is due to the coagulation of these proteins, cell death mediated by dry heat is probably the result of oxidative processes (6).

Heat in the form of saturated steam under pressure has proved to be the most efficient and reliable sterilising agent. The mechanism of its effect is thought to involve the breakage of intramolecular disulphide bonds to give free sulphydryl (—SH) groups in the cell's protein molecules (4). These reactive groups may subsequently recombine in a random manner, giving rise to polypeptides with an incorrect secondary and tertiary structure and frequently forming insoluble complexes by coagulation.

The effect of dry heat on cellular components is probably rather different. It has been suggested that, as the temperature increases and water is driven off from the cell, the activity of the various polar groups such as hydroxyl (—OH) and sulphydryl (—SH) on the polypeptide chains decreases markedly. Consequently, more energy is required to disrupt the native secondary and tertiary structure of those molecules so that higher temperatures and exposure times are required for their inactivation. Since conditions of 160°–180°C for 15 min may be required for adequate treatment, it is evident that this method is only applicable to materials that will not be adversely affected by such harsh oxidative conditions.

2.3.2. Ionising Radiations

The radiations commonly used for sterilisation purposes are gamma rays (which have largely superseded X-rays) and high-energy electrons produced by a linear accelerator. The lethal effects of these agents are largely attributable to the increased reactivity of the cellular macromolecules induced by ionisation due to transfer of energy from the radiation beam. In the case of the cell's DNA, irradiation is known to lead to various chemical changes including opening of purine and pyrimidine rings, dephosphorylation and scission of the polynucleotide chain (7).

Radiation may also affect the cell's genetic material in other, less direct, ways. For instance the activities of certain enzymes involved in the synthesis of nucleotide precursors of chromosomal DNA are known to be inhibited by irradiation with gamma or X-rays.

It has also been shown that irradiation can result in the disruption of intracellular structures, allowing release of membrane-bound enzymes, which can therefore no longer carry out their normal activities (2).

2.3.3. Toxic Chemicals

Many chemical preparations are currently used for disinfection and decontamination purposes, but only very few are of practical importance for sterilisation of cell culture equipment. The most important of these are the highly reactive compounds ethylene oxide and formaldehyde, which exert their effects mainly through their irreversible reactions with amino ($-NH_2$) groups on essential components of the cell.

Ethylene oxide, $(CH_2)_2O$, is a highly toxic chemical which is gaseous at normal ambient temperature, having a boiling point of 10.8°C. It readily alkylates the free amino groups which are to be found on both proteins and nucleic acids, producing modified macromolecules which are unable to fulfil their normal cellular functions. Basic proteins containing a high proportion of the amino acids lysine and arginine will be particularly affected by ethylene oxide and would be rendered inactive.

The purine and pyrimidine components of nucleic acids, which also contain free amino groups, will be alkylated in an analogous fashion to amino acids, and in the case of DNA the ability of the alkylated deoxyribonucleic acid chain to replicate during the cell proliferation cycle will be greatly impaired.

Formaldehyde, HCHO, is also highly toxic to living cells including bacterial spores and is potent against many viruses. It is a liquid at normal temperature but is volatile, emitting an extremely pungent vapour. It is most effectively used at a temperature of 50°–60°C with a relative humidity of 80–

90%, its action being analogous to that of ethylene oxide. Under these conditions it readily formylates free amino groups on proteins, nucleic acids and other cell constituents. However, its use as a sterilising agent is precluded by its relatively poor power of penetration, which greatly restricts its activity. Thus microorganisms protected by very thin films of organic material such as dried blood serum will be quite resistant to the biocidal effects of formaldehyde. Its only practical application in the area of cell culture technology is in the decontamination or disinfection of equipment, such as hollow fibre cell growth reactors, which cannot be exposed to more efficient sterilising agents such as high-temperature steam or hot air. For this type of operation, formaldehyde could be used as, for example, a 10% aqueous solution, but such treatment cannot be realistically considered as sterilisation. For this reason, the use of formaldehyde is not discussed further as a method of sterilisation.

3. METHODS OF STERILISATION

3.1. Sterilisation by Steam

Heat in the form of saturated steam under pressure is the most efficient and reliable means of sterilisation. It is universally the method of choice, provided that (1) a suitable vessel is available for carrying out the process and (2) the materials to be sterilised are unaffected by high temperature and humidity.

Saturated steam has been shown to effectively destroy all microorganisms, including thermophilic bacteria and bacterial spores, when used for a sufficient length of time at temperatures not greatly above 100°C (3,9). The efficacy of steam under these conditions is mainly due to its latent heat content. Thus at 100°C steam possesses approximately seven times more heat than an equimolar amount of water at the same temperature. This latent heat may be rapidly transferred to an object being sterilised, by the process of condensation, until the whole object reaches the same temperature as the steam. When this is attained the sterilising process begins, resulting in the death of all contaminating microorganisms, provided the steam is able to penetrate throughout the material. Problems of incomplete inactivation of contaminants will certainly arise if there is inadequate access of steam to the interior of, for example, an aspirator with various attachments including tubing, filters etc. Similarly, where there is air entrainment in porous materials or packages steam will not penetrate without prior removal of the air.

The rate of sterilisation by steam is very dependent on the temperature of the steam being used. It is therefore usual to use steam at an elevated

TABLE I Relationship between Temperature and Time for Sterilising by Saturated Steam under Pressure

Steam temperature (°C)	Sterilising time (minutes)	Steam pressure	
		lb/in.^{2}	kg/cm^2
115	30	10	0.70
121	15	15	1.05
126	10	20	1.40
134	3	30	2.10

temperature (i.e. under pressure) to effect sterilisation within an acceptable time. Data presented by Sykes (9) has shown a relationship between steam temperature and time for a wide range of bacterial spores, including thermophilic species, which can be used as a general guide for sterilisation of cell culture equipment (Table I).

A commonly used set of conditions for this purpose is 121°C (i.e. at 15 lb/in.^{2} or 1.05 kg/cm^2) for 15–20 min. However, there are species differences in the susceptibility of spores to the effects of steam and it may, after carrying out efficacy tests, be found necessary to increase the severity of the sterilising treatment. This would normally be accomplished by increasing the time of exposure of the load to 20 or possibly 30 min; it should rarely be necessary to increase the steam temperature, unless a shortened exposure time is essential.

It must be remembered that the sterilising time only begins when all parts of the load have attained the steam temperature. The pre-sterilisation time is dependent on several factors including the size and composition of the load and the size of the sterilising chamber.

3.1.1. Effect of Air

A major obstacle to achieving optimal conditions for steam sterilisation arises from the differences in the physical properties of steam and the air which is present in the sterilising chamber before the sterilisation process begins. When steam is introduced into the chamber, the air, due to its much higher density and lower temperature than steam, tends to form layers near the floor of the chamber. The poor miscibility of these two gases ensures that the temperature in this region is considerably lower than that of the pure steam and thus the desired sterilising temperature will not necessarily be attained throughout the chamber.

The obvious answer to this problem is to completely replace all air in the chamber by steam during the pre-sterilisation phase. This has been approached in two different ways, traditionally by using steam to displace air downward and out of the chamber via a drain line and more recently by use of a system for vacuum-assisted air removal, which has been incorporated into most modern autoclaves.

3.1.2. Autoclaves

The sterilising effect of steam is usually harnessed by means of an auto-clave. There are several different designs in common use, but basically the autoclave is a pressure vessel designed and constructed to specifications that allow its functions to be carried out safely and efficiently at a steam pressure of 20 lb/in.^{2} (1.4 kg/cm^2) or in some cases up to 30 lb/in.^{2} (2.1 kg/cm^2). The increased pressure, whilst essential for attaining temperatures above 100°C, is not required for the sterilising process itself, which depends on the com-bined effects of heat and moisture. Since many modern autoclaves incorpo-rate a high-vacuum cycle, these vessels must not allow leakage of air into the chamber under reduced pressure, which would compromise the efficiency of sterilisation.

The function of the autoclave, then, is to provide an environment in which the whole of the material to be sterilised is exposed to saturated steam under controlled conditions of temperature and time. Any factors that tend to detract from these conditions must be overcome by good autoclave design, which should be aimed at maximum efficiency and reproducibility of the sterlising process.

3.1.3. Downward Displacement System

The problem of incomplete air removal, leading to inefficient sterilisation, is not overcome in the downward displacement type of autoclave vessel, which in any case can only realistically be used for surface sterilisation of unwrapped items such as metal instruments, or bottles containing fluids, provided a long sterilisation time is used to ensure completion of the pro-cess. In this system, steam is fed into the chamber, usually at a point near the top, and displaces air downward and out via the chamber drain line. However, even though a baffle plate is used to disperse steam throughout the chamber, some air will inevitably remain in layers and pockets and will contribute to regions of lower temperature than that of the steam near the top of the chamber. Even when the autoclave chamber is loaded carefully to allow maximum access of steam around the items, this system will not cope with removal of air contained in, for instance, packs of wrapped tubing or empty, sealed vessels protected by an air filter. For these loads a more efficient system of air removal must be employed.

3.1.4. The High-vacuum Autoclave

Modern autoclaves usually incorporate a vacuum cycle to allow virtually complete removal of air from the chamber and its load before sterilisation begins. This is accomplished by means of a powerful vacuum pump which can reduce the chamber pressure to less than 20 mm of mercury, leaving only a small quantity of residual air (<3%) in the chamber. This process will normally ensure complete access of steam to all parts of the load. However, it is important to consider how the evacuation process may be employed optimally in relation to the size and nature of the load to be sterilised.

A typical autoclave serving a large-scale cell culture laboratory may combine two or three different cycles to cope with various types of load. In the author's laboratory three fully automatic cycles are incorporated. These are (1) a "downward displacement" cycle, which is mainly used for sterilisation of bottles containing heat-stable liquids for use in cell culture and fermentation processes. The temperature and holding time for the sterilising period are both adjustable by the operator but all other aspects of the process are fully automatic. The second cycle (2) is a vacuum-assisted cycle which begins with a single vacuum pulse followed by the admission of steam and then the sterilising period. This cycle is particularly useful for sterilising pieces of equipment such as empty bottles for use as inoculum vessels. These bottles are closed except for a narrow tube protected by an air filter, which allows access of steam after air has been withdrawn by the pre-sterilisation vacuum pulse. This cycle is probably the most useful for general purposes but there is also a third cycle (3) which is particularly useful for treatment of fabrics and other loads that may contain entrained air. Essential to this "porous load" cycle is the removal of air from the chamber and load by the use of a series of rapid alternating steam/vacuum pulses. Before the sterilising phase, steam under pressure is forced into the chamber, allowed to mix with the air and then withdrawn by vacuum. The pulsing is repeated a further five times, at which point virtually all the air entrained in the load will have been removed, allowing rapid penetration of steam immediately after the sterilising phase begins. At the end of the cycle a post-sterilisation vacuum is drawn as a single pulse to effect drying of the load as the autoclave cools down. This cycle is commonly used in hospital sterile service departments for the sterilisation of linen packs and instrument packs containing tubing which would normally entrap air. In the laboratory this cycle may be used for disposal of contaminated porous materials etc.

3.1.5. Causes of Sterilisation Failure

There are many reasons why, even when sterilising conditions known to completely inactive microorganisms are employed, a particular sterilisation treatment may fail. There are, of course, trivial reasons such as the possibility of incorrect instrumentation giving rise to insufficient temperature or time parameters; also the pre-sterilisation period may not be long enough or there may be incomplete removal of air, as discussed more fully in a previous section. Another common reason for failure is lack of cleanliness of materials to be sterilised; it is of paramount importance that items to be sterilised are as clean as possible, since very small amounts of adhering materials will contain large numbers of bacteria which will be well protected against all but the most rigorous sterilising conditions. In the author's laboratory it is axiomatic that any object that contained, for example, culture medium is thoroughly machine-washed using a very strong chemical agent containing sodium hydroxide before sterilisation.

A further cause of non-sterility of equipment after autoclaving may sometimes be traced to a faulty air filter on the vacuum release system. Thus at the end of a cycle, it is possible to recontaminate equipment by the access of unfiltered air into the sterilising chamber. This particular problem may go undetected due to the difficulty of *in situ* integrity testing of the air filter, so its prevention is best approached by regular changing of the filter according to a planned maintenance programme.

The importance of such a maintenance programme is emphasized by the space afforded its description in Health Technical Memorandum No. 10, Sterilisers (4a), which details all aspects of installation, maintenance and validation of sterilisation equipment with special reference to hospital requirements, where the regulations are particularly stringent. However, much detailed information and advice are also afforded in the Guide to Good Pharmaceutical Manufacturing Practice (4b), where the recommendations are directed towards the safe and reliable production of high-quality therapeutic materials. Since these are among the many end products of current processes in large-scale animal cell culture, it is extremely valuable to follow this advice.

3.1.6. Sterilisation of Cell Culture Equipment in Situ

The main use of *in situ* sterilisation is for items of equipment that either are too large to be autoclaved or are part of a system which needs to be sterilised frequently without the whole system being affected (as in a continuous fermentation system). Cell culture fermenters, together with process vessels, e.g. media and product holding vessels, are the main items to require sterilisation in this manner, and a typical air-lift fermenter system used for batch fermentation processes is shown in Fig. 1.

As previously stated, many toxic chemicals are available for decontamination processes but few are of practical importance for sterilisation of cell

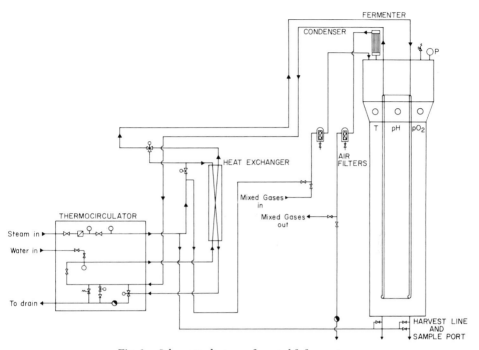

Fig. 1. Schematic diagram of an air-lift fermenter.

culture equipment and none are used routinely for fermenter sterilisation. Thus steam sterilisation is used for most process systems.

The main requirement for *in situ* steam sterilisation is that the vessel is suitably designed and manufactured to withstand the temperature (usually 121°C) and pressure (15 lb/in.^{2}) of the process. This limits the materials used in their manufacture to boro-silicate glass and stainless steel (usually grade 316). Many different types of fermenters have been used in animal cell culture but the same principles of sterilisation apply to them all.

The fermenter in Fig. 1 is an air-lift vessel of glass construction with stainless steel head plate, base plate and central collar which houses a pH probe, dissolved oxygen probe and temperature sensor. Passing through the base plate are sampling and harvest valve systems and attached through the head plate is a condenser with a gas/air outlet, an air/gas inlet and a steam/process water flow and return system. This fermenter contains a double-skinned draft tube through which the steam and process water flow for sterilisation and temperature control. Other systems may use an outer jacket surrounding the fermentation vessel through which the steam and process water flow.

Fermenter sterilisation involves the sterilisation of the vessel itself along with all attachments such as the probes, gas inlet and outlet filters and ancillary pipeware.

The pH probes and dissolved oxygen probes should ideally withstand *in situ* sterilisation, or alternatively retractable probes which are removed during the sterilisation process are used. During sterilisation of the empty vessel, the process water in the draft tube is replaced by steam via the thermo-circulator. Steam is also introduced to the vessel via the air/gas inlet filter line.

The air inlet and outlet filters must withstand the temperature and pressures of the sterilisation process and also must allow steam to pass through them without too great a pressure drop across the filter material. A suitable hydrophobic filter is required, usually of PTFE construction, although some types of filter fail to allow steam to pass through them when the surface becomes "wetted", i.e. covered with a surface layer of water, and this should be ascertained before use.

As steam is introduced into the vessel the air is vented out (as described previously for steam sterilisation by autoclaving) via the air outlet filter line. Cold water to the condenser system must be turned off during sterilisation as this would cause the steam passing through the air outlet filter to condense before reaching the filter housing.

During the initial stages of sterilisation some of the steam condenses due to the vessel being cool, and this must be drained continually otherwise a relatively cool condensate pool accumulates on the base plate. This is drained from the vessel via the harvest and sample valves. As the vessel temperature increases, less condensate is formed, and when the temperature reaches 100°C the air outlet line is closed along with the harvest and sample valves. This enables the temperature in the vessel to increase with the pressure. The steam condensate from the air outlet line is automatically removed from the system via the condensate trap. However, during the sterilisation process the condensate must be removed from the vessel at regular intervals (usually via the harvest and sample valves). The vessel is sterilised at 121°C (15 lb/in.^{2}) for a period of time which is usually calculated during the commissioning of the vessel. As it is essential that for the sterilisation process to occur effectively all parts of the vessel must reach 121°C and be maintained at that temperature for a minimum time of 15 min, one must consider the difficulty presented by metal components which lead to low surface temperatures. If it is necessary to incorporate large metal areas in the vessel, then sterilisation can be aided by applying external heat sources, e.g. electrical or steam jackets. The temperature/time required for the sterilisation of a particular vessel is usually calculated during its commissioning by means of thermocouples placed within the vessel. A sterilisation run is then carried out and is continued until all parts of the fermenter have reached sterilisation temperatures for the required time. This process may be tested by use of a biological indication system (usually *Bacillus stearothermophilus*), described in Section 4.

After the sterilisation time has elapsed the steam is turned off to the vessel and the vessel is allowed to cool from 121° to 100°C, when the valve to the condensate trap is closed and the valve in the air outlet line is opened. This enables air to be drawn in via the filter (sterile air) as the vessel cools. The vessel is normally allowed to cool slowly to its required temperature.

The valves used in this system are usually of a rubber diaphragm type. Another type used is a stainless steel ball valve, usually of a more complicated construction, which means it is harder to sterilise both the valve and attached pipeware. The diaphragm valve has several advantages:

1. It is of simple construction, basically a one-piece body which can be welded into the pipeline.

2. It has a single interface with the environment (round edge of the diaphragm).

3. The diaphragm is easily replaceable.

3.2. Sterilisation by Dry Heat

Steam is almost always preferred to dry heat for sterilisation of cell culture equipment and materials, primarily because of the efficacy of steam under much milder conditions of temperature than are required for dry heat. In practice, hot air may sometimes be a very convenient and effective method for sterilising metal or glass objects such as surgical instruments, syringes etc., and it is certainly more appropriate for pharmaceutical preparations including powders and ointments, which may be adversely affected by the moisture associated with steam sterilisation or may simply be impenetrable by steam. It is evident that dry heat is not widely applicable to sterilisation of cell culture equipment, which is virtually all sterilisable by steam, but might be considered as an alternative in case of emergency, e.g. following failure of an autoclave.

Dry heat sterilisation is normally carried out in a hot-air oven which can be controlled at temperatures up to 180°C and incorporates an automatic timer. Theoretically, as with steam sterilisation, the main requirement is to ensure that the whole of the load is brought to the required temperature and held for the correct predetermined time to effect sterilisation. In practice, due mainly to inadequate design of the hot-air oven itself, very large temperature gradients have been observed and an indicated temperature of 180°C may only produce 150°C in some parts of the oven. It is strongly recommended that, if hot-air sterilisation is to be used successfully, reference should be made to the relevant British Standard (BS3421) in order to assess the performance to be expected from a steriliser of this type. Full consideration must be given to the size, nature and distribution of the load and careful testing of the temperature attained throughout the load, using thermocouples, is a prerequisite.

Various data have been quoted on the conditions of time and temperature required for effective killing of a range of microorganisms including bacterial spores (3,9). The majority, including *Bacillus anthracis* and several *Clostridia* species, are killed at 170°C in less than 60 min and increasing the temperature to 180°C for 60 min proves lethal to virtually all species including heat-resistant soil spores. Thus 180°C for 1 hr is a commonly used set of conditions for hot-air sterilisation, though again it must be emphasized that the sterilising period only begins when all parts of the load reach the set temperature.

3.3. Sterilisation by Ionising Radiations

Gamma Rays

The use of X-irradiation for sterilisation purposes has now been largely superseded by gamma radiation, which has the advantage of greater power of penetration due to its shorter wavelength. However, as mentioned previously, due to the high cost of irradiation sources (usually cobalt-60 and caesium-137) and the necessary ancillary equipment, the use of this method is normally restricted to manufacturers of large numbers of small plastic items such as filters, syringes etc. Although most plastics can be irradiated without problem, some materials may suffer permanent damage after treatment with a dose of 2.5 Mrad (megarads), which is accepted widely as the standard treatment for medical disposable products. These include Teflon (PTFE), some rubbers and also glass materials, which become badly discoloured.

Although 2.5 Mrad has been shown to be more than adequate for killing a wide range of bacterial spores, moulds and yeasts (9), several unusually resistant organisms have been discovered, the most striking of these being *Micrococcus radiodurans*, which can survive doses up to 6 Mrad but is fortunately rarely encountered in the laboratory situation.

It is important to appreciate that although many bacterial toxins are inactivated by radiation by a similar dosage to spores, some enzymes that may be released from inactivated cells and also pyrogens released from the cell surface may require 10 or even 20 Mrad for complete inactivation (10). Also, although many eukaryotic viruses are readily inactivated by 2.5 Mrad, certain others such as polio virus and encephalitis exhibit resistance requiring up to 4.5 Mrad (5).

3.4. Sterilisation by Chemicals

Ethylene Oxide

The most common applications for ethylene oxide are in the industrial sterilisation of equipment and materials that would be adversely affected by steam or ionising radiation. It has also been widely used in hospitals for sterilising heat-sensitive medical apparatus, sometimes in combination with low-temperature steam at 80°C.

Under optimal conditions ethylene oxide diffuses through porous materials, including many plastics and rubber, effectively inactivating contaminating microorganisms. Treatment can be carried out in various ways by altering the parameters temperature, time and concentration. Thus where sterilisation of the interior of a closed container, such as a biological safety cabinet, is to be carried out *in situ* it may be convenient to use a low concentration (e.g. 10%) of ethylene oxide and allow exposure overnight (18 hr) at a temperature between 20° and 25°C. In other circumstances where a shorter exposure time is required, a higher concentration of the gas (80%) used at a temperature of 50°–60°C may be more appropriate (9). Another parameter to be considered is humidity, and it has been shown that for greatest efficacy, relative humidity greater than 30% and in some cases up to 90% may be required (9). As with all sterilisation methods a series of initial tests is essential to define optimal conditions for particular types of load.

It is relevant to point out some strong disadvantages of ethylene oxide for routine sterilisation, particulary in a laboratory situation. Apart from the fact that it forms a highly explosive mixture with air, which can be overcome by using a mixture of ethylene oxide with carbon dioxide or Freon, it is also an extremely toxic material and great care must be taken to ensure that operators are not exposed to it. This requires constant vigilance for leaks, which may develop unexpectedly during the use of equipment containing the gas, and automatically-operated low-level gas detectors are frequently employed for this purpose. Additionally, waste ethylene oxide after a sterilisation treatment must be safely disposed of, usually by passing through a "scrubber" system containing water, with which it readily reacts to form ethylene glycol.

In general, there does not appear to be great application for ethylene oxide in the cell culture area, where it has no apparent advantages except possibly in the *in situ* situation mentioned above for the decontamination of safety cabinets.

4. VALIDATION AND MONITORING OF STERILISATION

Whichever means of sterilisation is to be used, it is essential that its efficacy for the type of load to be processed be properly validated and regularly confirmed. Additionally, a reliable means of monitoring the process should be adopted and used to check each load to ensure that sterilisation conditions have actually been attained. It is imperative that records of all such validations and checks be kept for future reference.

Validation

Validation of the efficacy of sterilising conditions in an autoclave ultimately depends on determining that all parts of the load reach the desired temperature. This can only be accurately carried out by the use of thermocouple probes actually placed within the load at points where the temperature is likely to be lowest due to poor steam penetration. Thus the time taken for the load at these points to reach the sterilising temperature defines the pre-sterilisation period for that particular load and cycle.

The thermocouples used for this purpose are usually in the form of multiple needle probes or wires inserted into the autoclave chamber via a gland in the chamber wall. The probes can be inserted into the load before the cycle begins and connected to an accurate digital thermometer outside the autoclave. Thus the progress of temperature change during the process can be monitored and recorded simultaneously at several points in the load and chamber.

Having established the correct conditions of time and temperature required, re-validation should be carried out at frequent intervals, a period of 3 months being the maximum. This may be arranged to coincide with routine maintenance checks on the machine and it is often prudent to carry out a thorough re-validation immediately after such a service has been carried out.

Monitoring

Since it is inconvenient and impractical to use this direct temperature measurement for each load to be sterilised in a cell culture laboratory, other useful, but less reliable, means have been developed. They are chemical and biological indicators which can be inserted into the load prior to processing and examined for changes at the end of the sterilization cycle. However, it must be remembered that such tests are only useful in the negative sense and do not necessarily prove the success of a process, although they will indicate when sterilisation has failed.

Chemical indicators are available for heat, ethylene oxide and radiation sterilisation. The best known of these is the adhesive tape known as autoclave tape, which is simply used either to package a load, or placed at various points on the load. If there is no colour change on the tape at the end of a cycle it may be assumed that sterilisation conditions were not attained, although a colour change does not positively indicate sterility as such changes may take place before complete sterilisation conditions have been achieved. Another incidental use of this adhesive tape is that it allows ready distinction between items which have been processed and those which have not.

Other chemical indicators in common use are coloured patches or strips which can be attached to loads, and also there are small glass tubes (e.g. Browne's tubes) containing a coloured liquid, all of which change colour on heating. Browne's tubes have the advantage that they may be placed into a liquid to be sterilised rather than just on the exterior of the container.

Biological indicators are also commonly used to indicate the failure of a sterilisation procedure, and again these can be adapted for use with all types of sterilising agent. They consist of preparations of spores of resistant bacterial strains and are supplied either as strips of absorbent material impregnated with the spores, or as small glass tubes containing the spores. For monitoring of heat sterilisation processes the organism *B. stearothermophilus* is frequently used because of its inherently high heat resistance. As with chemical indicators, these strips or tubes may be attached to the load during the sterilisation process and examined subsequently for indications of failure of the process; the principle is to attempt to detect viable organisms in the indicator by incubating it in a suitable growth medium. Where spore-containing tubes are used, the tube is usually broken into the growth medium and incubated, growth being indicated by a colour change in the medium. In addition to the fact that these indications do not prove sterility, there is a significant danger of accidental transfer of viable bacterial spores to cell culture equipment. It is therefore necessary to take great care if such tests are to be used, and wherever possible it is preferable to use chemical indicators for this routine monitoring purpose.

To summarize:

1. Validation of the sterilisation process is of prime importance for the success of procedures in animal cell technology. Thermocouples must be used.

2. Re-validation must be carried out at frequent, regular intervals. Ideally it should be part of the planned maintenance programme, e.g. at 3-monthly service intervals for autoclaves and hot-air ovens.

3. An appropriate means of monitoring each load must be adopted. Ideally this should again be a temperature measurement using thermocouples, but biological or chemical indicators may be used so long as their limitations are fully understood.

ACKNOWLEDGMENTS

The authors wish to thank Lesley Anne Smyth for provision of the fermenter diagram and Jane Durham for preparation of the typescript.

REFERENCES

1. Anand, J. C. (1961). Heat resistance and shape of destruction rate curves of sporulating organisms. *J. Sci. Ind. Res., Sect. C* **20**, 295–298.
2. Bacq, A. N., and Alexander, P. (1961). "Fundamentals of Radiobiology," 2nd ed. Pergamon, Oxford.
3. Ernst, R. R. (1977). Sterilization by heat. *In* "Disinfection, Sterilization and Preservation" (S. S. Block, ed.), pp. 481–521. Lea & Febiger, Philadelphia, Pennsylvania.
4. Hansen, N. H., and Riemann, H. (1963). Factors affecting the heat resistance of nonsporing organisms. *J. Appl. Bacteriol.* **26**, 314–333.
4a. Her Majesty's Stationery Office (1980). "Sterilisers," Health Tech. Memo No. 10. HMSO, London.
4b. Her Majesty's Stationery Office (1983). "Guide to Good Pharmaceutical Manufacturing Practice." HMSO, London.
5. Jordan, R. T., and Kempe, L. L. (1956). Inactivation of some animal viruses with gamma-radiation from cobalt-60. *Proc. Soc. Exp. Biol. Med.* **91**, 212–215.
6. Rohn, O. (1945). Physical methods of sterilization of microorganisms. *Bacteriol. Rev.* **9**, 1–47.
7. Scholes, G., and Weiss, J. (1954). Chemical action of X-rays on nucleic acids and related substances in aqueous systems. *Biochem. J.* **56**, 65–72.
8. Sykes, G. (1965). "Disinfection and Sterilization," 2nd ed. Spon, London.
9. Sykes, G. (1969). Methods and equipment for sterilization of laboratory apparatus and media. *In* "Methods in Microbiology" (J. R. Norris and D. W. Ribbons, eds.), Vol. 1, pp. 77–121. Academic Press, New York.
10. Wagenaar, R. O., Dack, G. M., and Murrell, C. B. (1959). Purified type-A *Clostridium botulinum* toxin subjected to ultracentrifugation and irradiation. *Food Res.* **24**, 57–65.

6

Air Sterilization

G. J. HARPER*

Experimental Microbiology and Safety Reference Laboratory
Public Health Laboratory Service
Centre for Applied Microbiology and Research
Salisbury, Wiltshire
United Kingdom

1. INTRODUCTION

Air substantially free from living microorganisms is essential for the preparation, and subsequent handling, of cell cultures free from contamination.

*Present address: "S'Argamassa", Quidhampton, Salisbury, Wiltshire SP2 9AR, United Kingdom.

Animal Cell Biotechnology, Vol. 1

The protection of workers and the environment from airborne micro-organisms present in cell culture systems, such as viruses being propagated, or possibly carried in tumour cell lines requires that the effluent air is also rendered free of living microorganisms.

The presence of living microorganisms in air and methods for preventing the contamination of cultures by airborne microbes have been known since the early days of bacteriology. It was soon realised that contamination of cultures, a problem bedevilling early bacteriologists, could be prevented in a number of ways. The most familiar of these, and one still widely used in microbiology laboratories, is the cotton wool plug used to close flasks and tubes of liquid culture media. Although the cotton wool plug still suffices for static cultures, the need for flowing gas supplies and aeration in modern culture systems requires a higher level of protection than is provided by this simple method.

Sterility is often thought of as an absolute condition and is defined as "free of living microorganisms". Although it is common to refer to "sterile" air, and this chapter is so entitled, the application of this strict criterion to air is difficult to achieve and impossible to prove. A more practicable aim is to use a process that will reduce the microbial contamination of the air to a level that is unlikely to result in a contaminated culture or environment. In prac-tice the microbial content of air can be reduced to extremely low levels, below the threshold of sensitive test methods. The limitations of existing test methods led to the work of Elsworth et al. (19) in which they demonstrated that it is possible to estimate by extrapolation from laboratory data reduction factors of the order of 1×10^{15} ($10^{-13}\%$). This low level of penetration through an air treatment system is in accord with the value deduced by Gaden and Humphrey (21) for air treatment in mould fermentation plants in order that cultures should not be contaminated by extraneous microbes. This chapter reviews the methods currently available.

2. METHODS FOR REMOVING MICROORGANISMS FROM AIR

2.1. Ultraviolet Light

The ability of ultraviolet light, at wavelengths of around 2600 Å (260 nm) to kill bacteria and viruses rapidly has been used in various forms to reduce the numbers of microorganisms in air. This topic is discussed in detail by McCulloch (28), Morris (29) and Shechmeister (34). Jensen (24) demon-strated high killing rates for airborne viruses, using light intensities of not less than 0.03 W min ft^2 in airstreams flowing at 100–200 ft^3 min^{-1} (2.8–5.6

m^3 min^{-1}) with aerosol exposure times of 0.3–0.6 sec. Influenza and vaccinia viruses were reduced by >99.9%, whereas adeno 2, coxsackie B1 and Sinbis viruses were not completely inactivated. In general it is agreed that although useful for reducing air contamination, ultraviolet light cannot be relied upon to sterilize air. Fungal and bacterial spores (common air contaminants), and cells protected by protein require much higher light activities for inactivation, as do cells carried in large particles (25). It is of interest that British Standard 5726 (5) states, "The provision of an ultraviolet light source for disinfection purposes is not recommended". Ultraviolet lamps must be kept scrupulously clean to ensure maximum effectiveness and their output should be monitored at frequent intervals. There is some evidence that the lethal effect of ultraviolet light is influenced by both temperature and relative humidity (33).

2.2. Electrostatic Precipitation

This method is widely used in industry for the removal of small particles from airstreams, and is mentioned by White and Smith (36) as an adjunct to high-efficiency filtration. Although this principle has been used in air sampling equipment collecting at flow rates between 20 and 10,000 litres min^{-1} (30), it has not been successfully applied to processing air supplied or exhausted from microbial culture vessels. The main disadvantages are the need for high operating voltage and loss of efficiency in the event of a power failure.

2.3. Cyclone Separators

Removal of particles from air in cyclone separators is achieved by centrifugal force. Particles are deposited on the walls of a tapering cylinder, the air entering tangentially near the top of the cylinder and the deposited particles being removed from the base. The main cleaned airstream is extracted from the top of the cyclone. The efficiency of this system can be increased by introducing a stream of liquid into the airstream as it enters the cyclone. Decker et al. (10) examined a large-volume wet collection cyclone and found it to be about 70% efficient for the collection of microbial particles generated in a test chamber. Errington and Powell (20) reported similar collection efficiencies when sampling air with a wet cyclone system. Bartlett and Bainbridge (1) quote collection efficiencies of about 60% for 2-μm-diameter particles, increasing to around 100% for particles greater than 8 μm in diameter. Because of these relatively low collection efficiencies, cyclone separators cannot be considered for air treatment in the context of this chapter.

2.4. Heat

Hot-air sterilization was shown by Bourdillon *et al.* (*2*) to be a reliable and technically feasible method for reducing the microbial content of the air to below detectable levels. The theory of hot-air sterilization is discussed by Elsworth *et al.* (*18*). This principle was used successfully by Van Den Ende (*35*) to treat highly infectious air extracted from a safety cabinet. Many configurations of hot-air sterilizing systems have been described. For example, Elsworth *et al.* (*18*) give details of a system capable of handling 1700 litres min^{-1} (1.7 m^3 min^{-1}). They found that a temperature of 300°C with an exposure time of 1.6 sec could be relied on to produce nominally "sterile" air. In a more recent review on hot-air sterilization Elsworth (*17*) concluded that developments in filter theory and materials made filters technically competitive with heat treatment. As the running costs of an air filtration system are lower and the equipment less sophisticated than a heat treatment system, there now appears to be no advantage in the use of expensive energy to reduce the microbial content of air by heat. As with other systems directly dependent on power, a failure in the electrical supply can result in an unsafe condition. Elsworth considered that in some circumstances such as the decontamination of air extracted from pathogenic viral cultures there was still a case for the use of heat treatment. The modern practice of mounting high-efficiency air filters in series in areas requiring the highest level of containment is considered adequate to deal with effluent air containing the most dangerous viral pathogens (*13*).

2.5. Filtration

Due to the shortcomings of the methods briefly reviewed above, much effort over the past 25–30 years has gone into the design and manufacture of filters capable of producing particle-free air. A large impetus for the production of reliable high-efficiency particulate air (HEPA) filters arose from the needs of the nuclear, electronic and pharmaceutical industries, all of which require ultra-clean air. For a review of filter development and the theory of filtration, reference should be made to White and Smith (*36*).

A variety of filter materials have been used. These include porous ceramics, membranes, sintered glass or metal and a wide range of fibres. Slag wool, glass wool, cotton wool, cotton–asbestos and resin-coated wool have all been used to pack deep bed fibrous filters. Elsworth (*16*) gives details of the design and efficiencies of filters packed with slag wool and glass fibres.

The ability to manufacture uniform sheets of glass fibre paper using differing fibre diameters and paper thicknesses has been responsible for the development of modern HEPA filters operating at low pressure drops. HEPA filters with different levels of penetration are now readily available from a

TABLE I Properties of Air Treatment System

Property	Ultraviolet light	Electrostatic precipitation	Cyclone separation	Heat	Filtration
Efficiency for sub-micrometer particles	Good	Good	Poor	Good	Good
Efficiency for large particles	Poor	Good	Good	Good	Good
Loading	Not applicable	Good	Good	Good	Good
Running costs	Low	Low	Low	High	Low
Fail-safe	No	No	No	No	Yes

number of manufacturers. We are concerned only with HEPA filters of the highest grades, i.e. those showing the smallest penetration.

The advantages and disadvantages of the methods reviewed for reducing the microbial content of air, summarised in Table I, show that HEPA filtration is the method of choice for supplying and extracting air from culture vessels, and for containment areas such as microbiological safety cabinets. The remainder of this chapter will be devoted to the use of filtration methods.

3. FILTERS

3.1. Mode of Action

Membrane filters are frequently referred to as millipore filters. They consist of plastic films about 0.13 mm thick with uniform size pores making up about 80% of the filter volume. Commercially available filters have pore sizes ranging from 0.1 to 10 μm. When used to filter liquids, the pore size determines the size of particle the filter will retain; when used as air filters, membranes can capture particles smaller than their rated pore size because of impaction and static charges. The delicate mechanical structure of membrane filters limits their use on an industrial scale, though they have proved useful for low-flow-rate systems when suitably mounted. As these filters operate mainly as mechanical sieves, they can rapidly lose efficiency due to blockage of the pores when used in atmospheres heavily laden with particulates.

For most air sterilization applications use is made of fibre filters. The

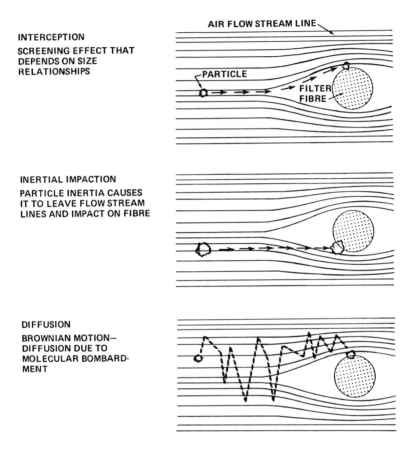

INTERCEPTION

SCREENING EFFECT THAT
DEPENDS ON SIZE
RELATIONSHIPS

INERTIAL IMPACTION

PARTICLE INERTIA CAUSES
IT TO LEAVE FLOW STREAM
LINES AND IMPACT ON FIBRE

DIFFUSION

BROWNIAN MOTION—
DIFFUSION DUE TO
MOLECULAR BOMBARD-
MENT

AIR FLOW STREAM LINE

PARTICLE

FILTER
FIBRE

SEDIMENTATION
ELECTROSTATIC ATTRACTION

Fig. 1.   Air filtration theory. Particle collection mechanisms.

theory of fibrous filtration, dealt with in detail by Dorman (15), is summarised below. In contrast to membrane filters, fibrous filters do not depend
on direct sieving for the retention of particles. Three mechanisms are principally involved: inertia, interception and diffusion (Fig. 1). *Inertia:* as a
particle of 5-μm diameter or larger approaches a fibre it will not follow the
flow line but because of its inertia will take a more direct path and will be
captured by collision with the fibre. *Interception:* particles between 0.5 and
5 μm in diameter tend to follow the flow line and are trapped by direct
contact with the fibre. *Diffusion:* particles smaller than about 0.3 μm in

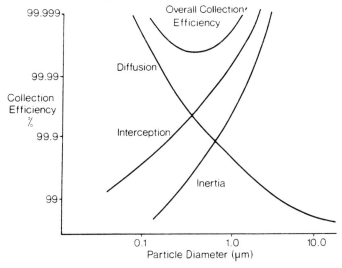

Fig. 2. Efficiency of particle removal by different mechanisms in fibrous filters.

diameter are subject to Brownian movement, which causes them to diffuse across the flow lines and so increases their chance of collision with a fibre. Gravitational settling and electrostatic attraction between fibres and particles can also make some contribution to particle removal.

With both inertia and interception, the larger the particle the higher the collection efficiency. With diffusion the reverse holds and the process is most efficient with the smallest particles. As all three mechanisms act together the overall efficiency of the filter depends upon a combination of differing filtration efficiencies. The relationship between the three principles of particle removal is shown in Fig. 2. From this it can be seen that the most difficult size of particle to remove from air by filtration is around 0.3 μm in diameter. This has caused some concern because most viruses are around, or below, this size. This concern is unnecessary for several reasons. First, the filtration efficiency does not decrease significantly for particles smaller than 0.3 μm in diameter; secondly, very small particles are seldom generated during laboratory operations: high energy is needed to produce small particles; and thirdly, even if single naked virus particles could be generated it is unlikely that they could survive the stresses involved in such rapid dehydration. Like most other microorganisms, viruses survive best when carried in large particles. Harstad and Filler (23) carried out an evaluation of HEPA filters manufactured from glass paper in which they measured (1) the penetration of purified bacteriophage particles with a number median diameter (NMD) of 0.12 μm, (2) bacterial spores (NMD 1 μm) and (3) dioctyl phtha-

TABLE II Effects of Aerosol Particle Size, Airflow, and Exposure to High Humidity on the Performance of Ultra-high-efficiency Filter Units[a]

| | | Geometric mean percent penetration[b] | | | |
| | | New filter unit | | After exposure to high humidity[d] | |
Filter unit manufacturer	Test aerosol[c]	7.5 cfm	30 cfm	7.5 cfm	30 cfm
1	Phage	0.00012	0.0020	0.00018	0.0025
	Spores	0.00023	0.00032	0.00056	0.00051
	DOP		0.005		0.005
2	Phage	0.00043	0.0017	0.0020	0.0021
	Spores	0.0023	0.000085	0.0034	0.00097
	DOP		0.013		0.009
3[e]	Phage	0.000035	0.00025	0.00019	0.00070
	Spores	0.000011	0.0000047	0.000043	0.000014
	DOP		<0.001		f
Overall geometric mean[f]	Phage	0.00012	0.00095	0.00041	0.0015
	Spores	0.000083	0.000050	0.00043	0.00019

[a] Reproduced by permission of the publisher of the *American Industrial Hygiene Association Journal*.

[b] Each value represents the geometric mean of three filter units evaluated once at each condition.

[c] DOP data by the manufacturer.

[d] Exposed for 5 hr at >95% relative humidity, then allowed to dry before retest.

[e] One unit was apparently defective upon receipt.

[f] Overall geometric means based on all nine filter units.

late (DOP), a standard test smoke used for filter testing. Tests were carried out at two airflow rates and were repeated after the filters had been exposed to high relative humidity for 5 hr. The results (Table II) led the authors to conclude that although aerosol particle size, airflow and exposure to high humidity have significant effects on filter performance, the filter units could be expected to provide excellent protection against sub-micrometer aerosols as well as against bacterial aerosols.

3.2. Characteristics of Filters

The glass paper medium used in HEPA filters shows a very high retention of small particles, with a low pressure drop. Most filters are marketed with a pressure drop of about 1 in. water gauge (250 Pa) at their rated flow—an important factor in energy conservation. The paper is non-flammable, does not begin to deteriorate until it reaches a temperature of about 500°C, and can tolerate relative humidities up to 100%. The chemical resistance of the paper is that of glass. Other glass fibre materials used in deep bed filters have similar characteristics.

For removal of the small particles of concern in reducing the microbial content of air to the levels needed to prevent contamination of cultures or the environment, filters of the highest grade are used. Filters are usually tested by manufacturers either with particles of sodium chloride, using the method given in British Standard (BS) 3928 (3), or by the use of DOP smoke (11). HEPA filters are marketed in several grades. A typical range will list three levels of sodium chloride penetration: 5%, 0.01% and <0.003%. Some manufacturers make filters suitable for repeated steam sterilization. Clearly, care must be taken to specify filters of the correct grade; for removal of microbial contaminants only use filters with a sodium chloride penetration of <0.003%. Some manufacturers scan the surface of filters, using a DOP smoke challenge, and these are usually marked as showing <0.01% penetration at any point on the filter surface. The sodium chloride test carried out according to BS 3928 with a penetration of <0.003% and the DOP scan test showing leakage of <0.01% are roughly equivalent tests of a filter's integrity.

HEPA filters are constructed by pleating the glass paper, which is about 0.8 mm thick, around spacers of corrugated metal, plastic or paper, so that large areas of the filter medium can be packed into conveniently sized containers of wood or metal. A filter of 12 × 12 × 6 in. (30 × 30 × 15 cm) containing about 30 ft^2 (3.1 m^2) of glass fibre paper is suitable for an airflow of 100 ft^3 min^{-1} (2.8 m^3 min^{-1}), at which rate the pressure drop will not exceed 1 in. water gauge (250 Pa). A recent development in which the corrugated separators are replaced with a thin plastic string cemented to the paper results in a smaller filter case for a given flow rate and pressure drop.

3.3. Physical Forms of Filters

Filters utilising beds of fibrous material such as slag wool or glass wool can be designed for any desired flow rate; Elsworth (16) gives packing details and operating rates for beds of grade AA glass fibre (Table III). For removal of the smaller particles, phages and viruses, the filter flow rate is set at around 5 cm sec^{-1}. Figure 3 shows a filter housing packed with glass fibre in use in the author's laboratory. These filters can be sterilised by autoclaving and can be used again unless there is visual evidence that the filter bed is disintegrating.

Commercially available in-line gas filters (Fig. 4), usually containing glass fibre or glass paper, are convenient and inexpensive for filtering at low flow rates of up to 20 litres min^{-1}. Autoclave vent filters (Fig. 5) with flow rates of

Fig. 3.   Filter housing for glass fibre filter.

Fig. 4. In-line gas filters. Reproduced by permission of Microflow Ltd.

up to 700 litres min^{-1} are used to protect the contents of autoclaves from contamination when air enters the chamber to relieve pressure or vacuum, or where air is passed through the chamber for a period to dry the contents.

Disposable filter cartridges are available in many configurations and flow rates. These are supplied as complete units (Fig. 6) or as cartridges suitable

TABLE III Packing Details and Operating Rates for Beds of Grade AA Glass Fibre[a,b]

Body diameter (in.)	Packing weight (g)	Packing depth (in.)	Air rates (litres min^{-1}) for removal of	
			Bacteria	Phages and viruses
1	2.5	2.5	3.0–4.5	1.5
1.5	6.0	2.5	6.7–10.0	3.4
2	10.0	2.5	12.0–18.0	6.0

[a] Reproduced by permission of the author and publisher from *Methods in Microbiology* (1969); copyright, Academic Press Inc., London.

[b] 0.6 to 6.0 μm in diameter (Johns-Manville Co. Ltd., London).

Fig. 5.   Autoclave vent filters. Reproduced by permission of Microflow Ltd.

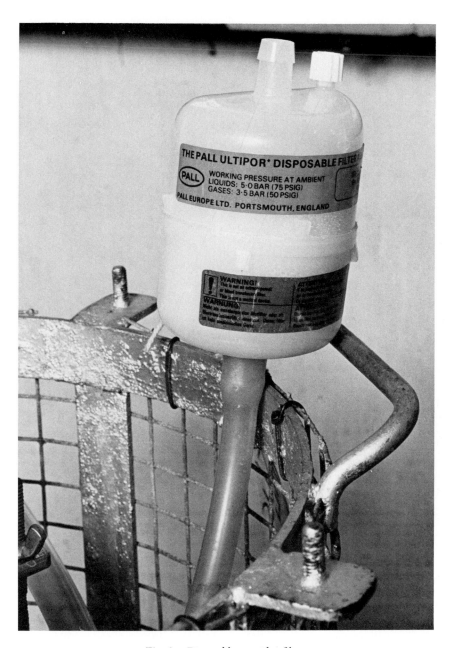

Fig. 6.   Disposable cartridge filter.

Fig. 7. Disposable filter cartridge unit for use in a permanent housing. Reproduced by permission of Microflow Ltd.

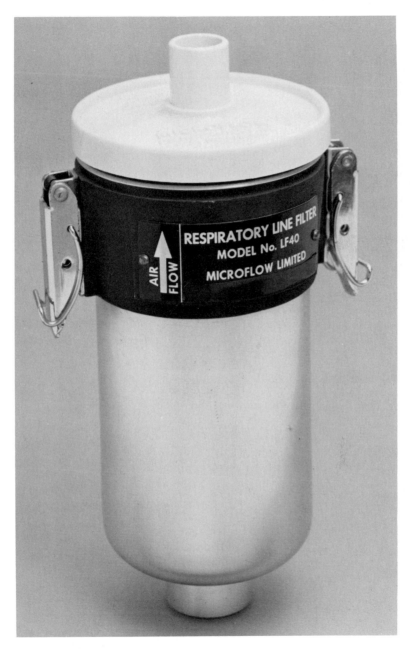

Fig. 8. Filter for use with patient respiratory apparatus. Reproduced by permission of Microflow Ltd.

Fig. 9. HEPA filter in air extract system from processing enclosure.

for enclosing in permanent housings (Fig. 7). Filters designed for use with patient respiratory apparatus (Fig. 8) with flow rates of up to 200 litres min^{-1} have been used to filter air in and out if isolators used for housing gnotobiotic animals. For higher flow rates HEPA filters mounted in wood or metal casings are available with capacities of 1000 ft^3 min^{-1} and more. Figure 9 shows a filter of this type used to extract air from an enclosure in which high concentrations of bacteria are processed. Where very high levels of contamination control are needed HEPA filters are often mounted in series. An air supply system using this arrangement is illustrated in Fig. 10. This illustration also shows the position of sampling points to enable *in situ* tests to be carried out.

There is now a wide choice of commercially available filters and these are usually manufactured to high standards. Individual filters are tested by the manufacturers and carry details of the test results and pressure drop at their rated flow rate on labels attached to the filter case. It is not cost effective for laboratories to make and test filters when reliable filters are so readily available.

Laboratory and pilot-plant-scale fermenters are also available from com-

Fig. 10.   HEPA filters mounted in series showing air sampling ports.

mercial sources. These systems are usually fitted with air filters with flow rates compatible with the capacity of the system and can be sterilized *in situ*.

3.4. Installation of Filters

It is a sound principle that filters should be fitted as close to the area to be protected as is physically possible to avoid unnecessary lengths of potentially contaminated ducts or pipework. The exception to this proviso is the need to treat effluent air from cultures to reduce its moisture content. The effluent air from cultures contains high levels of moisture; often the air is saturated with water and if this is passed through a filter condensation will take place within the filter, resulting in increased resistance to flow, loss of efficiency and in extreme cases microbial growth within or through the filter pack. Elsworth (*16*) describes a procedure suitable for use with laboratory-scale cultures (3 to 15 litres min^{-1}) in which a mains-water–cooled condenser and a catchpot are used to collect the condensate before passing the effluent air through a filter. For larger scale operations (100 litres min^{-1}) the arrangement shown in Fig. 11 is used. Greater security can be obtained by fitting a

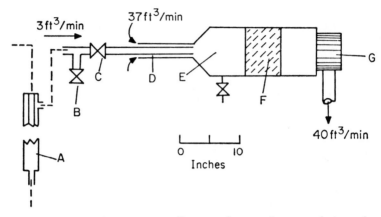

Fig. 11. Arrangement for processing effluent air from a culture vessel. A, catchpot on culture vessel; B and C, valves on effluent air line; D, inlet tube for laboratory air; E, expansion chamber; F, filter element; G, exhaust fan. Reproduced by permission of the author and publisher from *Methods in Microbiology* (1969); copyright, Academic Press Inc., London.

filter to the inlet tube D in Fig. 11. This will remove the risk of any back flow of contaminated air in the event of a power failure. Cameron and Favelle (6) and Pyle *et al.* (31) describe the fitting of heating devices to remove water from air prior to filtration.

Particular attention must be paid to the sealing of filters to ducting or pipeword. Most failures in air filtration arise from faulty sealing rather than from imperfect filters. When large HEPA filters are mounted in ducts the filter case should form part of the ductwork to avoid the possibility of leaks which could allow air to bypass a filter held within the duct. In Fig. 10 several unflanged HEPA filters are held in place by using compression exerted by tie-bars between the sections of the duct; the use of a silicone rubber cement on the gasket faces ensures a good seal. Filters with single or double flanges are available for direct attachment to ductwork.

As filtration efficiency is a function of the velocity through the filter medium it is essential that filters are not operated at flow rates in excess of their rated value (Fig. 12). Operating at higher flow rates will result in increased penetration through the filter medium. It is a good practice to under-run filters, i.e. to operate at less than their rated flow. This ensures high filtration efficiency and a longer operating life. To get the maximum life from HEPA filters used to process supply air drawn from the atmosphere it is usual to fit a coarse pre-filter before the HEPA. These are cheap disposable filters of low resistance and low efficiency for small particles, but are capable of trapping larger particles, thereby reducing the loading on the HEPA. British Standard 5726 (5) recommends using pre-filters with a gravimetric

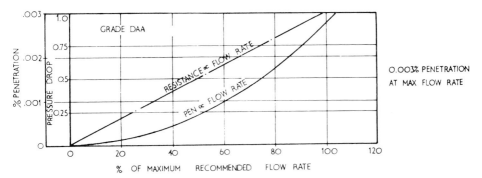

Fig. 12.   Relationship between filter flow rate, resistance and penetration. Pressure drop measured in inches water gauge.

efficiency of 95% against test dust no. 2, when tested at the designed flow rate in accordance with British Standard 2831 (4).

4. PROTECTIVE AIR ENCLOSURES

For the prevention of contamination during handling of cell cultures or suspensions in the laboratory during media changes, harvesting or inoculations, it is usual to work in some form of protective air enclosure. There are two types of enclosure: open cabinets using either a horizontal or vertical stream of HEPA-filtered air, or microbiological safety cabinets. The former are often incorrectly referred to as laminar-flow cabinets. As there is always some turbulence within the cabinets, a more correct description is "unidirectional" flow cabinets. Cabinets of this type are designed to protect the work being carried out within the cabinet. The air discharges from the cabinet directly into the surrounding room and offers no protection to the worker using the cabinet. Under no circumstances should this type of enclosure be used for handling tissue cultures. The use of unidirectional flow cabinets should be restricted to the preparation and dispensing of sterile materials prior to inoculation. If cabinets of this design are to be allowed in the laboratory they must be clearly labelled to restrict their use to clean operations only. To avoid possible misuse it is better not to use protective air enclosures, other than microbiological safety cabinets, in any laboratory handling cell cultures.

There are three classes of microbiological safety cabinets described in detail in British Standard 5726 (5). Class I and class III cabinets are mainly used for diagnostic microbiology when working with class A or B pathogens [Howie Code of Practice (1978)] or materials suspected of containing these

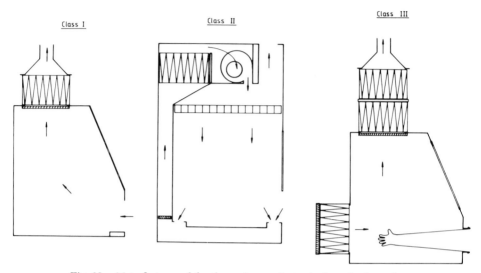

Fig. 13. Main features of the three classes of microbiological safety cabinets.

pathogens. The larger class II cabinets are much favoured in work with tissue cultures by virtue of their twin roles. Properly designed and correctly installed class II cabinets protect both the work and the operator. The main features of the three classes of microbiological safety cabinets are shown in Fig. 13.

Clark (7) and Clark and Goff (8) showed that a number of class II cabinets manufactured before the publication of BS 5726 failed to give adequate protection to the worker or the product, and in some cases failed to give adequate protection to either. Cabinets produced to comply with the standard did give the required protection. In addition to the British Standard there are other national standards and guidelines. These are summarised, along with guidelines on performance and approved test methods in a World Health Organisation document (37). Nevertheless, there are still some misgivings about the use of class II cabinets (14), and workers contemplating using this class of cabinet should consult these publications. It is essential to lay down a regular testing and maintenance schedule to ensure that the performance of any microbiological safety cabinet does not deteriorate with age or use. Recommended test procedures are described in the Howie Code of Practice (1978) for class I cabinets, and by the Department of Health and Human Services (12) for class II cabinets.

A recent paper by Collins and Yates (9) on the use of flexible film isolators suggests that these could be used for tissue culture work.

5. TESTING OF AIR FILTERS

Filters from commercial sources normally carry test results on a plate or label attached to the filter body. Provided there is no physical damage to the filter between the manufacturer's test and installation, the filters can be accepted as being as efficient as stated. However, it is necessary to check the efficiency after installation to establish the adequacy of the seals as well as the integrity of the filter. Tests can be carried out with a number of test materials (37). In the United Kingdom filters are usually tested with an aerosol of sodium chloride particles, using the procedure laid down in British Standard 3928 (3). An alternative test cloud, commonly used in the United States and Western Europe, consists of particles of DOP generated by vapourising DOP by means of heat (hot DOP) or with Laskin (27) nozzles (cold DOP). Recently (12) there has been some concern about potential health hazards from DOP aerosols. This has led to a search for substitute materials. Mineral oil and corn oil have been found to be satisfactory for use in place of DOP. The use of microbes for filter tests—either bacterial spores or small vegetative bacteria such as *Serratia marcescens*—is routine in the author's laboratory. Because of the different physical characteristics of these test aerosols, size of challenge particles and sensitivity of the detection systems used, it is not a simple matter to equate filter efficiencies measured by one method with those measured by another method. Elsworth (16) quotes Darlow, who compared the sodium chloride test method with bacterial spore challenges. A value of $10^{-3}\%$ by sodium chloride is equated with a value of $10^{-6}\%$ for monodispersed 1-μm particles of *Bacillus subtilis* spores. Recent work by Lach and Wright (26) did not show such a regular relationship between these two test methods, but they concluded that both test methods could be relied on for measuring the efficiency of HEPA filters. In an effort to attempt a better equation between the various test methods in common use, a WHO-sponsored international study is currently taking place.

For details of biological test methods see Restall (32) and BS 5726 (5). The method of choice will depend on the availability of test equipment and local experience.

6. DECONTAMINATION OF FILTERS

Filters need to be decontaminated before and between uses. Small in-line filters can be purchased already sterilized or they can be sterilized by autoclaving, either separately or already attached to culture vessels. In this laboratory such filters are resterilized up to 20 times before being discarded. Tests on small filters packed with glass fibre showed no penetration of either a bacterial aerosol ($<1 \times 10^{-6}\%$) or a sodium chloride cloud ($<1 \times 10^{-3}\%$) when retested after one, four and six cycles of autoclaving. Treatment with formaldehyde vapour is sometimes used to decontaminate large HEPA filters. Due to the poor penetration of formaldehyde this process cannot be relied upon to sterilize the whole depth of a large HEPA filter, but it is very effective for decontamination of the surface directly exposed to the vapour. With suitable precautions, i.e. the wearing of gloves and a gown and removal of the filter directly into a plastic bag, formaldehyde decontamination is suitable for treatment of filters prior to removal of potentially contaminated filters from a system. The use of formaldehyde does not cause any reduction in filtration efficiency. Glass wool and glass paper filters tested before and after treatment give similar results when challenged with a bacterial test aerosol.

ACKNOWLEDGMENTS

I am indebted to Mr. S. Clark for preparing many of the figures used in this chapter.

REFERENCES

1. Bartlett, J. T., and Bainbridge, A. (1978). Volumetric sampling of microorganisms in the atmosphere. In "Plant Disease Epidemiology" (J. Scott and A. Bainbridge, eds.), pp. 23–30. Butterworth, Oxford.
2. Bourdillon, R. B., Lidwell, O. M., and Raymond, W. F. (1948). Air disinfection by heat. A. Dry heat. Air sterilization in furnaces. Med. Res. Counc. (G.B.), Spec. Rep. Ser. 262, 190–203.
3. British Standards Institution (1969). "Method for Sodium Flame Test for Air Filters," Br. Stand. 3928. BSI, London.
4. British Standards Institution (1971). "Methods of Test for Air Filters used in Air Conditioning and General Ventilation," Br. Stand. 2831. BSI, London.
5. British Standards Institution (1979). "Specifications for Microbiological Safety Cabinets," Br. Stand. 5726. BSI, London.
6. Cameron, J., and Favelle, H. K. (1967). An effluent-air filter. J. Appl. Bacteriol. 30 (1), 261–263.
7. Clark, R. P. (1980). The evaluation of open-fronted biological safety cabinets. Lab. Pract. 29 (9), 926–929.
8. Clark, R. P., and Goff, M. R. (1981). The potassium iodide method for determining protection factors in open-fronted microbiological safety cabinets. J. Appl. Bacteriol. 51, 461–477.
9. Collins, C. H., and Yates, M. D. (1982). The use of a flexible film isolator in a diagnostic laboratory. Lab. Pract. Sept., 892–894.

10. Decker, H. M., Buchanan, L. M., Frisque, D. E., Filler, M. E., and Dahlgren, C. M. (1969). Advances in large volume air sampling. *Contam. Control* **8**, 13–17.
11. Defence Information Centre (1956). United States Military Standard 282. DIC, St. Mary Cray, Orpington, Kent.
12. Department of Health and Human Services (1982). "National Toxicity Program. Carcinogenesis Bioassay of Di(2-ethylhexyl) Phthalate," DHSS Publ. No. (NIH) 82-1773. Natl. Inst. Health, Bethesda, Maryland.
13. Department of Health and Social Security (1976). "Control of Laboratory Use in the United Kingdom of Pathogens Very Dangerous to Man." DHSS, London.
14. Department of Health and Social Security (1981). "Interim Advisory Committee on Safety in Clinical Laboratories," Bull. No. 2, pp. 14–15. DHSS, London.
15. Dorman, R. G. (1964). Theory of fibrous filtration. In "High-Efficiency Air Filtration" (P. A. F. White and S. E. Smith, eds.), pp. 67–99. Butterworth, London.
16. Elsworth, R. (1969). Treatment of process air for culture. In "Methods in Microbiology" (J. R. Norris and D. W. Ribbons, eds.), Vol. 1, pp. 123–136. Academic Press, London.
17. Elsworth, R. (1972). Notes on hot air sterilization. In "Safety in Microbiology" (D. A. Shapton and R. G. Board, eds.), pp. 61–71. Academic Press, London.
18. Elsworth, R., Telling, R. C., and Ford, J. W. S. (1955). Sterilization of air by heat. *J. Hyg.* **53**, 445–457.
19. Elsworth, R., Morris, E. J., and East, D. N. (1961). The heat sterilization of spore infected air. *Trans. Inst. Chem. Eng.* **39**, A47–A52.
20. Errington, F. P., and Powell, E. O. (1969). A cyclone separator for aerosol sampling in the field. *J. Hyg.* **67**, 387–399.
21. Gaden, E. L., Jr., and Humphrey, A. E. (1956). Fibrous filters for air sterilization. *Ind. Eng. Chem.* **48**, 2172–2176.
22. German Standards (1974). "Standards for HEPA Filters," Ger. Stand. DIN 24184. Deutschen Normenausschussers, Berlin, Federal Republic of Germany.
23. Harstad, J. B., and Filler, M. E. (1969). Evaluation of air filters with submicronic viral aerosols and bacterial aerosols. *Am. Ind. Hyg. Assoc. J.* **30**, 280–290.
24. Jensen, M. M. (1964). Inactivation of airborne virus by ultraviolet radiation. *Appl. Microbiol.* **12**, 418–420.
25. Kethley, T. W., and Branch, K. (1972). Ultraviolet lamps for room air disinfection: effect of sampling location and particle size of bacterial aerosol. *Arch. Environ. Health* **25**, 205–216.
26. Lach, V. H., and Wright, A. E. (1981). The testing of HEPA filters fitted to microbiological safety cabinets: A comparison of methods. *J. Hosp. Infect.* **2**, 385–388.
27. Laskin, S. (1948). "Submerged Aerosol Unit," AEC Proj. Q. Rep. UR-38. University of Rochester, Rochester, New York.
28. McCulloch, E. C. (1945). "Disinfection and Sterilization." Kimpton, London.
29. Morris, E. J. (1960). A survey of safety precautions in the microbiology laboratory. *Med. Lab. Technol.* **17**, 1–11.
30. Morris, E. J., Darlow, H. M., Peel, J. F. H., and Wright, W. C. (1961). The quantitative assay of mono-dispersed aerosols of bacteria and bacteriophage by electrostatic precipitation. *J. Hyg.* **59**, 487–496.
31. Pyle, P., Darlow, M., and Firman, J. E. (1969). A heated ultra high efficiency filter for mechanical ventilation. *Lancet* **1**, 136–137.
32. Restall, S. W. F. (1978). The adequacy and testing of biohazard cabinets. *Lab. Pract.* July, 496–557.
33. Riley, R. L., and Kaufman, J. E. (1972). Effect of relative humidity on the inactivation of *Serratia marcescens* by ultraviolet irradiation. *Appl. Microbiol.* **23**, 1113–1120.

34. Shechmeister, I. L. (1977). Sterilization by ultraviolet radiation. *In* "Disinfection, Steriliza-tion and Preservation" (S. S. Block, ed.), 2nd ed. Lea & Febiger, Philadelphia, Pennsylvania.
35. Van Den Ende, M. (1943). An apparatus for the safe inoculation of animals with dangerous pathogens. *J. Hyg.* **43**, 189–194.
36. White, P. A. F., and Smith, S. E. (1964). "High Efficiency Air Filtration." Butterworth, London.
37. World Health Organization (1981). "Guidelines for Biological Safety Cabinets," WHO CDS/SMM/81.21. World Health Organ., Geneva.

PART III

MASS CELL CULTIVATION

7

Mass Cultivation and Production of Animal Cells

H. Katinger
Institute of Applied Microbiology
University of Agriculture
Vienna, Austria

W. Scheirer
Sandoz Research Institute
Vienna, Austria

1. INTRODUCTION

Until recently mass cultivation systems have been technologically estab-lished only in a few exceptional cases, such as the production of foot-and-mouth disease (FMD) vaccines and of interferons at semicommercial scales. Traditional biologicals with broad application such as viral vaccines (for

Animal Cell Biotechnology, Vol. 1

human use) are produced very efficiently from anchorage-dependent prima-ry or diploid cells, and there is generally no reason to change the technology as long as the vaccine production remains restricted to primary or diploid host cell substrates.

Nontraditional biologicals from animal cells such as interferon, lympho-kines, enzymes, growth mediators and hormones are still in the state of research and development. Since most of these biologicals are coded by only one or at most a few genes they all are potential candidates for recombinant DNA technology and cheap production by microbial production systems. More complicated molecules requiring co- and post-translational modifica-tions during biosynthesis (glycosylation, thiolation), such as monoclonal anti-bodies, are produced by mass culture technology with animal cells.

TABLE I Aspects of Mass Production of Animal Cell Culture Products: Animal Cells Versus Microorganisms

Criteria for mass production	Animal cell (AC)	Genetically modified microorganism		Bacterial cells, animal cell relative factor AC = 1
		Prokaryotic	Eukaryotic	
Strain stability	Variable	Variable	?	1
Nutritional re-quirement	Complex	Simple	Simple	
Media (cost)	High (2–5)	Low (1)	Low (1)	2–5
Cell density (g·litre^{-1})	Low (0.05–0.5)	High (5)	High (20)	100
Productivity of biomass forma-tion (g·litre^{-1})	Low (0.2)	High (6)	High (6)	30
Cell specific product formation (g·g^{-1}·day^{-1})	Hybridoma-Ig (0.1–100)	Up to 5% enrich-ment (0.03)	?	0.0003–0.3
Product secretion	Yes	No	Variable	
Process control	Complex	Complex	Complex	1
Scale-up	Variable	Simple	Simple	<1
Biohazards	?	?	?	1
Product quality				
Homologous	Yes	No	?	?
Co- and post-translational processing	Yes	No	?	?
Product modifications	Variable	Possible	Possible	?
Downstream processing	Most simple	Mostly complex	Variable	?

The negative and positive aspects of animal cell technology compared to recombinant DNA technology with prokaryotes are shown in Table I. It is evident that the criteria of mass production clearly favour the application of recombinant DNA technology. However, as far as the quality of the biologicals produced by either alternative is concerned, the situation is not yet clear enough to draw final conclusions. Such considerations of quality and problems involved in gene expression and translation in bacterial recipients may have been responsible for the fact that DNA cloning in animal cell systems has become possible (7, 8, 18). If this continues, reliable mass cultivation systems for engineered animal cells will become obligatory.

In this chapter we will consider the aspects of mass cultivation of animal cells in theory and practice.

2. IDEALIZED FEATURES OF MASS CULTIVATION AND PRODUCTION

The interrelations of the factors involved in the mass cultivation and production of animal cell culture products are very complex. In Fig. 1 some idealized features are shown. The cells' activities and their requirements with respect to physical and chemical environmental stimuli *in vitro* are the central point of our considerations. A cell culture will display special characteristics (growth and/or product release) in response to a certain set of chemical and/or physicochemical factors. For simplicity we distinguish between physical and chemical environmental parameters, with the physical environment of a cell culture being largely affected by the configuration of a bioreactor and the power supplied, and the chemical input to a large extent determined by the composition of the nutrient medium. The resulting cell activity is fed back upon both levels of input, physically and chemically (e.g. by growth or cell aggregation and by secretion of substances). The overall result of these biological, chemical and physical interrelations defines the status of the cell population.

We judge the status of the culture by measuring state variables (on morphological, biological, ultrastructural, physical, chemical etc. levels), whose numerical value is usually of very limited value due to analytical restrictions (e.g. change of cell number versus time gives a simple growth curve).

Biotechnologists are interested in manipulating the culture in order to optimize growth and/or product formation. Optimization of the culture is achieved by interaction at different levels, by cell strain improvement (selection, mutation, genetic recombination) and by means such as improved parameter control, formulation of suitable nutrient media and stimuli (42), choice of proper techniques of cultivation (bioreactor configuration and reac-

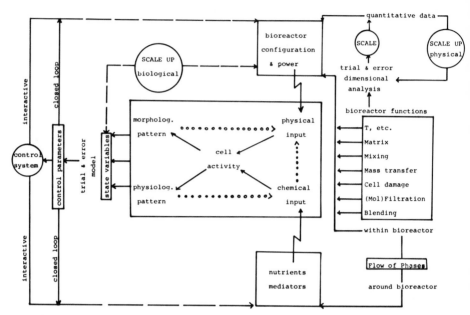

Fig. 1. Factors involved in the mass cultivation and production of animal cell culture products.

tor performance) and application of a suitable cultivation method (open or closed, homogeneous or heterogeneous, filtration and dialysis).

A control system consists of either simple closed control loops (set point control) such as temperature or pH capable of maintaining certain parameters at desired levels, or of more sophisticated systems (direct digital control, rarely applied in cell culture) which are able to interact at certain stages of culture development (by feeding some mediators or changing a physical parameter). Whether a measurable variable turns out to be a parameter which governs the process (i.e. a control parameter) or is just a variable suitable for indicating the state of the culture has to be delineated qualitatively and quantitatively by trial-and-error experimentation and process modelling (e.g. the correlation of lactic acid dehydrogenase activity with FMD virus infectivity status) (41).

Physical power in the bioreactor has functions such as fluid mixing, with some mass transfer between phases (liquid, gaseous and bio-phase) which may be more or less beneficial for a cell culture or even detrimental to cell growth and development (e.g. cell damage). These reactor functions are determined by the configuration and by the scale of a given reactor configuration. Dimensional analysis of physical reactor performance together with trial-and-error experiments can quantitatively translate defined bioreactor functions into characteristic numbers and coefficients, which define the reactor performance and make the choice of a proper reactor configuration possible.

From Fig. 1 it is also apparent that any change in the reactor size (scale up or scale down) will affect the interrelations between the different physical reactor functions. As one or more of these physical functions are changed due to physical scale up (or scale down) of the reactor, this will affect the development of the culture and complicate the scale up of the entire process. Because of the complex nature of all of the factors involved in mass cultivation and production, it may often happen in practice that one does not know why a process works properly. However, the precise determination of all the definable conditions constitutes the modus operandi of the process and is transferable under the heading of process "know-how".

2.1. Biological Aspects of Process Design

In technological terms the optimization of an *in vitro* cell culture system is possible on two general levels of manipulation, first by improvement of the genotypes of the cell strains and secondly by process design and control, which are aimed at bringing about the optimum phenotypic expression of the genetic potential of any production strain. The latter are considered here.

The make-up of a culture system is a composite of three basic elements:

1. Cultivation techniques, which include such aspects as reactor configuration, how to suspend, mix or immobilize a cell population etc.;
2. The choice of a proper cultivation process (batch, fed-batch, continuous culture etc.; i.e. flow of phases within or around a reactor); and
3. Sophisticated parameter estimation and control.

In designing the entire cultivation system and the engineering of a reactor, the techniques of suspending cells or supporting cells with solid or soft matrixes, the methodology of feeding cells with nutrients etc. must be adapted to the particular cellular requirements and their characteristics of growth and product formation. The various types of growth and metabolite formation are shown in Table II.

In practice, a wide variety of reactor constructions and techniques have been used for cell culture. In the following sections we will show that, with respect to their functions, only a few basic techniques and types of reactors are distinguishable.

TABLE II Cell Characters Important for the Design of *in Vitro* Cultivation and Production Systems

Growth type	Supporting matrix necessary
	Supporting matrix not necessary
	Sensitive type
	Generation number limited
	Generation number "unlimited"
	Cell density-dependent
Growth kinetics	Asynchronous—(synchronous)
	Exponential
	Rate-limited
	Stationary (non-propagating)
Metabolite (product)	Growth rate-related
formation kinetics	Growth phase-related
	Cell cycle-related
	Non-growth-related (permanent)
	Intracellular-secreted
Regulation of metabolite	Constitutive-induced (derepressed)
(product) formation	Catabolite-repressed (derepressed)
	Feedback repressed (gross regulation)
	Feedback inhibited (fine regulation)
	Mediator-related
	Host cell–virus interaction (various)
	Cell–cell interaction (complex, e.g. cross-feeding)

2.2. Influence of Reactor Design

From the engineering point of view, two general principles of cell cultivation are distinguishable: growth on gas/solid medium interphase (e.g. on nutrient agar surfaces) and the techniques of deep culture (i.e. cells surrounded by liquid nutrient). The principles of deep culture techniques are generally applied to both extremes of animal cell types, the anchorage-dependent and suspension cell types. There are only a few different techniques of deep culture; these are shown in Fig. 2.

In static culture the mass transfer between the cells (biophase), the nutrient medium (liquid phase) and the gas phase is limited by diffusional transport phenomena. The interfacial mass transfer rate (MTR) of any solute is described in its simplest form by the equation

$$\text{MTR} = K_L A (C^* - C_L) \tag{1}$$

where K_L is a rate coefficient for a given solute and a given superficial liquid motion

A represents the interfacial area

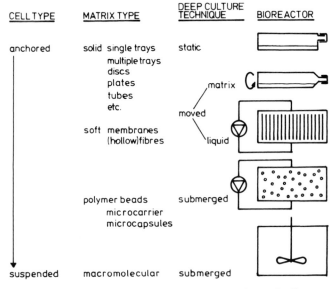

Fig. 2. Techniques used for deep culture of animal cells.

C^* describes the theoretical concentration of the solute at equilibrium conditions without consumption

C_L is the actual concentration seen by the cells, or at the site of consumption

From the simple relationship in Eq. (1) it is evident that in growing systems, spatial concentration gradients of solutes must exist.

In static deep culture vessels the interfacial mass transfer is predetermined by the geometric configuration and the fluid level, which defines the liquid depth and the area of interface for transfer. Static deep culture systems are generally used at the small scale for all types of animal cells. Their scale-up is easily achieved by keeping the relative interface area constant, a fact which leads to multiples of geometrically identical units (e.g. the development of the Multitray unit by S/A Nunc Denmark).

The scale-up of a process by increasing the number of units characterizes the process as a multiple process system, whereas when scale-up can be achieved by increasing the size of the unit, the process is defined as a unit process system. Most of the multiple systems enable the reduction of manual labor and allow simple handling of cell cultures on preparative scales (10^{10}–10^{11} cells) with decreased risk of contamination. However, the scale-up potential of static deep culture systems is generally limited. Increasing the fluid level causes increasing concentration gradients of reactants. Both improved MTR and reduced spatial concentration gradients are achieved in agitated reactors.

Three basic techniques for the construction of dynamic depth cultures can be distinguished (compare Fig. 2):

1. Surface (matrix) moved, liquid "static"; roller bottle type, rotating disk (15, 16, 28)

2. Surface (matrix) static, liquid moved; packed beds or columns, all types of supporting or protecting matrices possible (10, 26, 29, 30, 40, 50)

3. Matrix (and cells) moved, liquid moved; submerged deep culture in various reactor types (microcarrier culture)

Techniques 1. and 2.—the biological film reactors—are broadly established in animal cell technology, particularly for perfused immobilized cell culture systems of both the anchorage-dependent and suspension cell types. Cell systems which release their products in a more or less non-propagating physiological state are also handled by such techniques. Many types of cells and matrixes (plates, tubes, disks, filters, (hollow) fibres, beads, capsules etc.) are possible (42).

Submerged deep culture techniques are preferentially applied to "non-sensitive" suspension cell systems, to "sensitive" cell systems protected by matrixes or encapsuled in beads or to anchorage-dependent cells which are immobilized on beads, fibres etc. (e.g. microcarrier culture) (49).

How do we define the sensitive and non-sensitive suspension cell systems and how do matrixes interfere with growth? Many suspension cell systems can simply be propagated in conventional submerged deep culture reactors; others—the sensitive types—propagate more readily in static deep culture. Whether a suspension cell system belongs to the sensitive or non-sensitive type must be assessed by experiment. Hybridomas, which are non-adherent but stick to surfaces, very often turn out to be sensitive and may be grown and produced in Roux bottle culture or in multitrays better than in spinner flasks or any other submerged deep culture system. Complex interactions may lead to different growth results. Mixing phenomena, although not necessarily detrimental to the cells, have to be carefully considered in the design of any submerged deep culture system and whenever free cell surfaces are exposed to liquid or mechanical stress. It is obviously very complicated to combine both an optimum of mixing in a submerged reactor in order to achieve a reasonable MTR, with an appropriate degree of homogeneity in the liquid bulk and a minimum of effects that would be detrimental for the microenvironment of the cell surfaces. These two effects are at cross-purposes and leave a very small edge for the designer to go along, particularly when a given system has to be scaled up.

In this context matrix phenomena have to be considered for suspension cell systems because the necessity of a growth-supporting matrix is not as clearly assessed as for the anchorage-dependent cell type. There is, however, much experimental evidence that suspension cell systems, which differ with cell type and cultivation technique, also need a matrix which is macromolecular in nature and serves as a cell surface protector. Serum, for example, contributes more to the medium than a well-balanced mixture of nutrients; its macromolecules serve as important protectants in submerged systems. This macromolecular function is evident from experiments with media in which serum has been replaced. The addition of macromolecules such as PVP or Methocel (23, 37, 43) was essential for growth when submerged deep culture techniques were applied and was less important in static deep culture (4). If we consider the matter as a whole we can conclude that all suspension cells require certain protective matrixes. Both the type of matrix and the technique of deep culture have to be adapted to the particular requirements of the cell line. In some cases it was shown that aggregates of anchorage-dependent cells can substitute for a matrix (46). Normal human fibroblasts, considered to be entirely anchorage-dependent for proliferation, have been grown in methylcellulose medium without solid matrixes (34).

2.3. Influence of Operating Procedures

In the preceding section techniques were examined for their significance in the interfacial transfer of masses and other mixing phenomena within a bioreactor. Another important aspect in analyzing a bioreactor as well as the performance of an entire cultivation system is that of its bulk flow characteristics; i.e. the general regimen of the flow of phases (liquid bulk phase, biophase etc.) within and around a bioreactor. The connection between the flow regimen of a reactor (or reactor system) and the physiological patterns of a culture can be deduced for any given situation. It is, however, useful in such a consideration to think in idealized schemes and to look at their dynamic behaviour in extreme situations. Real systems will reflect, to a greater or lesser extent, the picture within these idealized extremes.

Bulk flow regimens within (throughout) a reactor will tell us whether the reactor is homogeneously (evenly) blended or whether the blending quality is heterogeneous. Bulk flow regimens around (or through) a reactor will determine whether a system is closed (batch) or open (continuous). Again thinking in idealized terms of fluid blending, either closed–homogeneous and closed–heterogeneous or open–homogeneous and open–heterogeneous systems must exist for submerged reactor types. The idealized regimens of fluid mixing in continuous bioreactors are shown in Fig. 3.

For any homogeneous reactor the blending (quality) is assumed to be perfect, which means that the concentration of any reactant solute entering

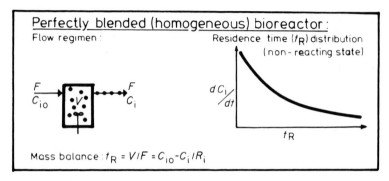

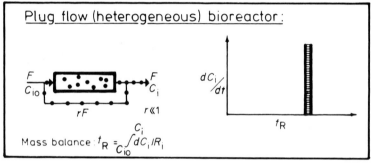

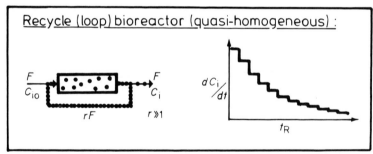

Fig. 3.   Ideal regimens of blending of continuous submerged bioreactors.

the reactor (C_{io}) is immediately distributed uniformly throughout the reactor and the reactant solute concentration (C_i) (or concentration of particles, if we assume perfect mixing) within the reactor and at the outlet is identical. The residence time (t_R) concept, which is illustrated for a non-reacting state in Fig. 3, shows that a given solute molecule (or a given particle if we include mixing quality in our considerations) may leave the reactor at time zero or later, or is in other words a statistically calculable mean value. In the plug flow reactor no blending is assumed, a situation which represents the extreme of heterogeneity with respect to blending quality. Thus t_R for any solute (or particle) entering the plug flow reactor is not subject to statistical deviations but is subject to an exact temporal and spatial regimen. The recycle (loop) reactor, which is a plug flow reactor having a very high ratio of recycling ($r>>1$), behaves in its overall performance similarly to the flow regimen of the perfectly blended reactor and approaches a "quasi-homogeneous" status.

The overall rates of biological reactions (R_i) are related to various environmental parameters such as the concentrations of reactants (C_i), pH and temperature and also the concentration and physiological state of cells:

$$R_i = f(C_i, T, \text{pH state and concentration of cells}) \qquad (2)$$

It is clear from Eq. (2) and from the mass balances that the different reactors shown in Fig. 2 must lead to different physiological patterns of growing cell systems, simply as a result of different mixing quality in or around a reactor.

So far, idealized regimens of flow (and mixing) have been considered and their relevance for the physiological pattern of a culture is easily assessed. However, real cultivation systems are more complex and much more complicated to overview for different reasons. In a stirred tank reactor (standard reactor) the limits to perfect mixing are clearly given by the deleterious effects of stress on the cells arising from the mixing and also by the reactor scale-up. Several standard reactors perfused in a serial cascade do, in practice, approach the characteristics of a plug flow reactor more closely than a tube-shaped one, etc.

2.4. Basic Cultivation Systems and Physiological Patterns

Cells behave differently to environmental stimuli and to concentration gradients of nutrients or mediators. In a closed (batch) culture the concentration of nutrients decreases and waste products accumulate in relation to cultivation time. This state of affairs is responsible for the variety of physiological responses observed during the course of a batch culture.

The open cultivation system with a continuous supply of fresh medium and withdrawal of exhausted medium, on the other hand, more closely approaches the situation *in vivo*. In the living animal an efficient circulatory system delivers nutrients to cells and removes their waste products and thus provides the cells with a stable environment. From this point of view the open system is the optimum since a steady-state environment is achieved. It is also worth mentioning that the (non-biological) temperature-dependent inactivation of "sensitive medium components" is also minimized in open cultivation systems, since these substances are constantly fed with the fresh medium into the culture at ideal concentration levels and are immediately used by the cells (if this were not the case, the storage of media in the cold would not make any sense). If we bear in mind all of these aspects, the relationship that exists between a particular cultivation system and a particular physiological pattern can be assessed.

An entire cultivation system is composed of purely technical elements providing mass transfer to submerged or immobilized, propagating or non-propagating cells and purely methodological elements (batch or continuous, homogeneous or heterogeneous culture). We have so far considered that the combination of particular techniques and methods in order to design a cultivation system will depend mainly on the type, kinetics and regulation of growth and product formation (see Table II) and also on some non-biological aspects. A condensed systematic evaluation of these aspects is shown in Fig. 4, where the basic types of deep culture systems and their inherent physiological patterns are summarized. The cell density (X), concentration of substrate (S) which is actually seen by the cells and cell-specific substrate turnover rate (Q_s) are used to describe typical physiological patterns.

In the batch reactor, the conventional routine cell culture system, the growing cells are exposed to a permanently changing environment. After inoculation into a medium that contains the right nutrients the cells grow, nutrients are consumed and waste products accumulate. In the early phases of the batch culture, the specific substrate turnover rates (Q_s) are artificially enhanced due to the fact that most cells (habituated to homeostatic concentrations *in vivo*) are not programmed to control uptake and turnover of substrates properly. Both enhanced Q_s and enhanced accumulation of waste products, as well as higher losses of sensitive materials in the medium due to inactivation at physiological temperatures, will impose limits on the culture development in later phases of batch culture. The physiological patterns of batch systems are therefore characterized in Fig. 4 by constantly changing S and changing Q_s during the batch time and by final low cell densities (X).

These physiological patterns can be considerably changed (or improved with respect to cell densities) when closed systems are changed successively to partially open or fully open cultivation systems. In fed-batch systems the nutrients are kept at the proper concentrations by controlled feeding; however, waste products and cells are not withdrawn. As a result Q_s and S are stabilized to constant levels and the overall cell densities and the cell yield per litre of medium tend to increase. In cultivation systems where the cells are retained by filters (or molecular sieves) and the culture is continuously perfused with fresh nutrients and waste products are removed, the effect on culture development is further improved and cells at very high densities can be kept viable (12, 45).

The various forms of perfusion culture, in which suspended cells are retained by filters or immobilized on beads, provide an *in vivo*-like environment. They are suitable for the cultivation of all types of cells and very good

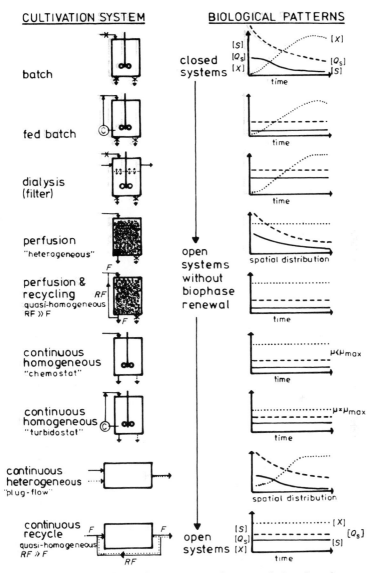

Fig. 4. Basic types of deep culture systems and associated physiological patterns.

growth results have been reported with various types of cells at various scales of operation (12, 50, 51). Perfusion systems are particularly advantageous for non-propagating (stationary phase) cells, when cells of limited life-span are used and when the kinetics of product formation are typically non-growth-related, such as with urokinase, monoclonal antibodies etc. Additionally, the combination of perfusion and external recycling of media offers the application of "quasi-homogeneous" cultivation as well as packed- and fluidized-bed reactors.

The various forms of fully open (continuous) cultivation systems, which also include continuous withdrawal of cells, are limited in application to permanent cell lines. The most common open system is the chemostat (see Chapter 8, this volume) in which the specific growth rate (u) of the culture is limited by the concentration of any reactant and is kept in a self-balancing steady-state equilibrium. A special form of homogeneous continuous culture is the principle of the turbidostat, in which the steady state is maintained by an external growth-related control loop and cell growth is permanently maintained at the maximum rate (u_{max}). The chemostat represents the most efficient system for cell mass production since cell nutrients can be supplied to the cells in low and well-balanced concentrations and non-biological inac- tivation of sensitive substrates is minimized (19).

The continuous heterogeneous reactor, the plug flow reactor, is, at first glance, of more theoretical interest. The "perfect inhomogeneity" in this reactor gives rise to physiological patterns which, with respect to their spa- tial distribution in the reactor, are comparable to those observed in the batch culture with respect to time. In other words, the plug flow reactor repre- sents the growth phases of the batch culture on a time-independent basis. Two aspects are clear from this comparison: firstly, whether the quality of mixing is of physiological relevance at all, and secondly, that in a serial cascade of several continuous homogeneous reactors, which will represent the realization of a plug flow reactor, all phases of a batch culture can be permanently reproduced in practice. Whenever a production system re- quires the continuous renewal of biomass and the product formation re- quires a separate production step (e.g. virus production, interferon produc- tion), the application of two-step or multistep continuous culture systems should be considered, since it combines the optimum condition of biomass production in the first reactor and the possibility of optimizing and condi- tioning the production phase in the following steps (e.g. by addition of inducers, virus). The heterogeneous continuous culture will probably be- come the system of choice when inducible promoters are cloned in animal recipient cell strains by recombinant DNA techniques. The latter philoso- phy, as expressed here, has already been verified to a large extent by experi- mental data published by Collins (8) and Hofschneider (18).

3. SUBMERGED DEEP CULTIVATION

Most of the submerged deep culture reactors currently in use have been designed for the mass cultivation of microorganisms and thereafter have been adapted to the specific requirements of animal cells. The criteria by which microbial reactors have been adapted to use for animal cell culture deserve detailed consideration since there are requirements common for both cellular and microbial purposes as well as quite contradictory ones. Reactor functions such as mixing quality, parameter detection and control, low contamination risk, simple and transparent behaviour on scale-up etc. are desired both for the cultivation of microorganisms and for animal cells. However, animal cells are generally more sensitive with respect to mechanical and liquid stress factors. It can also be seen from Table I that animal cell cultures are 1.5–2 orders of magnitude less productive of biomass due to lower cell densities and lower rates of propagation. Submerged deep culture reactors for animal cell cultures must therefore be designed for effective liquid mixing combined with low stress factors (mechanical and hydrodynamic) and for adequate interfacial mass transfer capacity.

3.1. Operational Reactors

In Fig. 3 we have shown an idealized picture of reactor performance. This idealized reactor performance is realized in practice by two general reactor designs: the standard reactor (Rushton type), in which mixing and mass transfer are achieved by mechanical power input (e.g. by impellers, marine propellers, vibro mixers) and pneumatic systems such as the air-lift type of reactor.

The standard reactor type, the stirred tank reactor, is the most often and widely used for animal cell cultivation, and belongs to the more difficult to understand types of reactor with respect to its performance. Usually the stirred tank reactor is never optimal but is most often reasonable. The physical functions of the standard reactor have been extensively analysed and described in hundreds of publications. A condensed evaluation of its usefulness for the cultivation of animal cells can be deduced from the trends shown in Table III, in which the alterations of physical reactor properties with scale-up are pointed out. From Tables III and IV it is clear that reactor functions such as specific liquid bulk flow, mechanical and hydraulic shear and oxygen transfer via the head space (parameters which we would like to have under control in a scale-up procedure) are dependent on undesired alterations in scaling up geometrically similar reactor configurations. The concept of geometric similarity in scale-up of standard reactors cannot therefore be recommended for animal cell culture. Alternatively, the concept of momentum factor published by Blakebrough and Sambamurthy (5) seems to be more reliable. The momentum introduced by stirrers into the liquid is scaled up in the latter case by using slowly rotating stirrers with very large impeller blade areas, resulting in high liquid bulk flow and reasonable mixing in the reactor. Increasing the impeller diameter (D_i) with reactor scale is

TABLE III Alteration of Physical Reactor Properties with Scale-up[a]

Properties		Bench-scale volume (1)			Production-scale volume (125)		
Biological	Physical						
Specific power dissipation (P/V)	$N^3 D_i^2$	1	1.0	25.0	0.2	0.0016	
Specific liquid bulk flow (Q/V)	N	1	0.3	1.0	0.2	0.04	
Mechanical shear at impeller tip (V-tip)	$N D_i$	1	1.7	5.0	1.0	0.2	
Hydraulic shear (Reynolds number)	$N D_i^2$	1	8.5	25.0	5.0	1.0	

[a] Geometric similarity, scale-up ratio 1 : 125. D_i, impeller diameter; N, rotation speed.

critical since both hydraulic and mechanical shear stress and the local power dissipation density at the tip of the impeller increase drastically with D_i. The application of these scale-up principles consequently leads to helix-shaped multiplate impellers similar to those recommended recently by Feder and Tolbert (*12*) for animal cell culture.

The principle of the vibromixer is also recommended instead of rotating impellers for animal cell cultivation (*13, 14, 31*). The power dissipated at the blades of a vibromixer is uniform with respect to its local dissipation density. This characteristic does not change during scale-up. Thus a certain condition of mixing, once established in a bench-scale reactor, can easily be scaled up.

The problem of head-space aeration (see Table IV) remains an unsolved question in scaling up. The oxygen transfer rate coefficient [K_L; see Eq. (1)] for head-space aeration has been estimated in our laboratory to be between 0.2 and 0.3 cm/min, corresponding to 20–30 $\mu g O_2/cm^2 \cdot hr$ (gas phase, air; liquid phase, aqueous). From these data, the gas–liquid head-space interface of a given reactor vessel and Eq. (1), the oxygen transfer via head-space can be calculated. If we assume that an average specific oxygen demand for

TABLE IV Oxygen Demand Versus Surface Aeration at Different Reactor Volumes

Culture volume (litre)	Head space area (cm^2)[a]	Oxygen supply by head space aeration (μg)	Volumetric oxygen demand (μg)
1	100	2,000	2,000
10	500	10,000	20,000
100	2,500	50,000	250,000

[a] Liquid height (H)/diameter (D) = 1.

growing cells of different kinds is generally in the order of 2–3 $\mu g/10^6$ cells per hour (*19*), head space aeration becomes critical for vessels of sizes in the range 10–100 litres; oxygen supply by air sparging or permeation of oxygen through membranes or any other technique of increased gas–liquid mass transfer becomes necessary (Table IV).

Direct sparging of gas is sometimes detrimental to cell growth. Whether oxygen supply by air sparging is possible has to be assessed for a particular cell system. When air sparging is acceptable the air-lift reactor can be used as an alternative to mechanically stirred reactors. The principles of the air-lift reactor and its application in large-scale cultivation of suspension cell systems have already been discussed (*19, 21, 22*). While mixing efficiency is somewhat lower than in a stirred tank reactor and there are additional problems in foam formation (both of which problems can be overcome by extending the draught tube to just below the surface and just above the bottom of the vessel), there is practically no technical limit to the scaling up of air-lift reactors.

3.2. Suspension Cell Cultivation Systems Currently in Use

Existing equipment and methods for cell cultivation are discussed here for suspension cells only, because there are reviews on anchorage-dependent cell techniques in this volume and in earlier reports (*42*). For suspension cells only a few new ideas of clear importance and future technical relevance have been reported. Most work has been done with equipment originally designed for microbial growth and modified for animal cell applications (*1, 6, 13, 14, 31, 38, 52*). All these systems later showed limitations, especially when cultivating sensitive cells. Reports on vessels for cell propagation larger than 1000 litres do not exist (compare oxygenation considerations above). In 1971 Telling and Radlett described (*44*) the necessity to minimize turbulence by low speed and used a special agitator design. They showed the dependence of animal cell growth on different chemical and physical parameters and reported the control of pH by sparging of CO_2 and air as required. The range of oxygen tensions most useful for cell yield and minimal glucose consumption for BHK cells was between 40 and 80 mm Hg. The same authors (in 1972) described with more sophisticated methods a sharp optimum of pO_2 at 80 mm Hg (*36*). The specific oxygen uptake was correlated with the growth rate of the cells (see also Fig. 7).

Zwerner *et al.* (*52*) reported results which indicate that the choice of the proper agitation system has to be carefully evaluated for any particular case. With different culture vessels of sizes up to 14 litres (spinner, vibro and marine impeller type) they found a marked influence of agitator system on the expression of surface antigens on mouse cells. Compared to an 8-litre spinner and a 14-litre marine impeller, the 3-litre Vibromixer showed comparable growth but 40% less expression of surface antigens.

Our own development is a reactor with extremely low power input and shear forces based on a bubble column (21, 22). We tested reactors with net volumes up to 700 litres and found good results with some lymphoblastoid cell lines and BHK 21. This air-lift–type reactor needs, contrary to more conventional types, less energy per volume in large vessels than in smaller ones. This is the reason that larger reactors work more effectively than small ones. Scale-down is restricted to sizes larger than 5 litres. The maximum local power dissipation density within the air-lift reactor is similar, irrespective of the reactor size, whereas in conventional reactors the local power dissipation density increases with reactor size as described above. The oxygen supply is not dependent on reactor size and can be managed with minute amounts of driving gas at any desired scale (less than 1 volume of gas per liquid volume per hour) (21).

In 1978 Pollard and Khosrovi (35) proposed a reactor design for insect cells and proposed a tubular flow reactor, the practicable form of the more idealized model of plug flow reactor described above. They designed a plant with a reactor of 185 m^3 total volume for continuous cell and virus propagation. It was proposed to do the virus multiplication in line in the rear part of the plug flow reactor. With a liquid velocity of approximately 1 m/hr they predict a production of *Autographa california* nuclear polyhedrosis virus of 7.5 tonnes per year, enough to treat about 10^6 acres with this biological pesticide.

Another interesting system is the spin filter culture described by Himmelfarb (17) and Tolbert et al. (12, 45). This is a typical perfusion reactor type where cells are kept in the reactor and only the medium is removed from the suspension by filtration through a rotating filter cartridge. This movement improves the problem of clogging by cell mass, which occurs during separation of animal cells from liquid by filters. The rotating filter cartridge generates a laminar liquid layer beneath the filter surface. The centrifugal force generated by the rotation forces the cells to move away from the filter and keeps the fluid on the filter surface free from cells. With this device it should be possible to achieve perfusion culture at close to the ideal culture conditions described above for extended periods.

Beside high cell densities it should also be possible to get better product generation. Cells are not wasted with the effluent but stay in the production process. The authors reported experiments at bench scale and in a 40-litre vessel where they sparged pure oxygen to meet the consumption of the cells. With a rat tumor cell line (Walker 256) they reached cell densities as high as 3×10^7 cells/ml at about 100% viability compared to 1.5×10^6 in a conventional spinner vessel. The medium consumption of a per-cell basis was about one-third of the conventional quantity.

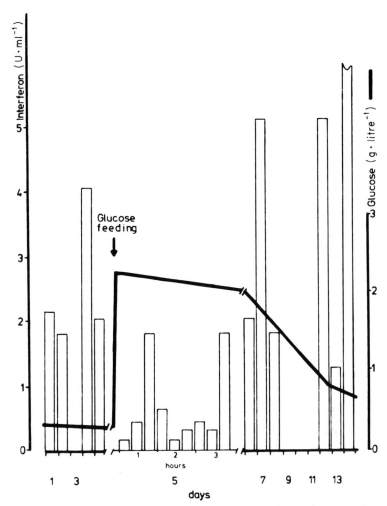

Fig. 5.   Interferon production from Namalwa cells grown under steady-sate conditions in continuous culture is decreased after Glucose feeding reversibly (induction by Sendai-virus under standard conditions in fresh medium RPMI 1640 without serum).

Publications on chemostatic cultures appeared as early as 1959 (3, 9). The first report on a long-time chemostatic culture with defined limiting substrate (glucose) was published in 1976 by Tovey and Brouty-Boyé (48). They found a good correlation of animal cell growth with the Monod equation and demonstrated that animal cell cultures show growth kinetics comparable to those of microorganisms. In 1980 Tovey (47) gave a paper on tumor cell growth in the chemostat and reported much better cell yields in the chemostat than in batch processes (about the eight-fold output), an important

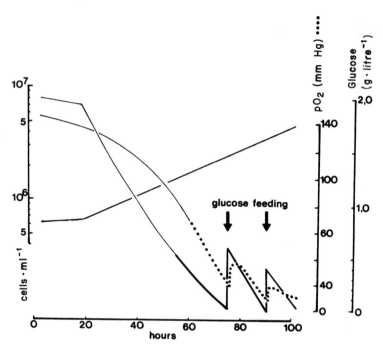

Fig. 6.   Batch culture of Namalwa cells in RPMI 1640 supplemented with 2% human serum. Oxygen consumption is reversibly inhibited by glucose feeding as shown by increase of pO_2 in an air-lift system with constant aeration and transfer rates.

finding for production processes, correlating well with the theoretical considerations above. Not only cell mass but also product titers may be improved by chemostatic fermentation. Details of the modern methodology and advantages of such culture systems may be obtained from Chapter 8 in this volume. Tovey's example of enhanced interferon yields in continuous culture was in good agreement with our own findings (24) as shown in Fig. 5. These and the following few examples may indicate that cell mass propagation and production processes need further improvement on all the levels sketched in Fig. 1.

We showed (cf. Fig. 6) that there is a direct interaction between glucose concentration and oxygen uptake which turned out to affect the adenylate energy concentration in the cells (24). Other phenomena resulting from oxygen tension (uptake) and glucose consumption have been described by Telling (44) and Taylor (43). In Fig. 7 the growth of BHK 21 cells in continuous culture at steady states with two different dilution rates in an excess of glucose is shown. At higher growth rates enhanced oxygen and glucose consumption and acid production (hydroxide consumption for neutralization)

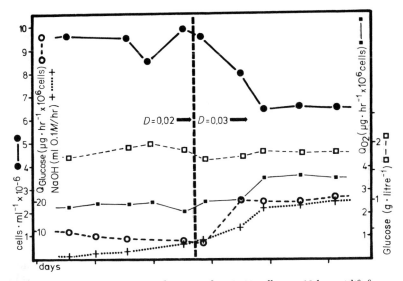

Fig. 7. Example of continuous cultivation of BHK 21 cells in a 10-litre airlift fermentor using Pirbright 7 medium. From a long time run the days before and after change of the dilution rate are shown. Decreasing cell density at roughly constant Glucose concentration is accompanied by increasing Glucose consumption, increased Oxygen uptake and increased acid production (NaOH consumption for neutralization).

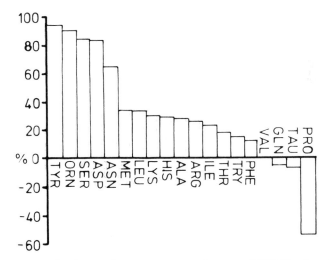

Fig. 8. Amino acid utilization during continuous cultivation of BHK 21 cells analyzed from the culture medium during steady state conditions at a dilution rate of 0.02 per hour. Bars give the percentage of use compared to fresh medium (Pirbright 7 M supplemented with 10% fetal calf serum). Negative values indicate an increase of this particular amino acid in the medium during cultivation.

are visible. These steady-state conditions allowed us to analyse the amino acid utilization during culture (25). We found very different use of particular amino acids in batch and continuous culture. As shown in Fig. 8, three amino acids were produced even during continuous cultivation. Under conditions of batch culture, however, the profile of amino acids consumed is quite different and depends on the phase of culture development. Similar observations were reported by Arathoon (2) and by Patterson (33). The improper composition of presently used cell culture media for mass cultivation can be inferred from these findings.

However, not only is medium composition a quite underdeveloped area in technical cell cultivation, but also reactor design is in a very early stage of development, particularly for the cultivation of sensitive cells such as hybridomas or anchorage-dependent cells on microcarriers. Hybridoma cells seem to be the most promising candidates for mass production in the near future (see also Chapter 3 in Volume 2). Their cultivation in standard fermenter units has worked up to now with only limited success. From our experience hybridomas in many cases do not grow and produce product at expected levels in submerged deep culture. For these types of sensitive cells a new approach for protecting fragile cells with a polymer coat with the use of existing standard fermenters was developed. The microcontainers thus achieved minimize the mechanical stress experienced by the cell surface. The polymer resins in which the cells are encapsulated have diameters of 50–500 μm, allowing the embedded cells to build up their own microenvironment, which is not destroyed by convection in the dynamic environment within the bioreactor.

Two methods were published in this field, one from Lim and Moss (27), who used polylysine beads, the other from Nilsson *et al.* (32), using agarose beads. The polylysine leads to microcontainers of small pore size, allowing nutrients to enter the capsule but retaining the secreted high-molecular-weight globulins inside. Antibody concentrations comparable to those of mouse ascites cultures have been reported after freeing the content of the capsules; this means about 100 times higher globulin titers than in conventional *in vitro* culture (11). This excellent production is obviously a result of both improved microenvironment and improved nutrient supply achieved by perfusion. To harvest the globulin the capsules must be opened by an enzymatic treatment. The disadvantage of this method may be the loss of cell material at harvest and the necessity for centrifugation and re-encapsulation.

The agarose method, which is used in our own laboratory, allows free diffusion of substances through the microspheres for molecules up to 300,000 daltons. Furthermore, it gives roughly the same or higher concentrations of antibody as bottle culture but requires fewer cells. In theory, the microspheres can be used indefinitely. For results see Figs 9 and 10. There is no doubt that the techniques of encapsulation of animal cells are in

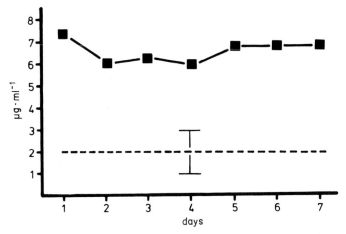

Fig. 9. Production of anti-Urokinase-IgM from a mouse–mouse hybridoma line. Cells were encapsulated at a density of 5.6 × 10⁶ cells/ml in 50- to 200-μm spheres consisting of 1% Agraosegel. 120 ml of these microspheres were cutivated in 400-ml culture medium (Dulbecco's modified MEM supplemented with 10% fetal calf serum). Medium was changed daily by separation of spheres through a 50-μm stainless steel sieve mounted to the roller flask. Production is shown as μg/ml IgM (■————■) compared to supernatants from stationary bottle cultures [(— — —), average and range].

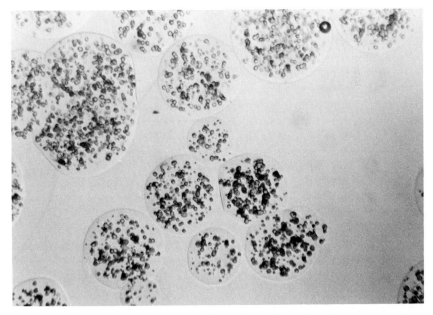

Fig. 10. Photomicrograph of encapsulated hybridoma cells one day after entrapment. Density of cells is 6.2 x 10⁶/ml gel (2.5% Agarose), sphere size between 80 and 200 μm diameter.

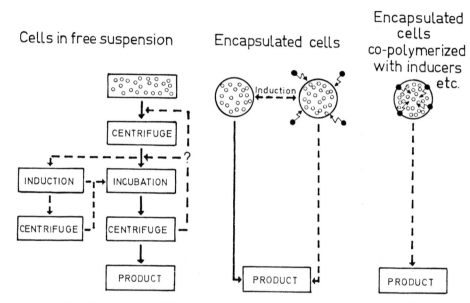

Fig. 11. Strategy for simplification of production processes by encapsulation of cells. Centrifugation steps are saved and entrapped cells may be used indefinitely. For processess where exogeneous inductors etc. are involved, these may obviously be immobilized within the encapsulation polymer to give the most simple form possible for the scheduled process.

their early stages and need further improvement. We assume that the techniques of microencapsulation of living cells, which are currently further developed and applied on a broad basis to plant cells and microorganisms, will stimulate research and promote their application with the more fragile and sensitive animal cells. A strategy for making microcontainers more functional with respect to their technological use is proposed in Fig. 11.

If we take into consideration that the encapsulation technique would allow the co-polymerization (or co-immobilization) of compounds such as inducers, enzymes, specific inhibitors and symbiotic systems together with the cell culture, another dimension of process improvement would be gained.

4. CONCLUSION

Taken together, the practical experience available on systems, reactors and methods with relevance to scaling up is poor. The paucity of the available data reflects the complexity of the matter in physical and chemical terms. Comparing hybridoma cultivation *in vivo* (ascites) and *in vitro*, it is clear that the presently used *in vitro* techniques in this case give yields in the range of only 1% of ascites fluid or even less. There is no logical reason for not achieving titres similar to those in ascites *in vitro*. This speculation may give an idea of how far technical cell cultivation is from optimal conditions. Even giant production plants for FMD vaccines with fermenter volumes of up to 4000 litres work suboptimally, as can be concluded from medium analysis and growth results (2). We hope that there will be more activity towards the optimization of technical cell culture methods so that these processes will have a chance to show their true potential and give realistic values compared with the results for microorganisms for which there are genetically modified organisms and well-established and optimized production processes.

REFERENCES

1. Acton, R. T., Barstad, P. A., Cox, P. M., Zwerner, R. K., Wise, K. S., and Lynn, S. D. (1977). Large-scale production of murine lymphoblastoid cell lines expressing differentiation alloantigens. *In* "Cell Culture and Its Application" (R. T. Acton and J. D. Lynn, eds.), pp. 129–161. Academic Press, New York.
2. Arathoon, W. R., and Telling, R. C. (1982). Uptake of amino acids and glucose by BHK 21 clone 13 suspension cells during cell growth. *Dev. Biol. Stand.* **50,** 145–154.
3. Averdunk, R., and Voss, H. (1968). Suspensionskultur-Verfahren für permanente Zellinien von Warmblütlern. *Zentralb. Bakteriol. Parasitenkd., Infektionstr. Hyg., Abt. 1: Orig.* **207,** 265–281.
4. Birch, J. R. (1980). The role of serum in the culture of a human lymphoblastoid cell line. *Dev. Biol. Stand.* **46,** 21–27.
5. Blakbrough, N., and Sambamurthy, K. (1966). Mass transfer and mixing rates. *Biotechnol. Bioeng.* **8,** 25–42.
6. Cherry, W. R., and Hull, R. N. (1960). Studies and observations on the growth of mammalian cells agitated in fluid medium. *J. Biochem. Microbiol. Technol. Eng.* **3,** 267–285.
7. Colbere-Garapin, F., Horodniceanu, F., Kourilsky, P., and Garapin, A. C. (1982). Construction of a dominant selective marker useful for gene transfer studies in animal cells. *Dev. Biol. Stand.* **50,** 323–326.
8. Collins, J. (1982). Gewinnung von Interferonen durch DNA-Rekombination. *Symp. Biotechnol. Arzneim.-Forsch. 1982, Bad-Schliersee, Federal Republic of Germany.*
9. Cooper, P. D., Wilson, J. N., and Burt, A. M. (1958). Critical effect of oxygen tension on rate of growth of animal cells in continuous suspended culture. *Nature (London)* **182,** 1508–1509; *J. Gen. Microbiol.* **21,** 702–720 (1959).
10. Corbell, M., Trudel, M., and Payment, P. (1979). Production of cells and viruses in a new multiple-tube tissue culture propagator. *J. Clin. Microbiol.* **10,** 91–95.
11. Editorial (1982). Method may boost monoclonal antibody output. *Chem. Eng. (N.Y.)* **60,** 22–23.
12. Feder, J., and Tolbert, W. R. (1983). The large scale cultivation of mammalian cells. *Sci. Am.* **248,** 36–43.
13. Fontanges, R., Deschaux, P., and Beaudry, Y. (1971). Apparatus for continuous, large-volume suspended cell culture. *Biotechnol. Bioeng.* **13,** 457–470.
14. Girard, H. C., Okay, G., and Kivilcim, Y. (1973). Use of the vibro fermentor for multiplication of BHK cells in suspension and for replication of FMD virus. *Bull. Off. Int. Epizoot.* **79,** 805–822.

15. Girard, H. C., Sütcü, M., Erdum, H., and Gürhan, I. (1978). Monolayer culture of animal cells with the gryogen equipped with tubes. *Dev. Biol. Stand.* **42**, 127–133.

16. Gullino, P. M., and Knazek, R. A. (1979). Tissue culture on artificial cappillaries. *In* "Methods in Enzymology" (W. B. Jakoby and I. H. Pastan, eds.), Vol. 58, pp. 178–184. Academic Press, New York.

17. Himmelfarb, P., Thayer, P. S., and Martin, H. E. (1969). Spin filter culture: The propagation of mammalian cells in suspension. *Science* **164**, 555–557.

18. Hofschneider, P. H. (1982). Gewinnung von Vakzinen durch DNA-Rekombination. *Symp. Biotechn. Arzneim.-Forsch.*, *1982,Bad–Schliersee, Federal Republic of Germany.*

19. Katinger, H. W. D., and Scheirer, W. (1982). Status and developments of animal cell technology using suspension culture techniques. *Acta Biotechnol.* **2**, 3–41.

20. Katinger, H. W. D., and Scheirer, W., unpublished results.

21. Katinger, H. W. D., Scheirer, W., and Krömer, E. (1979). Bubble column reactor for mass propagation of animal cells in suspension culture. *Ger. Chem. Eng. (Engl. Transl.)* **2**, 31–38.

22. Katinger, H. W. D., Scheirer, W., and Krömer, E. (1978). Der Blasensäulenfermentor für die Massensuspensionskultur tierischer Zellen. *Chem.-Ing.-Tech.* **50**, 472.

23. Keay, L. (1975). Autoclavable low cost serum-free cell culture media. The growth of L-cells and BHK-cells on peptones. *Biotechnol. Bioeng.* **17**, 745–764.

24. Krömer, E., and Katinger, H. W. D. (1982). Effects of glucose-feeding on glucose and oxygen-requirement, energy-charge and interferon-production of human lymphoid cells cultivated in vitro. *Dev. Biol. Stand.* **50**, 349–354.

25. Krömer, E. (1977). Diplomarbeit, Institut für angewandte Mikrobiologie der Universität für Bodenkultur, Wien.

26. Ku, K., Kuo, M. J., Delente, J., Wildi, B. S., and Feder, J. (1981). Development of a hollow-fiber system for large-scale culture of mammalian cells. *Biotechnol. Bioeng.* **23**, 79–95.

27. Lim, F., and Moss, R. D. (1981). Microencapsulation of living cells and tissues. *J. Therm. Sci.* **70**, 351–354.

28. Litwin, J. (1976). A new type of multi-surface rotating vessel for anchorage dependent cells. *Proc. Gen. Meet. Eur. Soc. Animal Cell Technol., 1st, 1976, Amsterdam* (R. E. Spier and A. L. van Wezel, eds.), pp. 49–55.

29. Mann, G. F., and de Mudea, J. (1977). Replication of polio virus in perfused cultures of MRC-5 diploid cells. *Dev. Biol. Stand.* **37**, 255–258.

30. Merk, A. M. (1982). Large scale production of human fibroblast interferon in cell fermenters. *Dev. Biol. Stand.* **50**, 137–140.

31. Moore, G. E., Hasenpusch, P., Gerner, R. E., and Burns, A. A. (1968). A pilot plant for mammalian cell culture. *Biotechnol. Bioeng.* **10**, 625–640.

32. Nilsson, K., Scheirer, W., Merten, O.-W., Östberg, L., Liehl, E., Katinger, H. W. D., and Mosbach, K. (1983). Entrapment of animal cells for the production of monoclonal antibodies and other biomolecules. *Nature (London)* **302**, 629–630.

33. Patterson, M. K., Jr. (1972). Uptake and utilization of amino acids by cell cultures. *In* "Growth, Nutrition and Metabolism of Cells in Culture" (G. H. Rothblat and V. J. Cristofalo, eds.), Vol. 1, pp. 171–209. Academic Press, New York.

34. Peehl, D. M., and Stanbridge, E. J. (1981). Anchorage independent growth of normal human fibroblasts. *Proc. Natl. Acad. Sci. U.S.A.* **78**, 3053–3057.

35. Pollard, R., and Khosrovi, B. (1978). Reactor design for fermentation of fragile tissue cells. *Process Biochem.* **13**, 31–37.

36. Radlett, P. J., Telling, R. C., Whiteside, J. P., and Maskell, M. A. (1972). The supply of oxygen to submerged cultures of BHK 21 cells. *Biotechnol. Bioeng.* **14**, 437–445.

37. Radlett, P. J., Telling, R. C., Stone, C. J., and Whiteside, J. P. (1971). Improvements in the growth of BHK 21 cells in submerged culture. *Appl. Microbiol.* **22**, 534–537.
38. Rightsel, W. A., McCalpin, H., and McLean, I. W. (1960). Studies on large-scale methods for propagation of animal cells. *J. Biochem. Microbiol. Technol. Eng.* **2**, 313–325.
39. Scheirer, W., unpublished results.
40. Schleicher, J. B., and Weiss, R. E. (1968). Application of a multiple surface tissue culture propagator for the production of cell monolayers, virus and biochemicals. *Biotechnol. Bioeng.* **10**, 617–624.
41. Spier, R. E. (1977). Determination of the time to harvest foot and mouth disease virus by measurement of the supernatant concentration of lactic dehydrogenase. *Biotechnol. Bioeng.* **20**, 929–932.
42. Spier, R. E. (1980). Recent developments in the large scale cultivation in animal cells in monolayers. *Adv. Biochem. Eng.* **14**, 119.
43. Taylor, G. W., Kondig, J. P., Nagle, S. C., Jr., and Higuchi, K. (1971). Growth and metabolism of L cells in a chemically defined medium in a controlled environment culture system. *Appl. Microbiol.* **21**, 928–933.
44. Telling, R. C., and Radlett, P. J. (1971). Large scale cultivation of mammalian cells. *Adv. Appl. Microbiol.* **13**, 91–119.
45. Tolbert, W. R., Peder, J., and Kimes, R. C. (1981). Large-scale rotating filter perfusion system for high density growth of mammalian suspension cultures. *In Vitro* **17**, 885–890.
46. Tolbert, W. R., Hitt, M. M., and Feder, J. (1980). Cell aggregate suspension culture for large-scale production of biomolecules. *In Vitro* **16**, 486–490.
47. Tovey, M. G. (1980). The cultivation of animal cells in the chemostat: Application to the study of tumor cell multiplication. *Adv. Cancer Res.* **33**, 1–37.
48. Tovey, M., and Brouty-Boyé, D. (1976). Characteristics of the chemostat culture of murine leukemia L1210-cells. *Exp. Cell Res.* **101**, 346–454.
49. van Wezel, A. L. This volume. Chapter 11.
50. Whiteside, J. P., and Spier, R. E. (1981). The scale-up from 0.1 to 100 liters of a unit process system based on 3 mm diameter glass spheres for the production of different strains of FMDV from BHK monolayers. *Biotechnol. Bioeng.* **23**, 551–565.
51. Whiteside, J. P., Whiting, B. R., and Spier, R. E. (1979). Development of a methodology for the production of foot and mouth disease virus from BHK 21-C13 monolayer cells grown in a 100 l (20 m²) glass sphere propagator. *Dev. Biol. Stand.* **42**, 113–119.
52. Zwerner, R. K., Runyan, C., Cox, R. M., Lynn, D., and Acton, R. T. (1975). An evaluation of suspension culture system for the growth of murine lymphoblastoid lines expressing TL and THY-1 alloantigens. *Biotechnol. Bioeng.* **17**, 629–657.

8

The Cultivation of Animal Cells in Continuous-flow Culture

MICHAEL G. TOVEY

Laboratory of Viral Oncology
Institut de Recherches Scientifiques sur le Cancer
Villejuif, France

1. INTRODUCTION

Although continuous-flow culture is now firmly established as an important technique in microbiology, most cell biologists are quite unfamiliar with the system and indeed many regard the traditional methods of cultivating animal cells as wholly adequate. Until recently there have been only a few isolated reports of attempts to cultivate animal cells in continuous-flow culture and in general this system has received little attention. The apparent complexity of continuous-flow culture has undoubtedly deterred many people from adopting the system. However, in actual fact the basic concept of continuous-flow culture is simple and the apparatus used is a good deal less

Animal Cell Biotechnology. Vol. 1

complicated than many apparatuses currently used in laboratories. Similarly, fears about the risks of contamination in long-running culture have also proved to be largely unfounded (1). I hope to show in this chapter that continuous-flow culture (2, 3) has numerous advantages for the cultivation of animal cells.

2. BATCH CULTURE

2.1. Limitations of Conventional Batch Culture

Traditionally animal cells have been cultivated in batch culture either as suspensions or monolayers. A batch culture is a closed system in which cells are inoculated into a quantity of nutrient medium contained in a suitable vessel. The culture is incubated at the required temperature and events are then allowed to run their course. As cell multiplication proceeds nutrients are consumed and metabolites accumulate, thereby changing the environment of the culture. These changes in turn affect cell metabolism and lead ultimately to cessation of cell multiplication.

2.2. Batch Cultures with Medium Feeds

A closed batch culture consists of a series of transient states difficult to define and even more difficult to control. Numerous attempts have been made to ameliorate the basic batch culture either by means of a substrate feed (intermittent or continuous) or by perfusing the culture with nutrient medium. The periodic replacement of a constant fraction of cell suspension by fresh medium, which is the most common means of subcultivating animal cells, is often purported to be continuous or semi-continuous culture. In fact, this type of culture, which has been termed solera culture (4), is really only a succession of batch cultures in which part of the old culture is used as inoculum. Solera culture has been used to produce Namalwa cells on a semi-industrial scale for the production of human interferon (5).

In fed batch culture (6) the culture is fed continuously with medium with a corresponding increase in the volume of the culture. Since the medium flow rate is maintained constant and the culture volume increases progressively, the dilution rate and consequently the growth rate of the culture will decrease progressively (7). If a portion of the culture is then withdrawn at intervals, the culture can be maintained more or less continuously. Fed batch culture would seem to be a good system for the production of substances such as interferon, the synthesis of which is maximal after cells have reached a stationary phase (8).

Several devices have been developed for perfusing cultures of animal cells with fresh medium; in such cultures the cells are retained by some mechanical means, such as a dialysis membrane or filter, which permits exchange of spent medium with fresh medium outside the culture (9–12). Although such methods undoubtedly allow high cell densities to be attained, they still suffer from the same basic limitation of batch culture, that is, they are *closed* systems (where cells, or cells together with nutrients, are retained) consisting of a series of transient states culminating eventually in the demise of the culture. Such difficulties can, however, be overcome by the use of an *open* system such as continuous-flow culture, in which there is input of substrate and output of cells and cell products. Such an open system allows a steady state to be obtained in which constant conditions can be maintained indefinitely.

3. CONTINUOUS-FLOW CULTURE

Chemostat continuous culture is based on the principle established by Monod (13) that at submaximal growth rates, the growth rate of an organism is determined by the concentration of a single growth-limiting substrate. The theory of continuous-flow culture was first established by Monod (14) and independently by Novick and Szilard (15), who introduced the term chemostat. In practice, a chemostat consists of a culture of fixed volume, into which nutrient medium is fed at a constant rate. Medium is mixed with cells (sufficiently to approximate to the ideal of perfect mixing) and cells together with medium leave the culture at the same rate. The culture is first started off batchwise and then, during the period of exponential multiplication, is fed with medium containing a single growth-limiting nutrient, all other nutrients being supplied in excess (Fig. 1). Cell concentration then increases to the value supported by the concentration of the growth-limiting nutrient, that is, provided the washout rate (i.e. the rate at which the culture is being diluted with fresh medium) is less than the maximum growth rate of the cells. The decrease in the concentration of the growth-limiting nutrient will then slow the growth rate of the cells until the growth rate equals the washout rate. A steady state will thus be established in which both cell density and substrate concentration remain constant. The steady-state cell concentration is controlled by the concentration of the single growth-limiting component of the medium, and cell growth rate is controlled by the rate of supply of this component to the culture.

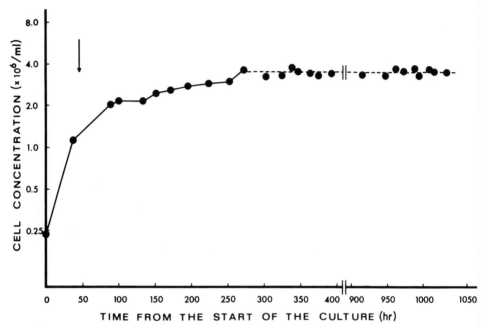

Fig. 1. Establishment of a steady-state chemostat culture of Namalwa cells at a dilution rate of 0.693 day^{-1}. The arrow indicates the start of continuous-flow culture. (- - - - -) Steady-state mean cell concentration ± standard deviation, 3.6 ± 0.2 × 10^6 cells/ml. From Tovey (39) by permission of Academic Press, New York.

During the steady state, the cell growth rate (μ) is equal to the dilution rate (D), that is, the quotient of the medium flow rate (f) and the culture volume (V):

$$\mu = D = \frac{f}{V}\text{day}^{-1}$$

Since the culture volume is maintained constant, cell growth rate can be changed simply by changing the medium flow rate. The doubling time (t_d) of the culture can be determined from the expression $t_d = \ln 2/D$.

The chemostat is a self-regulating system: a temporary decrease in the steady-state cell concentration will cause a corresponding increase in cell growth rate, which will act to restore steady-state conditions. An increase in the steady-state cell concentration will have the converse effect and will again act to restore steady-state conditions.

In addition to chemostat culture, in which cell density adjusts to a particular fixed flow rate, a further type of homogeneous continuous-flow culture known as the the turbidostat has been described (16). In the turbidostat the rate of medium supply is varied to maintain cell density at a particular fixed value. This is usually accomplished by means of a photoelectric cell which determines the turbidity of the culture and operates a pump to add fresh medium to the culture when cell density exceeds the chosen level. The advantage of this type of culture is that it allows cells to be cultivated at their maximum growth rate in an excess of substrate. Since cell growth rate is not fixed, there is a powerful pressure in such a system for the selection of faster growing organisms and this method has been used for the automatic selection of antibiotic-resistant bacteria (16). The main limitation of the turbidostat is that it operates satisfactorily only at growth rates near the maximum growth rate, where cell concentration varies rapidly with a change in the dilution rate. However, it is precisely under such conditions that the chemostat operates least satisfactorily since at high dilution rates, approaching the maximum specific growth rate, the cell population tends to wash out. Thus the two systems are to a certain extent complementary.

A number of variations of the turbidostat have been described in which growth control is based on the detection of a growth-linked product. Such systems have been termed productstats (17) and include systems based on the monitoring of pH, the phauxostat (18), and CO_2 production, the CO_2stat (19). The main application of such systems is in the large-scale production of a desired product, and with the advent of enzyme electrodes (20, 21) for the measurement of such substances as penicillin (22) and cholesterol (23), productstats should find increasing application for the industrial production of a variety of substances.

3.1. The Cultivation of Animal Cells in Continuous-flow Culture

A number of animal cell lines have been cultivated in the chemostat (Table I). In the majority of these studies, cells were cultivated in serum containing medium either under glucose limitation or under conditions where the growth-limiting nutrient was not identified. Limitation of glucose in a culture of animal cells cultivated in a complex medium containing serum and amino acids, and under conditions in which some of the glucose is incompletely oxidized to lactic acid (24), is not comparable to the limitation of the sole carbon source in a culture of microorganisms. Although the theory of continuous culture is based on cell multiplication being limited by a single nutrient, it has become apparent that under certain conditions the growth rate of an organism may be simultaneously limited by two or more substances (25). It is not clear whether the growth rate of the organism can be limited by only one substrate at a time or whether both substrates affect the overall growth rate of the organism. It has also been suggested that one

TABLE I Animal Cell Lines Cultivated in the Chemostat

Cell type	Species	Growth-limiting nutrient	Reference
L	Mouse	Glucose	(4)
P815Y	Mouse	Glucose	(24)
LS	Mouse	Glucose	(30)
		Unknown	(23a)
		Choline	(31)
L1210	Mouse	Glucose	(1)
ERK	Rabbit	Unknown	(42, 43)
		Glucose	(4)
BHK	Hamster	Phospate	(49)
		Unknown	(48, 49)
HeLa	Human	Unknown	(23b, 47)
KB	Human	Unknown	(23b)
Namalwa	Human	Glucose	(39)

nutrient may control the rate of cell multiplication while another may ultimately control the extent of cell multiplication (25). Mouse LS cells have, however, been cultivated in the chemostat in a chemically defined protein-free medium (26–29) under glucose (30) and choline (31) limitation. This is the first report of the cultivation of animal cells in a chemically defined medium in chemostat culture, and is to date the closest approximation to microbial systems in which growth is controlled by the limitation of a sole carbon source.

The majority of studies on the chemostat culture of animal cells have been carried out with mouse leukemia L1210 cells (1, 32–40). Mouse L1210 cells are ideally suited for chemostat studies, since they multiply in suspension culture, exhibit a high specific growth rate, and do not form clumps or attach to surfaces. Wall growth can indeed be a major problem in continuous-flow culture, since, apart from the risk of blocking medium lines or overflow pipes, wall growth can cause major deviations from the predicted performance of a chemostat (2, 3) owing to the continued reinoculation of cells from the walls of the culture vessel.

Mouse leukemia L1210 cells can be readily cultivated under glucose limitation in the chemostat provided that the cells are free from *Mycoplasma* contamination. We have been unable to obtain steady-state cultures from *Mycoplasma*-contaminated clones of L1210 cells (1). The L1210 cells respond well to glucose limitation, which has the added advantage of stabilizing the pH of the culture. In fact, the pH of the culture can be maintained at 7.4 ± 0.1 simply by gassing continuously the culture vessel and medium reservoir with 5% CO_2 in air.

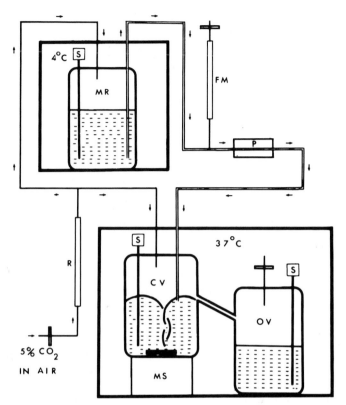

Fig. 2. Apparatus for the chemostat culture of mouse L1210 cells. CV, chemostat culture vessel; MS, magnetic stirrer; OV, overflow vessel; MR, medium reservoir; P, peristaltic pump; FM, medium flow meter; S, sampling port; R, rotameter. From Tovey and Brouty-Boyé (1), by permission of Academic Press, New York.

Animal cells can be cultivated either in suitably modified commercially available fermentation equipment such as the New Brunswick Bioflo or in a simple apparatus (1) constructed from readily available laboratory glassware, as represented diagrammatically in Fig. 2. A magnetically stirred (150–300 rpm) glass culture vessel is connected via a sidearm overflow to a collection vessel. The sidearm overflow is arranged so that the working volume (300 ml) of the vessel is approximately half the total volume. Nutrient medium stored in a refrigerated (4°C)-reservoir is supplied to the culture vessel at a constant rate via a variable-speed peristaltic pump (MHRE-2 Watson-Marlow, Falmouth, U.K.). The medium reservoir and overflow vessel can be constructed from 1.0-liter flasks with silicone seals and connected to the culture vessel with silicone tubing. All tubing and seals are made of either stainless steel or silicone rubber. The medium reservoir, culture vessel, and overflow vessel are equipped with sampling ports for the aseptic addition of medium and withdrawal of culture. The culture vessel can also be equipped with additional ports for pH and oxygen electrodes. Suitable electrodes and monitoring equipment are commercially available. The apparatus is fitted with autoclavable air filters (millipore) and can be sterilized as an ensemble.

A batch culture of L1210 cells is initiated in the chemostat culture vessel at 10^5 cells/ml in Eagle's minimal essential medium supplemented with 10% horse serum. During the period of exponential multiplication the culture is fed with nutrient medium containing a growth-limiting concentration of glucose (≤ 1.0 mg/ml) at the required dilution rate. Cell concentration then increases to the value supported by the growth-limiting concentration of glucose in the inflowing medium (Fig. 1).

Steady-state cultures of L1210 cells become established after a period of adjustment ranging from 100 to 450 hr of continuous operation of the chemostat. Adjustment times are proportional to the dilution rate of the chemostat (41). For example, if one assumes that no cell multiplication occurs, then a flow of medium equivalent to five volume changes is required to reduce cell concentration in the chemostat to 1% of its initial value. This would take a period of 5 days at a dilution rate of 1.0 day^{-1} and 50 days at a dilution rate of 0.1 day^{-1}. The long adjustment periods necessary for the establishment of steady-state cultures of L1210 cells at low dilution rates may explain in part the failure of some investigators to obtain stable steady-state cultures of animals cells in the chemostat (42, 43).

Steady-state cultures of mouse L1210 cells are characterized by constant values of a number of parameters (Table II). The steady states obtained can be considered true steady states since the standard deviation of the steady-state mean of each of these parameters is of the same magnitude as the standard deviation of their respective assays (Fig. 3). Cultures of mouse L1210 cells have been regularly maintained in the chemostat continuously for periods in excess of 1000 hr, and individual steady states have been maintained for up to 600 hr (1).

Stable steady-state cultures of L1210 cells have been established in the chemostat at dilution rates ranging from 0.1 day^{-1} ($t_d = 166.3$ hr) to 2.0 day^{-1} ($t_d = 8.3$ hr). The relationship between the steady-state cell concentration and dilution rate was found to be in good agreement with the theoretical curves of cell density and glucose concentration computed from the Monod equations (2, 3) using values for the constants determined experimentally for L1210 cells (1). However, at fast dilution rates substantial deviation from the theoretical curves was observed. For example, steady-state cultures of L1210 cells were established at growth rates greater than the maximum specific growth rate of L1210 cells (1.75 day^{-1} determined in batch culture). Similar

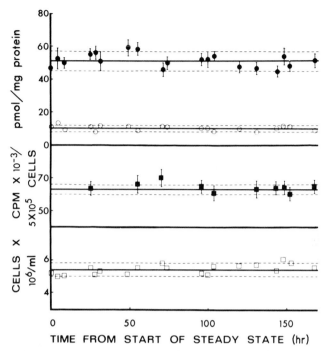

Fig. 3. A chemostat culture of L1210 cells under steady-state conditions at a dilution rate of 0.5 day^{-1} (t_d = 33.3 hr). Initial glucose concentration = 1.0 mg/ml. (●) Intracellular concentration of cyclic AMP (steady-state mean, 51.3 ± 6.3 pmol/mg protein). (○) Intracellular concentration of cyclic GMP (steady-state mean, 10.5 ± 1.8 pmol/mg protein). (□) Cell concentration (steady-state mean, 5.4 ± 0.4 × 10⁶ cells/ml). (■) Incorporation of [³H]thymidine into acid-precipitable material (steady-state mean, 63,382 ± 3,273 CPM/5 × 10⁵ cells). (╤) Steady-state mean value with standard deviation. The error bars represent the standard deviation of the replicates for a particular point. From Tovey *et al.* (38) by permission of the editors of the *Journal of Cellular Physiology.*

TABLE II Characterization of Steady-State Chemostat Cultures of Mouse L1210 Cells

Constant cell number	Constant rates of incorporation of:
Constant cell volume	³H-labelled thymidine
Constant intracellular concentrations of:	³H-labelled uridine
DNA	¹⁴C-labelled amino acids
RNA	³H-labelled 2-deoxy-D-glucose
Protein	
Adenosine 3′,5′-cyclic monophosphate	Constant percentage of cells
Guanosine 3′,5′-cyclic monophosphate	labelled by autoradiography
Constant concentration of L-lactate in the	
culture supernatant	Constant rate of heat production

effects have been observed in chemostat cultures of microorganisms and are probably due to imperfect mixing of the culture (2, 3).

In glucose-limited chemostat cultures of microorganisms at growth rates below about one-tenth of the maximum specific growth rate cell yield decreases, reflecting an increased maintenance energy requirement (44–46) (i.e., the energy required for specific cellular functions other than growth, such as transport processes and the turnover of cell materials). However, at comparable dilution rates the yield of L1210 cells remained constant (1). Although it is possible that a decreased yield of L1210 cells would be observed at dilution rates lower than those studied, it is perhaps not surprising that differences should be observed in the behavior of chemostat cultures of microorganisms and animal cells.

The output of L1210 cells from the chemostat was found to follow the general form of the theoretical curve derived from the Monod equations (1). A maximum output was obtained at a dilution rate of 1.3 day^{-1} (1). At higher dilution rates, even though flow rates were greater, the steady-state cell concentrations were considerably lower, and the output of L1210 cells fell sharply (1).

The results of our experiments on the cultivation of mouse L1210 tumor cells in the chemostat show that the principles of continuous culture, derived from the study of microorganisms, are to a large extent applicable to the cultivation of animal cells.

3.1.1. Production of Cells in the Chemostat

One of the principal advantages of chemostat culture is that it enables large quantities of animal cells to be produced under precise physiological conditions. Cells can be cultivated in the chemostat in a number of unique environments, under a variety of growth-limiting conditions, and at a wide range of growth rates.

Consider the production of L1210 cells in chemostat culture. The yield of cells, expressed either as the quantity of cells produced per milligram of glucose or as the quantity of cells produced per milliliter of medium, is markedly increased relative to batch culture (Table III). Thus, the use of the chemostat enables production costs to be reduced, especially if an expensive component of the medium such as serum or a growth factor is made growth-limiting. But it is in terms of productivity (i.e. cells produced per unit time) that the chemostat really excels. Thus in our studies the maximum output of a 300-ml chemostat was 66.4×10^6 cells/hr or 8.24 mg dry weight/hr. That is eightfold more L1210 cells were produced per unit time in the chemostat than in conventional batch culture (Table III). This is obviously of considerable imporatnce for the large-scale production of animal cells or animal cell products. Other advantages afforded by chemostat culture include the abil-

TABLE III Comparison of the Production of L1210 Cells in Batch and Chemostat Culture[a]

Type of culture	Volume (ml)	Yield ($\times 10^6$ cells/ mg glucose)	Yield (μg/mg glucose)	Yield ($\times 10^6$/ml medium)	Output (cells/week)
Batch[b]	300	3.5	524	2.3	1.38×10^9
Chemostat[c]	300	6.2	601	6.2	1.12×10^{10}

[a] From Tovey (39) by permission of Academic Press, New York.

[b] The batch culture took 3 days to attain a maximum population density of 2.5×10^6 cells/ml. When inoculated at 0.2×10^6 cells/ml a minimum of 1 day was required for "downtime" between cultures.

[c] Dilution rate, 0.3 day^{-1}.

ity to control culture conditions and to automate the process, reduction in size of the culture vessel required to produce a given quantity of cells, reduction in downtime (the time required for sterilization, inoculation, harvesting, etc.), and production of cells in the precise physiological state required.

3.1.2. Production of Cell Products in the Chemostat

Conditions that are optimal for the maximal rate of cell multiplication may not be optimal for the maximal production of a particular cell product. Furthermore, conditions optimal for product formation may be attained only transiently or perhaps not at all in conventional batch culture. In addition, a cell product may be inhibited by further production of that product or by production of another cell product or metabolite. A cell product may also inhibit further cell multiplication.

The advantage of the chemostat is that it allows a whole range of different environments to be established and to be maintained indefinitely. Cell products are continuously removed from the chemostat culture vessel, preventing inhibition of product formation or inhibition of cell multiplication. Furthermore, a two-stage chemostat may be employed in which conditions are optimal for the maximal output of cells in the first stage and for maximal product formation in the second stage.

The production of animal cell products in chemostat culture has to date been confined to studies on the production of viruses and interferon.

Production of Viruses. Chemostat culture offers the prospect of enhanced productivity and increased uniformity of product compared to the conventional methods of producing virus in batch culture. These are important advantages for the production of viruses for use either in the preparation of viral vaccines or as interferon inducers.

The choice of the continuous-flow culture system to be used depends upon the characteristics of the virus cell system to be employed. A virus which establishes a persistent infection without causing a cytopathic effect and without reducing cell growth rate too drastically could be propagated in cells cultivated in a single-stage chemostat. However, a virus that causes a lytic infection would have to be produced in the second stage of a two-stage chemostat, the host cells being produced in the first stage.

Holmström (47) reported the successful propagation of rubella virus in HeLa cells cultivated in a single-stage chemostat. Similarly, Van Hemert *et al.* (48) and Kilburn and Van Wezel (49) also reported this production of rubella virus in chemostat cultures of BHK cells and showed that virus titer, complement-fixing antigen, and antibody-inducing capacity were maintained in virus produced after 40 days of continuous operation of the chemostat. Kilburn and Van Wezel (49) also reported some increase in the rate of virus production with increasing cell growth rate in the chemostat, although it was not clear whether the rate of virus production per cell was increased. Gori (50) has described a quasi-chemostat termed the lysostat in which polio virus (which causes a lytic infection) was propagated in HeLa cells cultivated in a single-stage lysostat and adenovirus was produced in a two-stage lysostat (50). However, steady-state conditions were not attained in either system.

Production of Interferon.   Interferon for use in clinical trials in patients with neoplasia is currently being produced on an industrial scale from the human Burkitt lymphoma cell line Namalwa cultivated in large-scale batch culture (51). Chemostat culture, however, is a potentially more advantageous system for the large-scale production of human lymphoblastoid interferon, since it allows a high product output to be obtained from a relatively small culture vessel, thereby reducing loss of interferon titer in scale-up. Furthermore, the use of the chemostat offers the possibility of obtaining continuous interferon production. Interferon is an induced cellular protein, whose production ceases after a limited period, at which time cells are refractory to further stimulation (52). However, chemostat culture is a continuous-flow process which enables a continuous output of cells and cell products to be maintained more or less indefinitely.

Tovey *et al.* (30) showed that repeated induction of interferon could be obtained in chemostat cultures of mouse LS cells without the development of a refractory state. Higher titers of interferon were obtained from glucose-limited chemostat cultures of LS cells than from LS cells cultivated in the presence of excess glucose in batch culture. Furthermore, interferon preparations of relatively high specific activity and free of contaminating serum proteins were obtained by the use of a protein-free medium (53). Interferon production in the chemostat appeared to be independent of cell growth rate over the range of growth rates tested (30). Namalwa cells multiply well in the chemostat, establishing stable steady-state cultures at high cell densities (Fig. 1). I have shown that chemostat cultures of Namalwa cells can be induced with a potent virus inducer to produce levels of interferon in the chemostat similar to those obtained in batch culture (M. G. Tovey, unpublished results). Thus the production of interferon in chemostat continuous-flow culture should facilitate the large-scale production of human lymphoblastoid interferon.

4. FUTURE APPLICATIONS OF THE CONTINUOUS-FLOW CULTURE OF ANIMAL CELLS

Although chemostat culture is based upon the cultivation of cells in homogeneous suspension culture, anchorage-dependent cells can be cultivated in suspension culture if attached to a suitable microcarrier (54, 55). Since anchorage-dependent cells can detach from a microcarrier (56) and reattach to new beads, it should thus be possible to supply new beads to the culture and obtain some form of continuous-flow culture. Such a system might prove to be useful for the production of interferon-β or vaccine strains of certain viruses from human diploid fibroblasts.

The genes of a number of proteins of commercial interest have recently been cloned in bacteria. Although bacterial fermentation presents a number of advantages for the large-scale production of animal proteins, it may be advantageous to produce certain glycoproteins such as human interferon-γ in animal cells. The gene for human interferon-γ, unlike the genes for interferons α and β, contains introns and the protein also appears to be subject to post-translational processing (57). The introduction of amplified copies of the gene for human interferon-γ into animal cells would ensure correct processing, glycosylation, and secretion of the protein, which is not achieved in microbial cells. For the reasons outlined above chemostat culture would then become the system of choice for the production of such proteins from genetically modified cells.

REFERENCES

1. Tovey, M. G., and Brouty-Boyé, D. (1976). Characteristics of the chemostat culture of murine leukemia L1210 cells. Exp. Cell Res. 101, 346–354.
2. Herbert, D., Elsworth, R. E., and Telling, R. C. (1956). The continuous culture of bacteria: A theoretical and experimental study. J. Gen. Microbiol. 14, 601–622.
3. Pirt, S. J. (1975). "Principles of Microbe and Cell Cultivation." Blackwell, Oxford.

4. Pirt, S. J., and Callow, D. S. (1964). Continuous-flow culture of the ERK and L types of mammalian cells. *Exp. Cell Res.* **33**, 413–421.
5. Reuveny, S., Bino, T., Rosenberg, H., Traub, A., and Mizrahi, A. (1980). Pilot plant scale production of human lymphoblastoid interferon. *Dev. Biol. Stand.* **46**, 281–288.
6. Yoshida, F., Yamane, T., and Ken-Ichiro, N. (1973). Fed-batch hydrocarbon fermentation with colloidal emulsion feed. *Biotechnol. Bioeng.* **15**, 257–270.
7. Pirt, S. J. (1974). The theory of fed batch culture with reference to the penicillin fermentation. *J. Appl. Chem. Biotechnol.* **24**, 415–424.
8. Klein, F., Ricketts, R. T., Jones, W. I., Dearmon, I. A., Temple, M. J., Zoon, K. C., and Bridgen, P. J. (1979). *Antimicrob. Agents Chemother.* **15**, 420–427.
9. Graff, S., and McCarthy, K. S. (1957). Sustained cell culture. *Exp. Cell Res.* **13**, 348–357.
10. Kruse, P. F., Jr., Myhr, B. C., Johnson, J. E., and White, P. B. (1963). Perfusion system for replicate mammalian cell cultures in T-60 flasks. *J. Natl. Cancer Inst. (U.S.)* **31**, 109–123.
11. Himmelfarb, P., Thayer, P. S., and Martin, H. E. (1969). Spin filter culture: The propagation of mammalian cells in suspension. *Science* **164**, 555–557.
12. Thayer, P. S. (1973). Spin filter device for suspension cultures. *In* "Tissue Culture: Methods and Applications" (P. F. Kruse Jr. and M. K. Patterson, Jr., eds.), pp. 345–351. Academic Press, New York.
13. Monod, J. (1942). "Recherches sur la croissance des cultures bactériennes." Hermann, Paris.
14. Monod, J. (1950). La technique de culture continue; théorie et applications. *Ann. Inst. Pasteur, Paris* **79**, 390–410.
15. Novick, A., and Szilard, L. (1950). Description of the chemostat. *Science* **112**, 715–716.
16. Bryson, V. (1952). Turbidostat selector—a device for automatic isolation of bacterial variants. *Science* **116**, 45–51.
17. Silman, R. W., and Bagley, B. J. (1979). The viscostat: Productstat method of feed-rate control in continuous fermentations. *Biotechnol. Bioeng.* **21**, 173–179.
18. Martin, G. A., and Hempfling, W. P. (1976). A method for the regulation of microbial population density during continuous culture at high growth rates. *Arch. Microbiol.* **107**, 41–47.
19. Watson, T. G. (1969). Steady-state operation of a continuous culture at maximum growth rate by control of carbon dioxide production. *J. Gen. Microbiol.* **59**, 83–89.
20. Gough, D. A., and Andrade, J. D. (1973). Enzyme electrodes. *Science* **180**, 380–384.
21. Guilbault, G. G. (1976). Enzyme electrodes and solid surface fluorescence methods. *In* "Methods in Enzymology" (K. Mosbach, ed.), Vol. 44, pp. 579–632. Academic Press, New York.
22. Nilsson, H., and Mosbach, K. (1978). An enzyme electrode for measurement of penicillin in fermentation broth: A step towards the application of enzyme electrodes in fermentation control. *Biotechnol. Bioeng.* **20**, 527–539.
23. Satoh, I., Karube, I., and Suzuki, S. (1977). Enzyme electrode for free cholesterol. *Biotechnol. Bioeng.* **19**, 1095–1099.
23a. Griffiths, J. B., and Pirt, S. J. (1967). The uptake of amino acids by mouse cells (strain LS) during growth in batch culture and chemostat culture: The influence of cell growth rate. *Proc. R. Soc. London, Ser. B* **168**, 421–438.
23b. Cohen, E. P., and Eagle, H. (1961). A simplified chemostat for the growth of mammalian cells: Characteristic of cell growth in continuous culture. *J. Exp. Med.* **113**, 467–474.
24. Moser, H., and Vecchio, G. (1967). The production of stable steady-states in mouse ascites mast cell cultures maintained in the chemostat. *Experientia* **23**, 1–10.
25. Bader, F. G. (1978). Analysis of double-substrate limited growth. *Biotechnol. Bioeng.* **20**, 183–202.

26. Birch, J. R., and Pirt, S. J. (1969). The choline and serum protein requirements of mouse fibroblast cells (strain LS) in culture. *J. Cell Sci.* **5**, 135–142.
27. Birch, J. R., and Pirt, S. J. (1970). Improvements in a chemically defined medium for the growth of mouse cells (strain LS) in suspension. *J. Cell Sci.* **7**, 661–670.
28. Birch, J. R., and Pirt, S. J. (1971). The quantitative glucose and mineral requirements of mouse LS (suspension) cells in a chemically defined medium. *J. Cell Sci.* **8**, 693–700.
29. Blaker, G. J., and Pirt, S. J. (1971). The uptake of vitamins by mouse fibroblast cells (strain LS) during growth on a chemically defined medium. *J. Cell Sci.* **8**, 701–708.
30. Tovey, M. G., Mathison, G. E., and Pirt, S. J. (1973). The production of interferon by chemostat cultures of mouse LS-cells grown in chemically-defined, protein-free medium. *J. Gen. Virol.* **20**, 29–35.
31. Tovey, M. G. (1971). Interferon production by mouse L-cells in suspension culture. Ph.D. Thesis, University of London.
32. Tovey, M. G., Brouty-Boyé, D., and Gresser, I. (1975). Early effect of interferon on mouse leukemia cells cultivated in a chemostat. *Proc. Natl. Acad. Sci. U.S.A.* **72**, 2265–2269.
33. Tovey, M. G., and Brouty-Boyé, D. (1976). The cultivation of animal cells in a chemostat. *J. Appl. Chem. Biotechnol.* **26**, 345.
34. Brouty-Boyé, D., and Tovey, M. G. (1978). Inhibition of thymidine uptake in chemostat cultures of L1210 cells. *Interviorology* **9**, 243–252.
35. Tovey, M. G., and Brouty-Boyé, D. (1979). The use of the chemostat to study the relationship between cell growth rate, viability and the effect of interferon on L 1210 cells. *Exp. Cell Res.* **118**, 383–388.
36. Tovey, M. G., Rochette-Egly, C., and Castagna, M. (1979). Effect of interferon on concentrations of cyclic nucleotides in cultured cells. *Proc. Natl. Acad. Sci. U.S.A.* **76**, 3890–3893.
37. Tovey, M. G., and Rochette-Egly, C. (1980). The effect of interferon on cyclic nucleotides. *Ann. N.Y. Acad. Sci.* **350**, 266–278.
38. Tovey, M. G., Rochette-Egly, C., and Castagna, M. (1980). Correlation between growth rate, cell density, and intracellular concentrations of cyclic nucleotides in chemostat cultures of mouse L 1210 cells. *J. Cell Physiol.* **105**, 363–367.
39. Tovey, M. G. (1980). The cultivation of animal cells in the chemostat: Application to the study of tumor cell multiplication. *Adv. Cancer Res.* **33**, 1–37.
40. Tovey, M. G. (1981). Use of the chemostat culture for study of the effect of interferon on tumor cell multiplication. *In* "Methods in Enzymology" (S. Pestka, ed.), Vol. 79, pp. 391–404. Academic Press, New York.
41. Tempest, D. W., Herbert, D., and Phipps, P. J. (1967). Studies on the growth of *Aerobacter aerogenes* at low dilution rates in a chemostat. *In* "Microbial Physiology and Continuous Culture" (E. O. Powell, C. G. Evens, R. E. Strange, and D. W. Tempest, eds.), pp. 240–261. H. M. Stationery Office, London.
42. Cooper, P. D., Burt, A. M., and Wilson, J. N. (1958). Critical effect of oxygen tension on rate of growth of animal cells in continuous suspended culture. *Nature (London)* **182**, 1508–1509.
43. Cooper, P. D., Wilson, J. N., and Burt, A. M. (1959). The bulk growth of animal cells in continuous suspension culture. *J. Gen. Microbiol.* **21**, 702–720.
44. Herbert, D. (1959). Some principles of continuous culture. *In* "Recent Progress in Microbiology" (G. P. Tuneval, ed.), pp. 381–384. Almqvist & Wiksell, Stockholm.
45. Marr, A. G., Nilson, E. H., and Clark, D. J. (1963). The maintenance requirement of *Escherichia coli. Ann. N.Y. Acad. Sci.* **102**, 536–548.
46. Pirt, S. J. (1965). The maintenance energy of bacteria in growing cultures. *Proc. R. Soc. London, Ser. B* **163**, 224–231.

47. Holmström, B. (1968). Continuous flow culture of a HeLa cell line as a basis for a steady supply of rubella virus. *Biotechnol. Bioeng.* **10**, 373–384.
48. Van Hemert, P., Kilburn, D. G., and van Wezel, A. L. (1969). Homogeneous cultivation of animal cells for the production of virus and virus products. *Biotechnol. Bioeng.* **11**, 875–885.
49. Kilburn, D. G., and van Wezel, A. L. (1970). The effect of growth rate in continuous-flow cultures on the replication of rubella virus in BHK cells. *J. Gen. Virol.* **9**, 1–7.
50. Gori, G. B. (1965). Continuous cultivation of virus in cell suspensions by use of the lysostat. *Appl. Microbiol.* **13**, 909–917.
51. Finter, N. B., and Fantes, K. H. (1981). The purity and safety of interferons prepared for clinical use. The case for lymphoblastoid interferon. *In* "Interferon" (I. Gresser, ed.), Vol. 2, pp. 65–80. Academic Press, London.
52. Stringfellow, D. A. (1977). Production of the interferon protein: Hyporesponsiveness. *Tex. Rep. Biol. Med.* **35**, 126–131.
53. Mogensen, K. E., Tovey, M. G., Pirt, S. J., and Mathison, G. E. (1972). Induction of mouse interferon in a chemically defined system. *J. Gen. Virol.* **16**, 111–114.
54. van Wezel, A. L. (1967). Growth of cell strains and primary cells on microcarriers in homogeneous culture. *Nature (London)* **216**, 64–66.
55. Levine, D. W., Wong, J. S., Wang, D. I. C., and Thilly, W. G. (1977). Microcarrier cell culture: New methods for research scale application. *Somatic Cell. Genet.* **3**, 149–155.
56. Thilly, W. G., and Levine, D. W. (1979). Microcarrier culture: A homogeneous environment for studies of cellular biochemistry. *In* "Methods in Enzymology" (W. B. Jakoby and I. H. Pastan, eds.), Vol. 58, pp. 184–194. Academic Press, New York.
57. Gray, D. W., and Goeddel, D. V. (1982). Structure of the human immune interferon gene. *Nature (London)* **298**, 859–863.

9

Monolayer Growth Systems: Multiple Processes

GIAN FRANCO PANINA

Istituto Zooprofilattico Sperimentale
Brescia, Italy

1. INTRODUCTION

Animal cells grow *in vitro* in two different ways: attached to a suitable support or suspended in an appropriate medium.

Non-transformed cells are "anchorage-dependent", i.e. they grow in tissue culture in contact with a glass, plastic or metal surface, but not in

Animal Cell Biotechnology, Vol. 1

suspension. They develop *in vitro* according to an orderly pattern of growth and usually stop growing at the time when they reach the cell density of a confluent monolayer. Non-transformed cells were therefore defined as sensitive to a phenomenon called "contact inhibition" of growth or "density-dependent inhibition" of growth (89).

Transformed cells, on the other hand, clearly display a different pattern of growth *in vitro*. They are not contact-inhibited and usually they grow past the confluent monolayer stage and, by growing in multilayer, eventually reach a saturation density several times higher than that of a single monolayer. Theoretically, transformed cells could grow indefinitely in the same culture, the limiting factors being unfavourable environmental conditions, either in the medium or in the gaseous phase.

The transformation of the cells, besides preventing contact inhibition, reduces or annuls the role of the support as a conditioning factor of cell multiplication. Thus, some types of transformed cells grow in suspension, under conditions where the normal positive effect of the substrate is absent. (See Chapter 2, this volume.)

Other strains of transformed cells are in an intermediate state, i.e. they are able to grow either attached on a surface or in suspension. BHK 21-C13 cells are an example of a cell line in which there are strains growing in monolayer or in suspension or in both conditions.

Years ago, the phenomenon of contact inhibition was considered a sort of growth control of normal cells in tissue culture resulting from the activation of specific receptors on the cell surface upon contact with suitable sites on the surface of other cells. Dulbecco (*11*) stated in 1971 that the cell surface might exist in two states: a "growing state" or a "resting state". The growth of cells *in vitro* depends on the balance of positive and negative signals. The support provides the normal cells with positive signals, the negative ones being generated by intercellular contact.

At present, the term contact inhibition is no longer applied even to normal cells, in that a variety of factors have been described which are effective in promoting the overgrowth of confluent cells. Cell density would seem a variable which depends on the environmental conditions of the culture, and the cells which stop growing in a flask when a confluent monolayer is formed may be induced to continue to multiply exponentially beyond the monolayer stage, reaching a density of many confluence equivalents.

Overgrowth of density-inhibited cells may be stimulated by using a standard regime of medium changes (5, 6, 19), by a perfusion system (15, 41, 42), by the application of medium supplements such as serum (30), amino acids, glucose, phosphatase (31), steroids (94), various growth factors (59), ribonuclease, digitonin, hyaluronidase (101), polypeptides and hormones (32, 93) or by acting on additional growth-regulating factors such as the pH (9) or the

TABLE I Some Typical Physical and Biological Parameters of Monolayer Cell Cultures in Stationary Flat Containers and in Rolling Cylindrical Containers[a]

Containers	Stationary flat	Rolling cylindrical
Volume (cm^3)	1100	1200
Usable surface (cm^2)	210	620
Range of maintenance medium (ml)	100–70	100–10
Range of ratio of surface area (cm^2) to maintenance medium volume (ml)	2–3	6.2–62
Yield of cells × 10^6		
Calf kidney	40	150
BHK	110	500
Cells × 10^3/surface (cm^2)		
Calf kidney	190	240
BHK	520	800
Range of cells × 10^6/medium (ml)		
Calf kidney	0.4–0.6	1.5–15
BHK	1.1–1.6	5–50

[a] G. F. Panina and A. Ragni, unpublished results (1965).

electrostatic surface potential (83). It was experimentally demonstrated (83) that the simple aeration of a spent medium can be sufficient to "turn on" density-inhibited chick embryo fibroblasts. Similarly, a modification of the environmental conditions such as that obtained by rolling cultures in bottles can stimulate cell growth to give final populations of a characteristic degree of multilayering (74, 96) (Table I).

From what has been written, it is clear that the cell type used in a process determines the applicability of a particular production system. The cell type is one of the most important determinants on which the development of a particular technology is dependent.

2. CELL CULTURE TECHNOLOGIES

Primary cells and diploid cell lines which require a surface to grow upon are conventionally cultivated on the walls of bottles of different shape and composition, held in stationary racks or slowly rotated. New substrata of a different nature have recently been developed.

Established cell lines capable of growth for an indefinite number of passages in deep suspension have enabled technologists to scale up tissue culture systems.

Combined suspension–monolayer processes have been developed with cells of the intermediate state (98). In such cases, the cells for seeding the culture surfaces are obtained by "milking" a suspension culture. The main advantage of the process is the continuous availability of cell seeds, without having to perform trypsinization of fresh tissues or monolayers.

Doubtless, cell culture in deep suspension offers the best possibilities of scaling up. Suspension culture is a homogeneous system, easy to handle and control. Production plants involving suspension cultures of BHK cells in fermenters of 1000- to 5000-litre volumes have been realized for the production of foot-and-mouth disease (FMD) vaccines. However, there are biological limits to a broad extension of the suspension method.

Nobody has yet succeeded in growing diploid and primary cells in suspension and, according to the guidelines formulated in most countries, such cells are the only cells which may be used in generating products for human use. Established cell lines such as BHK can be used only for the production of veterinary vaccines.

A large cell suspension process using a human lymphoblastoid cell line has been developed for the production of interferon (39), but as far as we know the product has not yet been approved by regulatory agencies for sale to the public.

Several other considerations have influenced researchers to set up monolayer systems in spite of the advantages of suspension cell culture.

1. The monolayer system provides flexibility for culturing either primary cells or diploid and heteroploid cell lines, in contrast with the suspension system. In the author's laboratory, a large roller bottle plant produces alternatively, according to need, BHK cell line monolayers for FMD and Aujeszky's disease virus vaccines or calf kidney primary cell monolayers for infectious bovine rhinotracheitis (IBR) and parainfluenza 3 (PI$_3$) virus vaccines.

2. Monolayer cell cultures offer the possibility of easily removing the spent growth medium before infection of the cells. Growth medium usually contains serum. Traces of serum in the maintenance medium are responsible for anaphylactic reactions when the product is repeatedly injected in animals, as in the case of vaccinations (68). Moreover, residues of antibodies in the maintenance medium could interfere not only with the infectivity of the virus but also with its antigenicity (69). It is not easy to achieve a complete change of the medium in suspension cultures. It may be obtained by cell sedimentation, but the operation leaves behind a significant quantity of the growth medium. The author's laboratory is situated in an area where cow serum contains a high level of FMD antibodies because of annual compulsory vaccination of the cows with FMD vaccine. No problems arise from the use of such a serum when FMD virus is produced in rolling bottle cultures

because the simple change in the medium ensures the discharge of most of the antibody-containing serum. On the contrary, a negative influence on FMD virus infectivity is detected in suspension culture. In order to overcome such difficulties, cells in suspension are separated from the medium by continuous basket centrifugation before being infected. The operation is very efficient, but it is an additional step in the virus production cycle. It is not easy to perform and to maintain sterility and cell viability.

3. With rolling bottle cultures it is possible to produce highly concentrated FMD virus, without any difficulty in adapting new field isolates (70), and it has been demonstrated that potent vaccines against certain FMD virus strains can be produced to an acceptable immunogenic level only in BHK monolayer cells (88).

At present a great variety of products are made from animal cells grown attached to a surface. They are viral vaccines, hormones, enzymes, interferon and immunological reagents.

The aim of these operations is to obtain a product at the highest concentration. The concentration is related to the number of cells in the culture and to the volume of the medium used in the production phase (96). It is therefore advantageous to have in the culture system the highest number of cells possible and the flexibility of operating with as little medium as possible.

The number of cells in a culture system is related, within limits, to the surface area available for growth, and therefore technologists seek to realize systems able to contain a large surface area in a relatively small volume of medium.

In stationary flasks the possibility of realizing a convenient "surface/medium volume" ratio is prevented by the necessity of having a large volume of medium in order to cover the cell culture. A reduction of the volume of the medium can be obtained by rocking the flasks during incubation.

Furthermore, the rotating bottle gives the technologist the possibility of using a system which has a larger culture surface in comparison to the same volume in flat flasks. The amount of medium which is required in a rotating bottle such as for the production phase can also be very low, because of the rotation. High concentrations of cells per volume of medium can, therefore, be easily reached.

Different types of rotating bottles have been devised in order to improve the usable surface/medium volume ratio, and by using such a system many processes for the production of viral vaccines, interferon and other biological products have been successfully developed.

It is the general impression in the author's laboratory that, as Earle and Shilling (14) stated many years ago, with a very high level of cell population some type of circulation of the fluid nutrient gives superior results. In fact,

TABLE II Comparison of Glucose Metabolism in Calf Kidney and BHK Cell Cultures in
Stationary Flat Flasks and in Cylindrical Rolling Bottles[a]

	Calf kidney		BHK	
Test	Stationary flask	Rolling bottle	Stationary flask	Rolling bottle
Consumption of glucose	1.49	0.85	1.43	0.79
Production of lactic acid	1.20	0.77	1.37	0.45
Production of pyruvic acid	0.022	0.015	0.017	0.007

[a] Data are expressed as milligrams per 10^6 cells (G. F. Panina and A. Ragni, unpublished
results, 1965).

the cell density on a given area in a rotating culture is higher than that of an
identical area in a stationary culture (Table I). This is presumably due to a
different type of relationship of the medium to the cells. There is more
effective washing of nutrient over the cells and a continuous removal of the
possible toxic products of cell metabolism from the cell layer, which is
quickly diluted in the medium.

Due to the continuous rotation, the cells in the rolling system are exposed
alternately to the liquid and gaseous phases, and this results in more effective oxygenation of the cells. The utilization of glucose is more efficient than
in a stationary culture, where at the immediate cell–medium interface there
is a virtually anaerobic state. In the rotating cultures much more efficient
release of energy takes place and consequently, as introduced by Jensen
(38), cells can continue optimal growth with decreased amount of available
glucose.

Glucose metabolism in monolayer cell cultures was studied by the author,
and the figures presented in Table II indicate the metabolic difference between the stationary and rolling systems. MacMorine et al. (48) described
similar basic metabolic differences for monkey kidney cell cultures propagated in stationary bottles or in a rotating multi-surface cell propagator.

For the reasons above, the rotating system can be considered a significant
improvement in cell culture technology. However, MacMorine et al. (48)
stressed that caution must be used when a method of production is changed,
such as from stationary to rotating, because the quality of the product could
be affected by the altered cell metabolism.

There are, however, many disadvantages ascribed to the multiple process
systems. The most prominent are the handling of a large number of culture

units and the lack of monitoring and control of reactions which occur in tissue culture containers.

Monitoring and control can be achieved in each bottle of a multiple process only by using a perfusion control system (15, 41, 42) which permits the growth of multilayered tissue cultures. Hence, at present, efforts are being directed towards a transition from a multiple process to a unit process, i.e. consolidating the surface provided by many bottles in one vessel.

Various alternative systems have been proposed. They make use of plates, discs, tubes, spirals, hollow fibers, plastic bags and beads which mimic either the stationary flasks or the rotating bottles. Microcarriers, on the contrary, can be held in quasi-homogeneous suspension in a stirred tank reactor, which simulates a cell suspension culture.

It is clear that the unit process systems have many advantages resulting from lower capital and operating costs, better monitoring of the physical, chemical and biological parameters and improved standards of safety and working conditions (87). However, many of the expected advantages have not yet been fully achieved. Much remains to be done to make the unit process system more reliable and versatile, while systems such as those based on a multiplicity of bottles, tested through long experience, continue to be used successfully in current operations for the production of large quantities of vaccines.

The present chapter is a review of the monolayer cultures in multiple processes; the unit process systems will be discussed by others in this volume.

3. SUBSTRATA ON WHICH ANIMAL CELLS MULTIPLY

Cell attachment on a surface and cell spreading and growth are all dependent on the relationships between the cell and its supporting surface. Considering the nature of an artificial substratum, both chemical and physical factors have been identified as having a controlling influence. The macromolecular organization, the charge structure, the wettability and the physical form are considered of importance.

With regard to the macromolecular organization, Maroudas (54) has demonstrated that untransformed cells attach and grow on agarose gel, a polygalactose carbohydrate, only when the rate of cooling of the gel is carefully controlled in order to allow the galactose chains to form a regular fibrillar-like structure.

With regard to the charge structure, surfaces have been designated as "high" and "low" (20). The terms refer to the surface free energies present on different solid supports and the designations are based upon the known chemical composition of the surfaces. Pure surfaces of metals, metallic ox-

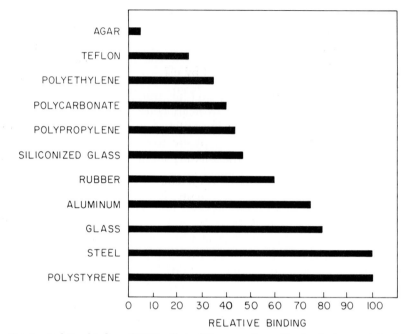

Fig. 1.   Relative binding of HTC cells to different surfaces. The incubations contained 10^6 cells and were carried out for 60 min. Each measurement was made at least four times. Polystyrene is Falcon treated polystyrene. (Grinnell *et al.* (20). With permission of Academic Press, Inc., New York.)

ides and glasses are high-energy surfaces, whereas organic surfaces are low-energy surfaces (85, 111).

The original surface can be altered by physical and chemical treatments. Thus materials such as polystyrene, having the properties of a low-energy surface, can, by appropriate treatment, acquire the properties of a high-energy surface.

It was demonstrated (20) that cells adhere better to high-energy than to low-energy surfaces (Fig. 1). Rappaport (77, 78) thought that a critical number of negative charges on the glass was required for cell adhesion, and Maroudas (52, 55) reported that for chemically etched substrata to be active they must contain at least five negatively charged groups per 100 Å^2. At present it is generally thought that cells can be cultivated on surfaces which possess either negative or positive charges (22, 34, 51, 52, 100) and that the basic factor in cell attachment and spreading is the density of the charges on the culture surface rather than the polarity of the charges.

Animal cells possess negative charges on their surface (7, 22) and traditional culture substrata such as glass or plastic are also negatively charged.

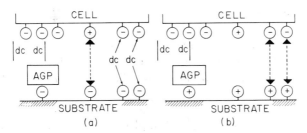

Fig. 2. A schematic representation of the essential elements involved when a cell attaches to and spreads on (a) negatively and (b) positively charged cell culture substrates. The process involves electrostatic interactions, divalent cations (dc) and an attachment glycoprotein (AGP). (Hirtenstein *et al.*, *(29)*. With permission of S. Karger AG., Basel.)

Thus, because of the electrostatic repulsion between the two negative charged surfaces, the initial cell–substratum contact must be mediated through ionic interaction and protein bridges *(22, 52)*.

In the case of positively charged substrata, the cell–substratum contact could be due solely to electrostatic forces *(22)*. However, the mechanism of attachment between cells and substrata is probably similar whether the substrata are positively or negatively charged *(29)*. In both cases the cell–substratum contact would be bridged by an amphoteric protein *(52)* such as fibronectin *(23, 33, 36)* and/or by divalent cations such as Ca^{2+} or Mg^{2+} *(29)* (Fig. 2).

The proteins involved in cell adhesion can be synthesized by certain cells, but they are also present in a tissue culture as components of the serum. Plasma proteins and serum included in the medium have profound effects on cellular adhesiveness, because the substratum, on exposure to protein solution, instantaneously adsorbs many of the proteins and thereby takes on their chemical characteristics *(27, 102)*. The rate of protein absorption and the thickness of the absorbed protein layer have been measured *(102)*. The latter appears to be a 50-Å electron-dense line between the cell and the substratum *(21, 92)*.

Protein absorption seems to be strictly related to the substratum wettability, whose favourable role in cell adhesion has been described *(4, 64, 103)*. Hydrophobic substrata do not generally support cell adhesion. They become more wettable following protein absorption and therefore the influence of substratum wettability on biological adhesion can be related to the absorption of particular proteins *(27, 102)*.

Regarding the physical form, it was observed that cells attached to a variety of surfaces spread in the direction of least curvature of the substratum. The mechanism appears to involve the cytoskeleton. This cell structure does not bend readily; therefore the direction of least curvature of the substratum becomes the direction of least resistance to cell spreading (*10, 26, 37, 80, 81, 105–107*).

Importance was also attached to the extent of the substratum. The circumference of glass bead substrata or the length of glass fibre substrata must be more than the normal length of the cells, in order to permit cell spreading and growth (*49–51*).

The materials commonly used for monolayer cell culture containers used in the multiple processes are glass and plastic, although cells grow on many other substrata such as metal and rubber.

3.1. Glassware

The composition of glass varies considerably and the different compositions contribute to differences in cell attachment to the substratum. Moreover, substances highly toxic to cells were found in some types of glass. Due to the alkaline environment, these substances are slowly brought into the liquid medium.

Glass is mainly divided into soda lime glass (normal glass, window glass) and aluminum borosilicate glass. Cells can grow on soda lime glass. However, significant quantities of alkali are brought into solution, especially at the high temperatures used for sterilization, and cell toxicity can be related to this source.

Aluminum borosilicate glass is better for growing cells. Its composition differs from that of soda lime glass because of its high B_2O_3 and Al_2O_3 contents and because of its very low content of alkaline oxides. It has a high hydrolytic resistance. It releases up to 10 times less alkali than soda lime glass.

The hydrolytic resistance of glass may be tested by the method described in the European Pharmacopoeia (*16*). The test is particularly severe and simulates a long period of storage.

Lyle (*47*) found that in sterilizing glass bottles full of distilled water at 121°C for 1 hr, the quantity of substances brought into solution corresponds to that released from the glass in 5 years at 20°C. With an alkaline medium such as that used for tissue culture, the quantity of substances brought into solution is higher than in distilled water.

Examples of aluminum borosilicate glass are the very well known Pyrex glass and the Neubor plasma bottles, round bottles which we have been using in our institute for many years and which appear to be quite satisfactory for rolling cell cultures. The 1-litre Neubor bottle is a high-quality glass of type I (*16*). Its hydrolytic resistance, expressed in millilitres of 0.01 N hydrochloric acid per 100 ml of test liquid, corresponds to 0.14 ml (*8*), a value lower than the limit prescribed for type I glass.

We have often noted that new bottles are rather unsatisfactory as far as cell attachment and spreading is concerned. After they have been used two or three times they give more satisfactory results, and after repeated use and washing with soda at 80°C cell growth becomes quite satisfactory.

The reason for this is obscure. There could be a slow loss of substances that are toxic for cells or that reduce cell adhesiveness. Metallic ions such as arsenic, lead, zinc or mercury may have been introduced into the glass as impurities in the raw materials. They are toxic for cells, but usually in the highest quality glass the traces of the above metals are absolutely negligible.

Used bottles could have a surface charge density different from that of new ones. The importance of the charge density in supporting cell adhesion has already been discussed, and Fig. 1 shows how easy it is to reduce the original energy of a glass surface by treatment with chemicals such as silicon (20). Other substances are known to be modulating factors of cell adhesiveness when added in varying thicknesses on the surfaces. The application of increasing concentrations of poly (2-hydroxyethyl methacrilate) (poly (HEMA)) results in a gradual decrease of cell adhesion (17).

Finally, the phenomenon observed with new bottles could be due to a simple physical modification induced on the glass surface by repeated treatment at high temperatures and with soda.

In spite of the above considerations, it is necessary to point out that brands of glass which were found to be toxic on some occasions were also found to be quite satisfactory on others. Therefore, in spite of any previous test, it is suggested that you find empirically a glass suitable for a particular purpose.

Moreover, apart from the flasks especially made for laboratory use, other glass containers on the market were found to be satisfactory for tissue cultures. Various types of bottles for medicaments, liquor, milk or oil have been used and some of them have been satisfactory.

3.2. Plasticware

We include under the general term "plastic", because of the similarity of their physical properties, both hydrocarbon polymers and carbohydrate polymers.

Cells adhere to plastic surfaces of different chemical composition, such as polystyrene, polyethylene, polycarbonate, polypropylene, Perspex, polyvinyl chloride, Teflon, cellophane, cellulose acetate and others. Some of them have the advantage of being transparent and cell growth can be visualized with a microscope. Otherwise growth can be visualized only by staining the cell layer with crystal violet.

Some of the commercially available plastic supports are unsuitable for cell culture because cells will not adhere to them. This is the case for most "bacterial grade" polystyrene dishes.

It is essential for cell growth that the plastic surfaces be wettable and carry a net charge. Plastic substrata of low wettability and of neutral charge do not generally support cell adhesion and are not suitable for routine tissue culture applications. They can be converted to wettable and charged substrata for tissue culture by appropriate techniques.

Plastic carriers of "tissue culture grade" are now commercially available. They principally stimulate the attachment of anchorage-dependent cell cultures through their ion-regulating surface, i.e. the ion arrangement has been grouped (35).

Maroudas (53) and Andrade et al. (3) have demonstrated that the disposable plastic material for cell culture carries a negative charge. Maroudas (52, 55), as previously reported, has indicated the minimal negative charge density required for cell spreading and has concluded that a substratum for cell culture need not be biochemically specific, provided it is physico-chemically polar, rigid and dense.

There is much speculation about the process whereby plastic substrata are made compatible with cell adhesion and growth. The subject covers one of the "magic" areas in tissue culture and the details are considered company secrets. In some instances they have been patented (1).

The producers of tissue culture plastics basically use a process which increases the electrostatic charge of the surface manyfold by chemical (56, 60, 82) or physical techniques (2, 61). In general, chemical processes consist of treating the surface layers with oxidizing agents ($KMnO_4$) or strong acids (sulphuric, chromic, nitric) (87). The physical processes most widely used at present involve ultra-violet light (87), "electret" formation (61) and the glow discharge (2) between two electrodes at a high voltage.

In principle, as Grinnell (22) stated, "bombardment of the plastic with high-energy electrons breaks the polymer structure, resulting in the formation of negatively charged carbonyl groups as well as unsaturated groups and possibly peroxides, depending on the atmosphere in which the bombardment occurs".

The negative surface charge on the dishes may be assayed by crystal violet binding (52). The dye bound to the surface is eluted and determined colourimetrically. The number of molecules of dye bound to the surface is assumed to be equal to the number of fixed negative charges.

By electron microscopy, it is possible to visualize the result of the surface treatment, which appears as a continuous layer of densely stained material (ruthenium red) (28).

The different sources of raw plastic materials are of importance since they cause different results in the activation process and failures in standard effect. The choice of raw material is therefore of importance, and in addition the speed, care of the moulding process and choice of the mould releasing agent, i.e. the substance which stimulates release from the mould, have an influence on the final functional quality (35).

The processes of surface activation are very efficient, and it can be concluded that many cells display better attachment to treated polystyrene surfaces than to glass (20, 46).

4. CONTAINERS FOR THE CULTURE OF ANIMAL CELLS IN MONOLAYER

Various sizes and shapes of containers, depending on specific requirements, are used for the culture of cell monolayers on a large scale.

When choosing a container, apart from the quality of the materials (see Section 3), one must remember the following factors:

1. The ratio between the growth surface and the volume of the bottle determines the growth surface which can be realized in a given incubator. It is advantageous to have the maximum growth surface per cubic centimetre of volume.

2. The ratio between the growth surface and the volume of the growth medium. This factor, in rolling bottles, influences the ratio between the growth surface in the liquid phase and the growth surface in the gaseous phase. This ratio is important during cell adhesion and the spreading of the culture.

3. The ratio between the growth surface and the volume of the maintenance medium. In order to obtain the best possible concentration of the product it is convenient to have the maximum growth surface per millilitre of maintenance medium (Table I).

4. Handling. Bottles with a smaller volume are easier to handle and enable the use of mechanical devices to facilitate operations such as washing and filling. The risk of breaking small containers is lower, even if one has to handle a larger number of them.

5. Containers available on the market are usually cheaper than those especially produced for tissue culture purposes. Moreover, as the volume of the bottle increases, the cost increases at a faster rate than the volume of the bottle.

A large-scale multiple process employs either containers with flat surfaces which are stationary during incubation or cylindrical containers which are incubated on a roller apparatus.

4.1. Flat Containers

Culture of cells in stationary flasks represents the traditional method and has been applied in the majority of laboratories for more than 20 years. The typical container is the 1-litre Roux flask with a usable surface of approximately 200 cm². Stationary containers having a smiliar shape but with a larger volume are also on the market.

An inconvenience of stationary flasks is that the usable surface is often not perfectly flat, and where the convexity is more pronounced the cell monolayer is not always fully confluent and the total cell number is reduced. In addition, because of the convexity of the surface, the volume of the maintenance medium cannot be reduced below certain levels, which results in higher dilution of the final product.

4.2. Cylindrical Containers

The evolution of the methodology of roller bottles was based on the simple observation that the concentration of a product in a cell culture is a function of the number of cells in the container and of its ratio to the volume of the fluid covering the cells during the liberation of virus from the cells (71, 72, 96, 110).

As stated before, when choosing a rotating bottle it is necessary to obtain the largest surface with the least volume and an optimum ratio between surface and volume of the medium. To do that it is necessary to remember a basic rule of geometry, i.e. by doubling the height of a cylinder the volume is doubled, but by doubling its diameter the volume is increased four times. In both cases the usable surface is doubled; however, in the latter case the capacity of the incubator is notably reduced. The volume of the medium must be doubled in both cases to keep the ratio of usable surface to medium constant. However, if the ratio between culture in the liquid phase and culture in the gaseous phase must also be kept constant, this can be done only by doubling the height of the bottles. When the doubling of the surface is achieved by increasing the diameter, the volume of the medium must be increased four times if the above ratio is to be kept constant. In such a case the ratio between usable surface and volume of medium is altered.

One-litre cylindrical Neubor bottles have been used in the author's laboratory for many years with satisfactory results. They have a usable surface of 620 cm². In larger containers the ratio of usable surface to total volume is not always advantageous.

Table I shows the concentration effect obtainable in the rolling bottles. There are two reasons for this effect. The first is the ratio of usable surface per millimetre of medium, which is higher in rotating bottles than in stationary flasks. The second is the multilayer that is obtainable in the rotating bottles, which gives more cells per unit of usable surface.

5. PROCESS PARAMETERS

In the monolayer cell systems a number of different kinds of parameters can be identified. Some of them, such as material, shape and size of the containers, are not changeable during a culture cycle.

Some parameters such as temperature, speed of rotation, kind of medium, volume of medium, cell type and size of inoculum can be held constant, whereas others such as pH, pO_2, CO_2 content, redox, glucose and amino acid concentrations are modified during the growth cycle.

Monitoring and control of the latter parameters are practicable only to a limited extent in the multiple processes because little or no influence can be exerted over the environment within each bottle. An improvement of the system is obtained by continuously adding to and withdrawing the medium. Such a perfusion system (15, 41, 42) allows one to monitor and control the growth parameters but it is not suitable for scaling up.

Criteria commonly employed for monitoring and controlling some of the above parameters with precision in the multiple processes are now discussed. Care must be taken in interpreting the results, as growth variabilities develop in single culture units and therefore monitoring must be performed on a certain number of cultures in order to obtain a statistically significant value.

5.1. Temperature

The rate of cell growth increases with increasing temperature until a maximum is reached and thereafter declines due to cell death. The optimum temperature varies from species to species.

For mammalian and avian tissues it is between $+36°$ and $+37.5°C$. For fish tissues it varies considerably according to the different fish species. Trout cells grow at $+20°$ to $+22°C$, whereas cells of some aquarium fish grow at $+33°C$. The range of temperature for amphibian tissues is $+20°$ to $+25°C$, for reptile tissues is $+18°$ to $+37°C$ and for arthropod tissues is $+25°$ to $+28°C$.

Temperatures lower than the optimum are better tolerated than the higher ones.

Control of temperature is usually obtained by placing the culture containers in incubators whose air temperature is controlled.

In industrial systems with very large incubators temperature gradients of

2°C or more may occur across the enclosure. In order to avoid such variations, which lead to quite large differences in cell yield, a uniform distribution of temperature is usually obtained by forcing the air circulation.

Overheating can be observed in enclosures having inside sources of heat such as motors for rolling mills. It may be prevented by installing a heat exchanger fed with cold water along the air distribution channel.

5.2. pH

The pH of a culture is considered a determinant of growth either for normal or transformed cells (9).

The optimum pH for mammalian cells usually lies between 7.2 and 7.4, a pH similar to that of the body fluids. Growth diminishes as the pH diverges from the optimum value. However, most mammalian cells survive in a range of pH between 6.6 and 7.8. Moreover, there are cases of cells growing at pH values beyond the above limits and cases of cells surviving at pH 8.5 and 5.5.

The pH in monolayer tissue cultures is controlled by different buffering systems. The most common is modelled on the naturally occurring CO_2–$NaHCO_3$ system present in the blood. Sodium bicarbonate is added to the medium and the pH value depends on the carbon dioxide tension in the medium and in the gaseous phase. The CO_2 tension can be controlled during the process by incubating the cultures in open vessels in an incubator with a CO_2-enriched atmosphere. The method is not applicable to the large-scale multiple monolayer processes. In such cases, the problem is overcome by tightly sealing the culture vessels. Carbon dioxide, provided by cell respiration, is retained and is normally sufficient to maintain the pH at satisfactory levels.

The CO_2–$NaHCO_3$ system is used in the majority of culture media. However, it has a number of disadvantages. The first is that the CO_2 content is not constant.

Variations in pH result from the metabolism of the cells; pH rises considerably during the initiation of a culture, reaching values near 8.0 in 24 hr, and decreases below 6.8 when the monolayer is formed.

Attempts have been made to use buffering systems other than sodium bicarbonate. The most successful have been tris(hydroxymethyl)aminomethane (tris) and glycoglycine (91), L-arginine (43) and HEPES (18, 86, 108). Some of them have the advantage that cells can grow in vessels open to the atmosphere. Buffer combinations for mammalian cell cultures have been studied by Eagle (12).

There are three parameters which can influence pH regulation in tightly sealed monolayer cultures. These are the volume of the vessel, the volume of the medium and the number of cells.

The aim of the large-scale monolayer process is to obtain cultures with the highest number of cells in the smallest volume of medium. Such a situation corresponds to a very high metabolic activity and to a reduced buffering capacity. Moreover, the volume of the gaseous phase is increased. Difficulties in buffering can arise from this situation.

In the author's laboratory problems related to the pH control of large-scale BHK cell cultures in rolling bottles were circumvented, as suggested by Polatnick and Bachrach (74), by buffering the medium with 0.02 M tris and by reducing the sodium bicarbonate to as low a level as possible as substrate for the cells. This medium, without being changed, sustains cell growth for 5–6 days, yielding confluent layers.

Tris buffers are particularly useful in limiting the rise of the pH during the initiation of the culture to acceptable values. The pH rise, for the first day, was found much more detrimental to BHK cell viability than the following drop below 7.0, due to cell growth.

Using tris buffers, the pH pattern is unchanged when culture bottles are sealed with loose caps (74).

Phenol red at 0.001% concentration is usually added to the medium as a pH indicator. The colour varies during the growth of the cells and careful observation of the several shades of colour, from violet to yellow, is the method used to monitor the pH in the multiple monolayer systems.

5.3. Cell Number

Estimation of the cell number in a monolayer process system is usually performed (1) on a cell suspension obtained from fresh tissue or from a monolayer cell culture used to prepare an inoculum containing an appropriate number of cells or (2) on a monolayer cell culture to estimate the cell growth on a given substratum.

Methods commonly employed make use of a direct count of the cells or of an indirect determination on the basis of known ratios between cell number and other parameters.

In a cell suspension, a direct count of the cells can easily be performed with a blood haematocytometer, such as the Fuchs Rosenthal or Thoma chamber.

Accurate cell counts can be obtained provided the cells are monodispersed; the count is repeated on three separate samples and at least 100 cells per sample are counted.

In order to help distinguish cells from other material which may be present, staining solutions such as crystal violet can be used. The count is much easier, but the method does not distinguish between living cells and dead cells.

One of the best estimates of cell viability is obtained from the plating efficiency. This consists in determining the proportion of cells in a population which give rise to cell colonies (75, 90) when the cell suspension is plated on a substratum. The test is not easy to perform, it requires very well standardized conditions and results are not obtained for several days.

The resistance of viable cells to staining affords an opportunity of differentiating viable from non-viable cells. The diffusion into the cells of dyes such as trypan blue, eosin, nigrosin, erythrosine or amethyst violet is used as a criterion of cell injury, because the dyes readily stain injured cells but do not stain uninjured cells. The validity of such methodologies may be influenced by variables such as serum concentration, dye concentration and time of exposure before counting (24). Living cells are stained by neutral red. For technical details see Melnick and Opton (58), Mayr et al. (57), and Paul (73).

At present, the cell number can also be estimated with electronic counters especially developed for blood cell counting (25, 65). These counters enable counts to be performed much more quickly and with higher accuracy. However, electronic counters do not distinguish between living and dead cells, nor between single cells, clumps or other material. Moreover, electronic counters are very expensive.

Having previously estimated the ratios cell/weight or cell/volume, the total number of cells in a suspension may be determined indirectly by measuring the weight or the volume of the packed cell sediment after centrifugation under standard conditions. The method furnishes very crude estimates but it is useful when large amounts of tissue are handled.

A rapid quantitative haematocrit method was described by Waymouth (104).

Cell concentration may be deduced from the turbidity of a cell suspension measured with a photometer at 590–650 nm. Cell number is obtained from a standard curve (40, 76, 109).

For the evaluation of the number of cells of a monolayer culture, cells are brought into suspension by scraping them from the glass or by using a proteolytic enzyme or a chelating agent. Cell count is then performed as previously described.

A count of nuclei was suggested as a more accurate method. Nuclei are separated from the cytoplasm by treating the cell monolayer with citric acid (84, 110). The procedure is a little more complicated and it does not distinguish between living and dead cells, but in the author's experience it makes it possible to distinguish big, elliptic and stained nuclei from unstained cell residues or other materials very easily.

The cell number of a monolayer can be indirectly estimated by a determination of the total protein (66). The washed cells attached to the substratum are first dissolved with the Lowry alkaline solution and then the protein content is calculated using the Folin–Ciocalteau reagent on the basis of the optical density of the blue colour at 660 nm. The author has found a very good relationship between these two parameters when testing proliferation curves of BHK cells in culture. The method is very useful if large numbers of cultures are to be tested. The cell number is easily estimated from a standard curve of the optical density plotted against the previously determined cell number.

5.4. Inoculum Size and Cell Yield

It has been shown (79) that the growth of a monolayer culture is determined largely by its initial cell concentration per unit area; below some minimum concentrations little or no growth occurs in the culture (13).

Cellular yield versus inoculum size determinations are therefore necessary to establish conditions which would give undegenerating maximal populations of different cell types at the expected time.

Without a medium change, primary cells from embryonal tissues usually reach a cell monolayer in 2 days, whereas those from adult animal tissues require as long as 5–6 days to reach the same population.

Lines of actively proliferating cells usually require the shortest time. Varying the cell concentrations gives the possibility of modulating the rate of growth, according to the age at which they will be used for a certain production.

An increase of the inoculum size usually shortens the time required for maximal growth of a cell line (74). This is not the case in primary cultures, where an optimum inoculum size must be chosen to avoid the detrimental effect of the large quantity of tissue debris, cell aggregates and dead cells that are usually present in the inoculum.

Figure 3 shows the growth curves of calf kidney and BHK cells in stationary flasks and in rotating bottles up to 8 days after seeding. The figure shows that a higher number of cells per square centimeter of usable surface is required with primary cells than with cell lines. Moreover, it shows a quite different situation as far as cell attachment and growth are concerned when using a cell line or primary cells. With BHK cells, 100% of the cells attach to the surface and the number of cells growing on the glass after 1 day surpasses the number initially added. By contrast, only a small percentage of the initial cells are found on the glass surface at 3 days when primary calf kidney cells are used. This is an obvious consequence of a different degree of viability of the cells in the two inocula.

The proportion of seeded primary cells which adhere to the glass is lower in rotating bottles (13%) than in stationary flasks (23%). Rotation therefore seems slightly deleterious for attachment to glass of the cells contained in the

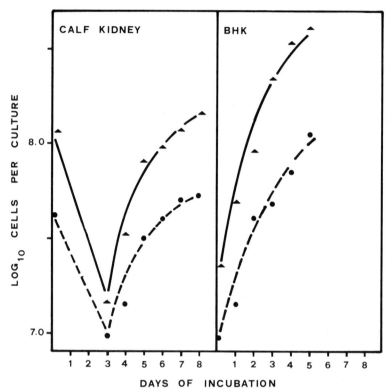

Fig. 3.   Growth curves of calf kidney and BHK cells in 1-litre stationary (●) and rotating (▲) containers (67, 96).

unhomogeneous suspensions obtained by trypsinization of animal tissues (96).

Figure 3 also gives the cell concentration of the population in stationary flasks and rolling bottles. Values indicate an overgrowth of BHK cells in rolling bottles. In such containers the cell number exceeds that expected when based on the ratio between the usable surfaces of the two types of containers (Table I).

5.5. Volume of Medium

Earle *et al.* (13) found that the volume of medium to be utilized by growing cells may be a critical limiting factor in determining the number of cells which proliferate in cultures.

Spier (87) studied the relationship of medium volume to culture proliferation and concluded that the number of cells growing in stationary flat con-

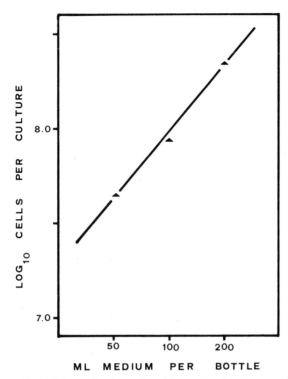

Fig. 4.   BHK cell yield obtained at 3 days from 1-litre rolling bottles planted with $10^{7.5}$ cells in different volumes of medium. Data from three experiments were averaged (67).

tainers is related to the volume of the medium used, provided the surface is not limiting.

The same relationship was found by the author, using BHK cultures in rolling bottles (Fig. 4).

5.6. Operational Ratios

It was already stated that the maximum concentration of a product derived from tissue culture is strongly influenced by the cell density per millilitre of medium. Such a ratio depends mainly on the size and shape of the culture containers.

The values shown in Table I represent a comparison of flat and cylindrical containers and clearly show the possiblity of obtaining highly concentrated products by using cylindrical bottles.

Very small volumes of maintenance medium may be used in rolling bottles. However, the author met difficulties in producing FMD antigens with monolayer cultures of BHK cells in 1-litre rolling bottles when surface area/medium volume ratios higher than 30 were used, because of the quick drop in the pH and the lability of the product at such a low pH.

6. SCALE-UP AND AUTOMATED VERSIONS

Monolayer systems based on a multiplicity of containers for the cultivation of cells are often criticized because the operations involved in the process, especially the handling of thousands of bottles, are cumbersome and clearly an obstacle to scaling up the process.

Culture containers are usually handled as single units or in racks or on trucks (62, 95).

Rotation of cylindrical bottles may be realized by simple devices having roller mills of different lengths on which the single bottles are rotated. A modular apparatus for bench-scale cultures is on the market. Its capacity may be easily expanded by adding additional roller tiers.

Large-scale processes were realized with special devices such as the wheel described by Leunen *et al.* (44, 45) or the rotating racks described by Ubertini *et al.* (96) in Italy and by Polatnick and Bachrach (74) in the United States.

Rotating racks have contributed significantly in rendering the operational features of plants using thousands of bottles acceptable, because the bottles remain in the racks during the whole operational cycle, dramatically reducing the number of units to be handled.

Cylindrical wire racks may be rotated on "pilot roll" units or on "mega roll" industrial units (96, 97). Both are on the market and several such units coupled together to reach the required production capacity have been adopted by various laboratories throughout the world (Figs 5 and 6).

Four mega roll industrial units having a total capacity of 28,800 culture bottles of 1 litre each were installed in 1966 in a large plant which is still operating at the Istituto Zooprofilattico in Brescia, Italy (63, 70, 99). For at least 19 years the plant has produced cell cultures to make billions of doses of FMD, IBR, PI_3 and Aujeszky virus vaccines. High-quality antigens have been obtained with a ratio of culture surface (cm^2) to volume of maintenance medium (ml) of about 10. This corresponds to 8×10^6 BHK cells or 2.4×10^6 calf kidney cells per millilitre of maintenance medium.

The Brescia plant is an example of an highly automated multiple process. The mechanical devices which were designed to enable it to operate are reviewed below in detail. All of them are on the market.

Fig. 5. Three-tiered "pilot roll" units for rotating the wire racks. (Veterinary Research Institute, Onderstepoort, South Africa. Courtesy of Dr. A. Pini.)

Racks. The 1-litre bottles in groups of 18 each are held in cylindrical wire racks (Fig. 7). The bottles are washed, sterilized, seeded with cell suspension, infected with virus, incubated and harvested of virus while held in the racks. They are never removed from the racks during the whole operational cycle.

Closure System. To avoid the impracticable individual stoppering of bottles, a sterile rubber sheet is used. A metal lid with foam rubber bonded to the underside forces the rubber sheet tightly against the lip of each bottle (Fig. 7).

Dispenser. A manually controlled volumetric dispensing device, working by gravity without a pump, provides a simple means for dispensing cells suspended in the growth medium (Fig. 7). A stainless steel plate with 18 discharge points fits directly on the top of the rack and 18 bottles are filled at the same time. Four thousand bottles per hour may be seeded in this way. To infect the cultures, maintenance medium and seed virus are dispensed with a similar device.

Fig. 6. One five-tiered "mega roll" unit holding 400 wire racks, i.e. 7200 individual roller bottles. (Istituto Zooprofilattico Sperimentale, Brescia, Italy.)

Fig. 7. Roller conveyors, cylindrical wire racks holding 18 1-litre Neubor bottles, metal lid with foam rubber, rack inverter for discarding the spent medium, volumetric dispenser for filling 18 bottles at the same time. (Istituto Zooprofilattico Sperimentale, Brescia, Italy.)

Conveyors. The transport of the bottle racks during the whole production cycle is effected with the aid of a roller–conveyor system (Fig. 7). A special movable belt conveyor moves the bottle racks directly to the different tiers of the incubator.

Incubators. Cultures are incubated in four incubators; each measures 18 m by 2.75 m and is 5.45 m high (Fig. 6). The capacity of the four incubators is 1600 racks, i.e. 28,800 bottles. Heating is obtained from 14-kW electric heaters for each incubator and a uniform distribution of temperature is ensured by forced air circulation, obtained through a centrifugal ventilator. A heat exchanger fed with cold water prevents overheating.

Rolling Unit. Four mega roll industrial units are installed in the incubators. Each unit has a loading capacity of 400 racks, corresponding to 7200 1-litre bottles. The mega roll consists of two five-tiered scaffoldings. Each scaffolding is 18 m long (Fig. 6).

Each tier is provided with a rolling device consisting of two shafts, each 17.50 m long. The distance between shafts is 220 mm. The shafts rotate at a speed which allows the basket to carry out four rotations per hour. In the author's experience such a speed is considered an optimum both for cell growth and for virus multiplication.

Two channelled chains are bolted at both sides of the shafts for transfer of the racks in the incubator. The racks are transported with the bottles held vertically; when they reach the correct position for rotation they are turned manually to bring the bottles to the horizontal position.

Rack Inverter.   Before infection, with the aid of a two-headed inverting apparatus, the spent growth medium is poured off from the 18 bottles held in the rack. The inverter is inserted along the line of the conveyors (Fig. 7).

Washing Machine.   The racks containing the bottles fit directly onto the header of an industrial tunnel washing machine, which includes a rack inverter for collecting virus fluid.

The inverted racks move slowly on a chain in the washing tunnel and the bottles placed in a vertical position are carried successively to different internal sectors of the machine, where they are squirted with various liquids for the appropriate length of time.

In detail, the washing process takes place as follows:

1. Washing at 80°C with an alkaline detergent
2. First rinsing with hot tap water
3. Second rinsing with cold tap water
4. Third rinsing with de-ionized water

The duration of the process, including virus harvesting, is 4 min. The machine furnishes a washed rack every 16 sec, a speed which corresponds to 4050 bottles/hr.

Sterilizer.   Sterilization of the bottles is carried out at 150°C in a dry oven, with two air-tight doors. The inside of the oven measures 11 m by 2.3 m and is 2.2 m high. The capacity is 480 racks, i.e. 8640 1 litre bottles.

Air circulation in the oven is forced by a centrifugal ventilator. Heating is performed by two steam heat exchangers with a maximum consumption of 500 kg/hr. Supplementary heating with a maximum consumption of 100 kW is installed for emergency only, because of the higher cost of electricity compared with steam.

Rapid cooling after sterilization is obtained by using a cold-water heat exchanger with a consumption of 8000 litres/hr. During such cooling sterile air is introduced into the oven. Air filtration is obtained with an absolute filter.

Positive pressure is maintained inside the oven during the complete cycle. A two-way moving base facilitates loading and unloading of the racks.

ACKNOWLEDGMENTS

The author is indebted to Miss G. Tolotti, Mr. L. Ragni, Dr. A. Pini and Dr. R. Spier for various kinds of help and assistance.

REFERENCES

1. Ambroski, L. E. (1966). Method of production non-fogging films. U.S. Patent 3,274,090.
2. Amstein, C. F., and Hartman, P. A. (1975). Adaptation of plastic surfaces for tissue culture on glow discharge. *J. Clin. Microbiol.* **2**, 46–54.
3. Andrade, J. D., Iwamoto, G. K., McNeill, B., and King, R. N. (1976). XPS Studies of polymer surfaces for biochemical applications. *Pap., 171st Meet., Am. Chem. Soc. (Div. Org. Coat. Plast. Chem.)*, 1976, Vol. 36, No. 1.
4. Baier, R. E. (1972). The role of surface energy in thrombogenesis. *Bull. N.Y. Acad. Med.* [2] **48**, 257–272.
5. Bard, J., and Elsdale, T. (1971). Specific Growth regulation in early subcultures of human diploid fibroblasts. *In* "Growth Control in Cell Cultures" (G. E. W. Wolstenholme and J. Knight, eds.), pp. 187–197. Churchill-Livingstone, Edinburgh and London.
6. Baserga, R., Rovera, G., and Farber, J. (1971). Control of cellular proliferation in human diploid fibroblasts. *In Vitro* **7**, 80–87.
7. Borysenko, J. Z., and Woods, W. (1979). Density, distribution and mobility of surface anions on a normal/transformed cell pair. *Exp. Cell Res.* **118**, 215–227.
8. Branchi, G., and Mecarelli, E. (1975). Il contenitore di vetro e la sua inerzia chimica a contatto con il farmaco. *Farmaco* **30**, 605–614.
9. Ceccarini, C., and Eagle, H. (1971). pH as a determinant of cellular growth and contact inhibition. *Proc. Natl. Acad. Sci. U.S.A.* **68**, 229–233.
10. Curtis, A. S. G., and Varde, M. (1964). Control of cell behavior: Topological factors. *J. Natl. Cancer Inst. (U.S.)* **33**, 15–23.
11. Dulbecco, R. (1971). Regulation of cell multiplication in tissue culture. *In* "Growth Control in Cell Cultures" (G. E. W. Wolstenholme and J. Knight, eds.), pp. 71–76. Churchill-Livingstone, Edinburgh and London.
12. Eagle, H. (1971). Buffer combinations for mammalian cell culture. *Science* **174**, 500–503.
13. Earle, W. R., Sanford, K. K., Evans, V. J., Waltz, H. K., and Shannon, J. E., Jr. (1951). The influence of inoculum size on proliferation in tissue cultures. *J. Natl. Cancer Inst. (U.S.)* **12**, 133–153.
14. Earle, W. R., and Schilling, E. L. (1954). Influence of tube rotation velocity on proliferation of strain L cells in surface substrate roller-tube cultures. *J. Natl. Cancer Inst. (U.S.)* **14**, 853–864.
15. Elsworth, R. (1970). Multi-layer perfusion tissue culture. *Process Biochem.* **5**, 21–22.
16. European Pharmacopoeia (1971). (Council of Europe, eds.), Vol. II, pp. 65–71. Maisonneuve S.A., Sainte Ruffine, France.
17. Folkman, J., and Moscona, A. (1978). Role of cell shape in growth control. *Nature (London)* **273**, 345–349.
18. Good, N. E., Winget, G. D., Winter, W., Conally, T. N., Izawa, S., and Sing, R. M. (1966). Hydrogen ion buffers for biological research. *Biochemistry* **5**, 467–477.

19. Griffiths, J. B. (1971). The effect of medium changes on the growth and metabolism of the human diploid cell, W1-38. *J. Cell Sci.* **8**, 43–52.
20. Grinnell, F., Milam, M., and Srere, P. A. (1972). Studies on cell adhesion. II. Adhesion of cells to surfaces of diverse chemical composition and inhibition of adhesion by sulphydryl binding reagents. *Arch. Biochem. Biophys.* **153**, 193–198.
21. Grinnell, F., Tobleman, M. Q., and Hackenbrock, C. R. (1976). Initial attachment of baby hamster kidney cells to an epoxy substratum. *J. Cell Biol.* **70**, 707–713.
22. Grinnell, F. (1978). Cellular adhesiveness and extracellular substrata. *Int. Rev. Cytol.* **53**, 65–144.
23. Grinnell, F., and Feld, M. K. (1979). Initial adhesion of human fibroblasts in serum-free medium: Possible role of secreted fibronectin. *Cell* **17**, 117–129.
24. Hanks, J. H., and Wallace, J. H. (1958). Determination of cell viability. *Proc. Soc. Exp. Biol. Med.* **98**, 188–192.
25. Harris, M. (1959). Growth measurements on monolayer cultures with an electronic cell counter. *Cancer Res.* **19**, 1020.
26. Harrison, R. G. (1914). The reaction of embryonic cells to solid structures. *J. Exp. Zool.* **17**, 521–544.
27. Häyry, P., Myllylä, G., Saxén, E., and Penttinen, K. (1966). The inhibition mechanism of serum on the attachment of HeLa cells on glass. *Ann. Med. Exp. Biol. Fenn.* **44**, 166–170.
28. Heckman, C. A., Vroman, L., and Pitlick, A. (1977). The nature of substrate-attached materials in human fibroblast cultures: Localization of cell and fetal calf serum components. *Tissue Cell* **9**, 317–334.
29. Hirtenstein, M., Clark, J., Lindgren, G., and Vretblad, P. (1980). Microcarriers for animal cell culture: A brief review of theory and practice. *Dev. Biol. Stand.* **46**, 109–116.
30. Holley, R. W., and Kiernan, J. A. (1968). "Contact inhibition" of cell division in 3T3 cells. *Proc. Natl. Acad. Sci. U.S.A.* **60**, 300–304.
31. Holley, R. W., Armour, R., and Baldwin, J. H. (1978). Density-dependent regulation of growth of BSC-1 cells in cell culture: Control of growth by low molecular weight nutrients. *Proc. Natl. Acad. Sci. U.S.A.* **75**, 339–341.
32. Holley, R. W. (1975). Control of growth of mammalian cells in cell culture. *Nature (London)* **258**, 487–490.
33. Höök, M., Rubin, K., Oldberg, A., Öbrink, B., and Vaheri, A. (1977). Cold-insoluble globulin mediates the adhesion of rat liver cells to plastic petri dishes. *Biochem. Biophys. Res. Commun.* **79**, 726–733.
34. Horng, C. B., and McLimans, W. (1975). Primary suspension culture of calf anterior pituitary cells on a microcarrier surface. *Biotechnol. Bioeng.* **17**, 713–732.
35. Houdkamp, J. (1982). Costar Europe Ltd. (personal communication).
36. Hughes, R. C., Pena, S. D., Clark, J., and Dourmashkin, R. R. (1979). Molecular requirements for the adhesion and spreading of hamster fibroblasts. *Exp. Cell Res.* **121**, 307–314.
37. Ivanova, O. Y., and Margolis, L. B. (1973). The use of phospholipid film for shaping cell cultures. *Nature (London)* **242**, 200–201.
38. Jensen, M. D. (1979). The application of environmental control to continuous culture and vaccine production. *In* "Practical Tissue Culture Applications" (K. Maramorosch and H. Hirumi, eds.), pp. 115–136. Academic Press, New York.
39. Johnston, M. D., Christofinis, G., Ball, G. D., Fantes, K. H., and Finter, N. B. (1979). A culture system for producing large amounts of human lymphoblastoid interferon. *Dev. Biol. Stand.* **42**, 189–192.
40. Kraus, W., and Schäfer, W. (1963). Apparatur zur Herstellung von primären Kulturen animaler Zellen. *Zentralbl. Veterinaermed., Reihe B* **10**, 394–399.
41. Kruse, P. F., Jr., Myhr, B. C., Johnson, J. E., and White, P. B. (1963). Perfusion system

for replicate mammalian cell cultures in T-60 flasks. *J. Natl. Cancer Inst. (U.S.)* **31**, 109–120.

42. Kruse, P. F., Jr., and Miedema, E. (1965). Production and characterization of multiple-layered populations of animal cells. *J. Cell Biol.* **27**, 273–279.

43. Leibovitz, A. (1963). The growth and maintenance of tissue-cell cultures in free gas exchange with the atmosphere. *Am. J. Hyg.* **78**, 173–180.

44. Leunen, J., Strobbe, R., and Mammerickx, M. (1962). Notice technique sur un appareil à rouler des flacons pour cultures de cellules rénales. *Bull. Off. Int. Epizoot.* **57**, 615–617.

45. Leunen, J., Mammerickx, M., and Strobbe, R. (1963). La préparation industrielle des cultures cellulaires à l'Institut National de Recherches Vétérinaires de Bruxelles par la technique des flacons roulants. *Bull. Off. Int. Epizoot.* **59**, 1019–1036.

46. Litwin, J. (1973). Titanium disks. *In* "Tissue Culture: Methods and Applications" (P. F. Kruse, Jr. and M. K. Patterson, eds.), p. 385. Academic Press, New York.

47. Lyle, A. K. (1943). Theoretical aspects of chemical attack of glass by water. *J. Am. Ceram. Soc.* **26**, 201–204.

48. MacMorine, H. G., Laurence, G. D., Parisius, W., and Cucakovich, N. B. (1977). Large-scale cultivation of cells in closed systems: Stationary and moving. Joint WHO/IABS Symposium on the standardization of cell substrates for the production of virus vaccines, Geneva, Dec. 1976. *Dev. Biol. Standard.* **37**, 139–142.

49. Maroudas, N. G. (1972). Anchorage dependence: Correlation between amount of growth and diameter of bead, for single cells grown on individual glass beads. *Exp. Cell Res.* **74**, 337–342.

50. Maroudas, N. G. (1973a). Growth of fibroblasts on linear and planar anchorages of limiting dimensions. *Exp. Cell Res.* **81**, 104–110.

51. Maroudas, N. G. (1973b). Chemical and mechanical requirements for fibroblast adhesion. *Nature (London)* **244**, 353–354.

52. Maroudas, N. G. (1975). Adhesion and spreading of cells on charged surfaces. *J. Theor. Biol.* **49**, 417–424.

53. Maroudas, N. G. (1976). Sulphonated polystyrene as an optimal substratum for the adhesion and spreading of mesenchymal cells in monovalent and divalent saline solutions. *J. Cell. Physiol.* **90**, 511–520.

54. Maroudas, N. G. (1977a). Polymer aggregation and cell adhesion. *In* "Cell Shape and Surface Architecture" (J. P. Revel, V. Henning, and C. F. Fox, eds.), pp. 511–520. Alan R. Liss, Inc., New York.

55. Maroudas, N. G. (1977b). Sulphonated polystyrene as an optimal substratum for the adhesion and spreading of mesenchymal cells in monovalent and divalent saline solutions. *J. Cell. Physiol.* **90**, 511–520.

56. Martin, G. R., and Rubin, H. (1974). Effects of cell adhesion to the substratum on the growth of chick embryo fibroblasts. *Exp. Cell Res.* **85**, 319–333.

57. Mayr, A., Bachmann, P. A., Bibrack, B., and Wittmann, G. (1974). Zellkulturen. *In* "Virologische Arbeitsmethoden," pp. 43–171. Fischer, Jena.

58. Melnick, J. L., and Opton, E. M. (1956). Assay of poliomyelitis neutralizing antibody in disposable plastic panels. *Bull. W. H. O.* **14**, 129–146.

59. Mierzejewski, K., and Rosengurt, E. (1977). Density-dependent inhibition of fibroblast growth is overcome by pure mitogenic factors. *Nature (London)* **269**, 155–156.

60. Munder, P. G., Modolell, M., and Wallach, D. F. H. (1971). Cell propagation on films of polymeric fluorocarbon as a means to regulate pericellular pH and pO_2 in cultured monolayers. *FEBS Lett.* **15**, 191–196.

61. Murphy, P. V., LaCroix, A., Merchant, S., and Bernhard, W. (1971). Quoted from Grinnell (22).

62. Nardelli, L., Ubertini, B., Dal Prato, A., Panina, G. F., and Santero, G. (1960). Weitere Erfahrungen über die Massenherstellung des MKS-Virus in Kälbernierenzellkulturen. *Arch. Exp. Veterinaermed.* **14,** 276–282.

63. Nardelli, L., and Panina, G. F. (1977). 10-years experience with a 28,800 roller bottle plant for FMD vaccine production. Joint WHO/IABS Symposium on the standardization of cell substrates for the production of virus vaccines, Geneva, Dec. 1976. *Dev. Biol. Stand.* **37,** 133–138.

64. Olsen, D. A., and Kletschka, H. D. (1973). Quoted from Grinnell (22).

65. Ørstavik, J., and Myhrvold, V. (1967). Electronic cell counting in cell culture studies. *Acta Pathol. Microbiol. Scand.* **70,** 341–348.

66. Oyama, V. I., and Eagle, H. (1956). Measurement of cell growth in tissue culture with a phenol reagent (Folin–Ciocalteau). *Proc. Soc. Exp. Biol. Med.* **91,** 305–307.

67. Panina, G. F., and Ragni, A. (1965). Unpublished results.

68. Panina, G. F., and De Simone, F. (1973). Immunological activity of foot-and-mouth disease virus purified by polyethylene glycol precipitation. *Zentralbl. Veterinaermed., Reihe B* **20,** 773–782.

69. Panina, G. F. (1975). Media and serum for the large-scale production of BHK cells. *Rep. Meet. FMD Eur. Comm.,* pp. 60–65. F.A.O., Rome.

70. Panina, G. F. (1976). Animal cell culture and FMD virus production in a large scale industrial "roller bottle" plant. *Proc. Gen. Meet. Eur. Soc. Animal Cell Technol. 1st,* (R. Spier and A. L. van Wezel, eds.), pp. 43–47. Rijks Inst. Volksgezondheid, Bilthoven, Netherlands.

71. Patty, R. E., and May, H. Y. (1961). The production of high concentrations of foot-and-mouth disease virus in cultures of cells on glass. *Am. J. Vet. Res.* **22,** 926–931.

72. Patty, R. E., and Norcross, N. L. (1961). Production of foot-and-mouth disease virus with high complement-fixing antigenicity. *Am. J. Vet. Res.* **22,** 775–778.

73. Paul, J. (1975). "Cell and Tissue Culture." Churchill-Livingstone, Edinburgh and London.

74. Polatnick, J., and Bachrach, H. L. (1964). Production and purification of milligram amounts of foot-and-mouth disease virus from baby hamster kidney cell cultures. *Appl. Microbiol.* **12,** 368–373.

75. Puck, T. T., Marcus, P. I., and Cieciura. S. J. (1956). Clonal growth of mammalian cells in vitro. Growth characteristics of colonies from single HeLa cells with and without a "feeder" layer. *J. Exp. Med.* **103,** 273–284.

76. Rappaport, C. (1957). Colorimetric method for estimating number of cells in monolayer cultures without physiological damage. *Proc. Soc. Exp. Biol. Med.* **96,** 309–316.

77. Rappaport, C., Poole, J. P., and Rappaport, H. P. (1960). Studies on properties of surfaces required for growth of mammalian cells in synthetic medium. I. The HeLa cell. *Exp. Cell Res.* **20,** 465–510.

78. Rappaport, C. (1972). Some aspects of the growth of mammalian cells on glass surfaces. *In* "Chemistry of Biosurfaces" (M. L. Hair, ed.), Vol. 2, p. 449. Dekker, New York.

79. Rein, A., and Rubin, H. (1968). Effects of local cell concentrations upon the growth of chick embryo cells in tissue culture. *Exp. Cell Res.* **49,** 666–678.

80. Rosenberg, M. D. (1963). Cell guidance by alterations in monomolecular films. *Science* **139,** 411–412.

81. Rovensky, Y. A., Slavnaja, I. L., and Vasiliev, J. M. (1971). Behavior of fibroblast-like cells on grooved surfaces. *Exp. Cell Res.* **65,** 193–201.

82. Rubin, H. (1966). Altering bacteriological plastic petri dishes for tissue culture use. *Public Health Rep.* **81,** 843–844.

83. Rubin, H. (1971). Growth regulation in cultures of chick embryo fibroblasts. *In* "Growth

Control in Cell Cultures" (G. E. W. Wolstenholme and J. Knight, eds.), pp. 127–145. Churchill-Livingstone, Edinburgh and London.

84. Sanford, K. K., Earle, W. R., Evans, V. J., Waltz, H. K., and Shannon, J. E. (1951). The measurement of proliferation in tissue cultures by enumeration of cell nuclei. *J. Natl. Cancer Inst. (U.S.)* 11, 773–795.

85. Sharp, L. H., and Schonhorn, H. (1964). In "Contact Angle, Wettability and Adhesive Joints" (R. F. Gould, ed.), pp. 189–201. Am. Chem. Soc., Washington, D.C.

86. Shipman, C. (1969). Evaluation of 4-(2-hydroxyethyl)-1-piperazineethanesulphonic acid (HEPES) as a tissue culture buffer. *Proc. Soc. Exp. Biol. Med.* 130, 305–310.

87. Spier, R. E. (1980). Recent developments in the large scale cultivation in animal cells in monolayers. *Adv. Biochem. Eng.* 14, 119–162.

88. Spier, R. E., and Whiteside, J. P. (1976). The production of foot-and-mouth disease virus from BHK 21 C 13 cells grown on the surface of glass spheres. *Biotechnol. Bioeng.* 18, 649–657.

89. Stoker, M., and Rubin, H. (1967). Microbiology. Density dependent inhibition of cell growth in culture. *Nature (London)* 215, 171–172.

90. Stulberg, C. S., Peterson, W. D., and Berman, L. (1962). Quantitative and qualitative preservation of cell-strain characteristics. *Natl. Cancer Inst. Monogr.* 7, 17–31.

91. Swim, H. E., and Parker, R. F. (1955). Nonbicarbonate buffers in cell culture media. *Science* 122, 466.

92. Taylor, A. C. (1970). Adhesion of cells to surfaces. In "Adhesion in Biological Systems" (R. S. Manly, ed.), pp. 51–71. Academic Press, New York.

93. Temin, H. (1967). Studies on carcinogenesis by avian sarcoma viruses. VI. Differential multiplication of uninfected and of converted cells in response to insulin. *J. Cell. Physiol.* 69, 377–384.

94. Thrash, C. R., Ho, T., and Cunningham, D. D. (1974). Structural features of steroids which initiate proliferation of density-inhibited 3T3 mouse fibroblasts. *J. Biol. Chem.* 249, 6099–6103.

95. Ubertini, B., Nardelli, L., Santero, G., and Panina, G. F. (1960). Process report: Large-scale production of foot-and-mouth disease virus. *J. Biochem. Microbiol. Technol. Eng.* 2, 327–338.

96. Ubertini, B., Nardelli, L., Dal Prato, A., Panina, G. F., and Santero, G. (1963). Large scale cultivation of foot-and-mouth disease virus on calf kidney cell monolayers in rolling bottles. *Zentralbl. Veterinaermed., Reihe, B* 10, 93–101.

97. Ubertini, B., Nardelli, L., Dal Prato, A., Panina, G. F., and Barei, S. (1967a). BHK 21 cell cultures for the large-scale production of foot-and-mouth disease virus. *Zentralbl. Veterinaermed., Reihe, B* 14, 432–441.

98. Ubertini, B., Nardelli, L., Panina, G. F., and Lodetti, E. (1967b). Some notes on techniques of foot-and-mouth disease virus production used in Brescia. *Rep. Meet. FMD Eur. Comm.* pp. 29–40. F.A.O., Rome.

99. Ubertini, B., Nardelli, L., Barei, S., Panina, G. F., and Lodetti, E. (1969). Production of foot-and-mouth disease (FMD) vaccine at the Istituto Zooprofilattico at Brescia. *Rep. Meet. FMD Euro. Comm.* pp. 16–31. F.A.O., Rome.

100. van Wezel, A. L. (1977). The large scale cultivation of diploid cell strains in microcarrier culture. Improvement of microcarriers. *Dev. Biol. Stand.* 37, 143–147.

101. Vasiliev, Yu. M., Gelfand, I. M., Guelstein, V. I., and Fetisova, E. K. (1970). Stimulation of DNA synthesis in culture of mouse embryo fibroblast-like cells. *J. Cell. Physiol.* 75, 305–314.

102. Vroman, L. (1967). Surface activity in blood coagulation. In "Blood Clotting Enzymology" (W. H. Seegers, ed.), pp. 279–322. Academic Press, New York.

103. Vroman, L. (1972). What factors determine thrombogenicity? *Bull. N.Y. Acad. Med. [2]* **48**, 302–310.
104. Waymouth, C. (1956). A rapid quantitative hematocrit method for measuring increase in cell population of strain L (Earle) cells cultivated in serum-free nutrient solutions. *J. Natl. Cancer Inst. (U.S.)* **17**, 305–313.
105. Weiss, P. A. (1945). Experiments on cell and axon orientation in vitro: The role of colloidal exudates in tissue organization. *J. Exp. Zool.* **100**, 353–386.
106. Weiss, P. A. (1961). Guiding principles in cell locomotion and cell aggregation. *Exp. Cell Res., Suppl.* **8**, 260–281.
107. Weiss, P. A. (1962). Cells and their environment, including other cells. *In* "Biological Interactions in Normal and Neoplastic Growth" (M. J. Brennan and W. L. Simpson, eds.), pp. 3–20. Little, Brown, Boston, Massachusetts.
108. Williamson, J. D., and Cox, P. (1968). Use of a new buffer in the culture of animal cells. *J. Gen. Virol.* **2**, 309–312.
109. Youngner, J. S. (1954). Monolayer tissue cultures. I. Preparation and standardization of suspensions of trypsin-dispersed monkey kidney cells. *Proc. Soc. Exp. Biol. Med.* **85**, 202–205.
110. Youngner, J. S. (1954). Monolayer tissue cultures. III. Propagation of poliomyelitis viruses in cultures of trypsin-dispersed monkey kidney cells. *J. Immunol.* **73**, 392–396.
111. Zisman, W. A. (1964). Relation of the equilibrium contact angle to liquid and solid constitution. *In* "Contact Angle, Wettability and Adhesive Joints" (R. F. Gould, ed.), pp. 1–51. Washington, D.C.

10

Monolayer Growth Systems: Heterogeneous Unit Processes

R. E. SPIER

Department of Microbiology
University of Surrey
Guildford, Surrey, United Kingdom

1. INTRODUCTION

Whilst it is well recognised that multiple process systems (Section 1.1) have served and still serve animal cell biotechnologists well (cf. Chapter 9, this volume), they are also subject to a number of severe limitations. The efforts of many investigators seeking to overcome such restrictions have led to a large number of alternative technologies within the unit process framework. For the purposes of this book, the unit process approach has been divided into two chapters, the first of which will deal with systems wherein

Animal Cell Biotechnology, Vol. 1
Copyright © 1985 by Academic Press Inc. (London) Ltd.

the solid substratum on, or within, which the cells grow is not evenly distributed throughout the culture apparatus, while the second will consider the homogeneous distribution of the solid phase throughout the apparatus. The distinction can be held to be artificial as either system can be forced into the alternative mode; however, it does serve to group the wide variety of monolayer systems into a form in which they can be readily described and discussed (cf. Chapter 11, this volume).

1.1. Disadvantages of Multiple Process Systems

It is clear from the preceding chapter that multiple process systems can be defined as those systems which are scaled up by increasing the number of units in operation in roughly the same proportion as the increase in scale. There are, however, a number of disadvantages in such a procedure. These have been summarised in Table I. Although most of the products which are derived from cultured animal cells are high-value, small (relatively) volume materials and the production cost can be readily taken up in the selling price, there is still much to be gained by operating with greater consistency, with higher yields and maximum product quality.

In contrast to the multiple process system, the scale-up of the unit process system is achieved by an increase in the size of the culture equipment

TABLE I Disadvantages of Multiple Process Systems

Higher capital costs	Larger floor area needed
	High equipment costs if using automated handling machinery
	Need for wide variety of complex equipment
Higher operating costs	More labour
	Less efficient use of materials and services
	Higher maintenance costs
	Higher rates of loss (breakages)
Less controllable	Lower yields
	Poorer quality
	Loss of time
	Fewer records
Poorer operating conditions	Unregulated liquid flows (from breakages, spillages)
	More noise
	Larger distances for operators to cover

without a substantial increase in the number of vessels. The important advantage inherent in the use of a unit process system is that it is practicable to monitor and control a range of process parameters (normally temperature, pH, dissolved oxygen and/or redox, glucose, lactic acid dehydrogenase etc.), which permits the process managers to generate and keep records of process conditions. This enables them to (1) understand their processes better, which often results in higher yields, and (2) provide for their administrators and regulatory agencies the data which justify their contention that there are no significant differences between the process under investigation and the process which led to a licensed product some time before.

1.2. Inhomogeneous and Homogeneous Unit Process Systems

There is little doubt that systems which are homogeneous can offer significant advantages over systems which are inhomogeneous. A sample of the vessel contents of a homogeneous system is representative of the whole vessel contents and measurements done on such a sample reflect the state of all the system components. With microcarrier systems it is possible to withdraw a representative sample of the solid substratum which supports the growing or producing cells. Simple microscopic techniques enable process operators to monitor the performance of the system by assessing the "state" of the cells—attached, rounded, flattened out, granular, exhibiting cytopathology, syncytia, overgrowth, confluency and nodule formation. A second reason to favour homogeneity is that all the cells throughout the system experience an identical environment in terms of the concentration of metabolites such as oxygen, lactic acid, ammonia and hydrogen ions. This means that once the most favourable conditions for growth and productivity have been defined, all the cells in the system can be made to function at the limits of their capability. The scale-up of a homogeneous system should also present fewer problems because there is no need to mimic the inevitable gradients which are inherent in the inhomogeneous systems. It is often held that the homogeneous microcarrier system is also space or volume saving in that the homogeneously distributed particles can be made to occupy a volume considerably less than in other systems (67). However, most investigators have found that the volume of the system is defined by the number of cells required, which, in turn, defines the volume of medium to be used (63). This latter volume determines the system volume. (Note: The number of cells produced per unit volume of medium is assumed to be maximal for these considerations.)

While the above advantages can be achieved with microcarrier cultures, many investigators have experienced considerable difficulties when using particular cell lines at the larger (greater than 10 litres) scales of operation. Although in some systems such as the baby hamster kidney monolayer cell (BHK-MC) system, the cells attach to the beads when the latter are in motion (59), other cells require that in the initial stages of the culture the microcarriers should be stationary (53). As this initial attachment and "flattening out" phase is critical to the subsequent development of the cells, the conditions for optimal performance are often sharply defined. This means that small changes in the cell seed, serum composition or hydrodynamic environment cause relatively large changes in system performance. Additional reasons favouring the use of alternative systems are (1) the cost of the substratum, (2) the substratum afforded by most microcarrier systems is different from that to which the cells have been accustomed, (3) the systems have proved to be difficult to aerate without generating a foam in which the microcarriers collect and (4) the removal of spent growth fluids or chemicals which control gene expression is difficult as the void volume of the sedimented carriers is significant and in some cases may have to be diluted out several times to remove traces of unwanted materials. Such considerations lead to the conclusion that where it is practicable, homogeneous systems are advantageous; however, in more critical situations or where the control of all the input materials cannot be achieved to a sufficiently high standard or where materials have to be removed and replaced rapidly and thoroughly, then one or other of the inhomogeneous systems described below could be a preferred method of proceeding to product generation.

1.3. Assessment of Unit Process Systems

It is essential for a thorough understanding of the massive and available literature to be able to appreciate the value of the units which are used to express productivities. In industry productivity is assessed in relation to the cost incurred. For laboratory systems this may be approximated to the volume of medium used and the cost per unit volume of that medium (7). Other ways of expressing productivities such as cells per unit area of surface or cells per unit of equipment volume may have a place under special circumstances, the former where one is engaged in considerations of contact inhibition or the latter when one wants to operate with a high cell concentration during one phase of the process. It is also important to relate the results obtained in propagators to standard bottle control systems run under optimal conditions. With both control and experimental results it is of great value to get an impression of the extent of the variance from which an average was computed and also the number of completely independent replicates measured.

2. DIFFERENT TYPES OF INHOMOGENEOUS UNIT PROCESS SYSTEMS

Three of the unit process system types designed for the cultivation of anchorage-dependent animal cells take their form from the conventional multiple systems commonly used and described above (Chapter 9, this volume) (Fig. 1). Thus the static bottle system exemplified by the Roux, Brock-

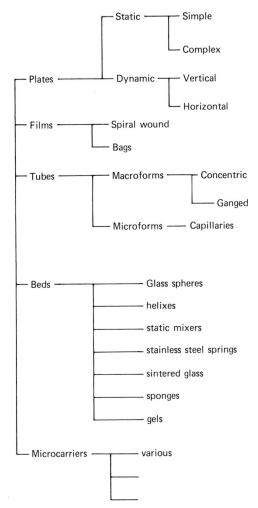

Fig. 1. Unit process systems used for cultivating anchorage-dependent animal cells. For information about microcarriers, see Chapter 11, this volume.

way, Carrel and Povitsky bottle has been developed into a number of different units which may be described as systems containing a plurality of parallel
stacked plates. Such plates may be either static or kept in motion (dynamic).
The roller bottle system, which has also been much used, has been developed into a number of propagators based on tubes or collections, groups or
gangs of tubes held together by various means but acting in a similar manner
to the roller bottle progenitor. A third system based on growing cells on
films of plastic can, under some circumstances, be likened to a continuous
flat plate or, if rolled up in a spiral, more like a continuous surface roller
bottle. Two additional systems may be defined. The one is based on a design
which mimics the organisation of cells within the animal body. It requires
that cells are grown on the outer or inner surfaces of hollow microfilaments
300 ± 50 μm in outside diameter (artificial capillaries) (56). The other takes
its basic form from the successful solid/gas or solid/liquid contacting systems
commonly used in the chemical industry. These packed beds, or more recently fluidised beds, are made up of an array of elements whose size, shape
and material of construction can be varied extensively. Such elements may
be held immobile in a bed or they may be made to move by inducing a flow
in the surrounding medium. In the latter mode of operation they approximate to the more homogeneous systems: *macro*carriers.

2.1. Plates

Flat or surface developed plates have been used in a number of configurations in cell propagators. The materials used for the plates have been varied.
Many of the systems described have been based on glass plates (*31, 34, 51,
68*), while others have been predominantly stainless steel (*2, 4*). Treated
polystyrene (*30, 50, 57*) and titanium (*43*) are also commonly used materials,
while polycarbonate, polyethylene, tetraphthalate and methyl pentene (*23*)
have also found a place. The surface area of the plates has been augmented
(developed) by specially shaped corrugations (*2*) or by using fused metal
spheres of titanium or aluminium (*39*). In this section a number of variants of
these plate systems will be examined. They can be broadly divided into
systems where the plates are in motion or where they are static. Alternative
groupings can be based on plate orientation (horizontal, vertical or variable)
and on whether the systems are designed to run in the perfused mode.

2.1.1. Static Plate Units

The four types of static plate units which dominate this technology are
presented in Fig. 2. The unit which performs most like a set of bottles is the
multitray equipment and its variants (2 in Fig. 2) (*23, 48, 57*). The depth of

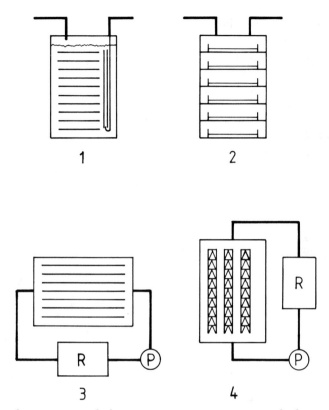

Fig. 2.   Alternative types of inhomogeneous unit process systems. The four types of static plate units. R, reservoir containing medium; P, pumping mechanism.

liquid over the cells is similar to that used for bottle cultures (typically 5 mm). While the techniques of handling banks of such multitray units have been developed to a considerable extent (50), the cumbersome manipulations and the complexity of the automated equipment which handles the units militate against the facile spread of this technology. Other factors such as the reusability and cost of the preformed, presterilised units can be a disadvantage, yet as such factors are offset by improved performance and decreased labour costs, efficient and useful systems can result.

Units formed from a multiplicity of stacked parallel plates held in a cage-like arrangement (68) have been scaled up to a medium volume of 200 litres (1 in Fig. 2). These units are inefficient in that only half of the surface can be used for cell growth and the methods used to circulate the medium are inadequate. The air-lift pump enhanced by a magnetic stirrer at the bottom of a stack of plates which have been "quartered" to aid mixing and prevent glass breakage during sterilisation does not improve performance sufficiently to make these units something other than a demonstration of the feasibility of unit process system potential.

The later systems based on static plates (4, 34) have increased the surface/volume ratio in a cell growth vessel and have held the required volume of medium in a second chamber (3 and 4 in Fig. 2). Medium circulation through the system is via a pumping mechanism. The problems with such systems stem from their complexity, with multiplicities of plates separated by gaskets, and in diffusion-controlled systems the distribution of medium over the cells cannot be controlled. While such units may be turned through 180° to make use of the surface on both sides of the plate, the mechanisms involved and the requirement for either flexible pipework or sterilisable connections make such units cumbersome, especially at scales of operation in excess of 100 litres of medium volume. A further disadvantage is the cost of the units, as mechanical complexity inevitably adds to production charges. There is, however, little doubt that such systems can be made to do an effective job, yet their limitations of cost, complexity and ability to trypsinise cells from such units may limit their potential as substitutes for multiple bottle processes. However, systems based on heat exchangers (4) offer some advantages in terms of temperature control as cells are grown in alternative chambers within the exchangers and the uninhabited chambers are used to circulate a temperature control medium. This feature helps in cooling equipment after sterilisation, a process which can be effected in a multitude of ways in other types of propagator.

2.1.2. *Dynamic Multiplate Propagators*

To overcome some of the disadvantages of the static plate propagators, multiplate propagators were developed. The primary advance evidenced by the dynamic systems was the generation of a more thoroughly mixed system. Concentration gradients of metabolites, excretory products or hydrogen ions were virtually eliminated. In addition, the vertical plate systems offered a higher surface area/medium volume ratio as only half of the plates needed to be submerged at any time (5, 6 and 8 in Fig. 3). Inevitably, the mechanical complexity of the system is increased by the addition of motors, bearings, supports and shafts. However, to compensate it is possible to remove the cells from supporting surfaces by increasing rotational speeds or using more complex mechanical wiping devices (21). A basic problem in the scale-up of such units is that as the volume of the system is expanded and the shaft is extended lengthwise to accommodate an increased number of plates of unincreased diameter, then the bending moment of the shaft becomes a significant mechanical problem which requires special treatment (hollow shaft, larger diameter, midpoint support bearing, counterweighting system out-

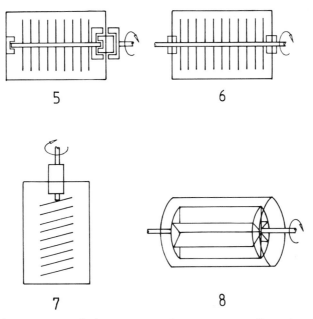

Fig. 3.   Alternative types of inhomogeneous unit process systems. Dynamic multiplate cell propagators.

side vessel). Alternatively, where the volume increase is accompanied by an increase in diameter, then the hydrodynamic characteristics of the fluid flow in relation to the plates is changed, a situation which is not completely remedied by changes in rotational velocity.

The operation whereby the cells are attached to the substratum presents a problem. One solution is to rotate the stationary plates through 180° and to recharge the apparatus with an additional 'dose' of cells. If, however, it is considered desirable to attach the cells to plates in motion, then the diagonally oriented system (7 in Fig. 3) permits this occurrence. However, even with the latter system it is necessary to turn the vessel in the reverse direction to obtain cells on both sides of the glass plates (*31*). The larger scale versions of such systems are awkward to use, difficult to manipulate in a steriliser and expensive to build; there are alternatives.

2.2. Film-based Systems

It is a logical progression from the plate systems to consider the infinitely extended unit plate convoluted in a way which enables it to fit into a conveniently sized chamber or vessel (Fig. 4). A unit based on a spiral of treated polystyrene film or sheet has been widely investigated (*16, 46, 65*) and is

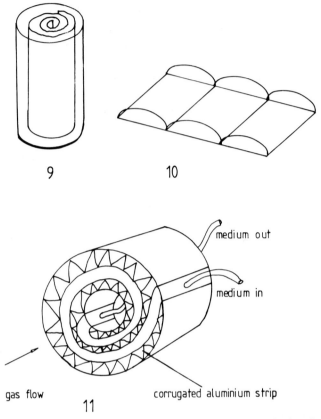

9   10

medium out

medium in

gas flow   corrugated aluminium strip

11

Fig. 4. Alternative types of inhomogeneous unit process systems. Film-based systems for cell propagation.

available as a presterilised packaged apparatus (9 in Fig. 4) (64). Other spiral systems have also been developed and used, at least on the small (less than 2 litres) scale, with some success; melinex polyester (ICI, United Kingdom) has been used by the author, while the application of a polycarbonate spiral system has also been recorded (11). Although it is possible to condense a large surface into a small volume, the problems inherent in such structures such as mixing, aeration and containment of medium for the requisite volume of cells have not been fully overcome, a feature which probably explains why such units, though successful at the 2-litre scale, have not been fully described in the literature (17) at larger scales of operation.

One approach to improving such spiral systems requires the control of medium flow and gas exchange throughout the bulk of the apparatus. This can be achieved by growing the cells in plastic bags which are permeable to the gases O_2, N_2 and CO_2 but impermeable to fluids (10 in Fig. 4) (*24*, *37*, *38*, *45*). The bags may be made from sealing together two strips of a Teflon (polymeric fluorocarbon) material. This concept has been embodied in the commercially available IL 410 system (*22*). The system consists of a long Teflon tube (typically 5.7 cm wide and about 900 cm long) which is wound in a spiral with a layer of corrugated aluminium between successive turns. Tubes are attached to each end of this long bag for the movement of fluids and the whole is mounted on a reel which is automatically rotated through 360° every hour, reversing direction after each revolution; this enables the top and bottom inner surfaces of the bag to be used for cell growth (11 in Fig. 4).

This system features some interesting properties. The cells can be detached from the substratum by mechanically stretching the unreeled bag. Also, as the bags are permeable to air, this offers a simple method of bag sterilisation and permits bag reuse. It also enables the cell sheet to be observed via a window cut into the retaining reel. Although this makes a high surface/volume ratio for such equipment, the absolute volume is better defined in terms of the total volume of medium required to propagate a defined number of cells. Also, the apparatus which holds the rotating reels is both costly and complex as each of 20 reels has to be attached to two rubber tubes. Thus, although this is a practicable approach to a unit process system, there are features which militate against its large-scale deployment.

2.3. Tube Propagators

There is a wide variety of tube-based unit process systems for cultivating monolayer cells. This approach follows naturally from growing cells in rolling tubes, a methodology which dates back to 1925 (*33*). The scale-up of the single rolling tube involved placing a number of independent tubes into a common holder and rotating the holder (*12*). In this way "massive" tissue culture was demonstrated in 1933. Two tube systems may be differentiated: systems based on large diameter tubes, typically much larger than 5 mm, while other systems depend on capillary tubes of diameter about 0.5 mm.

2.3.1. Macroform Tube Systems

In Fig. 5, five different macroform tube systems are represented schematically. Historically, the housing of a number of individually sealed tubes in a single chamber may be thought to be a first step on the path to a unit process system (13 in Fig. 5). This system has in somewhat different configuration been represented as a single bottle with a number of concentric tubes within it (12 in Fig. 5) (*26*). With a volume of 1.2 litres the bottle can be made to contain about 5000 cm^2 of surface area for cell growth on the assumption that

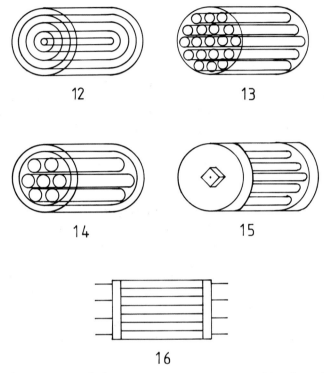

Fig. 5. Alternative types of inhomogeneous unit process systems. Macroform tube systems for cultivating monolayer cells.

both sides of the cylinders are made available for cell growth. However, it is more likely that only one such side would be used in practice. Such a system does not scale up readily as a unit process apparatus. Rather it is intended that a single such bottle may substitute for five or so standard roller bottles. Whether the extra cost of such specially prepared bottles and operating procedures can be paid back by space savings depends on the user's special circumstances.

A further simple expansion of the multiple tube concept is found in the propagator which consists of a number of open-ended tubes held between two end plates which have been designed to act as distributors for process fluids (15 in Fig. 5) (35). The sets of tubes may be of different lengths, thus allowing operational flexibility. The problem with such a unit occurs during the washing and reassembling phase, where, if the tubes for each of several propagators become mixed up, the slightly different lengths which are combined into a set lead to leakages. A more complex development of this concept is found in the "rotary column" method (16 in Fig. 5) (55). In this system each tube is operable as a separate vessel as a result of the valving at each end. However, the whole apparatus may be rotated about the centre of the tubes, which allows for steam sterilisation and medium filling, while the whole ganged tube system reverting to the horizontal position can be rotated about the long axis to ensure the use of all the inner surface. This equipment has been built to accommodate about 50 litres of medium over some 11 m² of surface area. It is an awkward and expensive piece of equipment which can be made to yield satisfactory results if patience, application and capability are on hand. It does not seem to be the robust and easy to use system which can find wide application in production situations.

The most recent development is that of the "Gyrogen" (14 in Fig. 5) (*13, 14*). This consists of a gang of open-ended tubes which are made to rotate within an outer vessel which has been filled to the halfway point with culture medium. Again, there is little doubt that such a system will produce useful results; however, the mechanical complexity (which determines cost and ease of use) and the inflexibility in the medium volume/surface area ratio (necessitated by having to keep the unit half-full) counteracts some of the advantages derived from the unit design (pH and dissolved oxygen control).

2.3.2. Microform Tube Systems

While large-scale cell culture has been used primarily for the production of vaccines and more recently for the manufacture of the interferons, there is also a need for animal cell culture systems in which the cell is made to perform as if it were still part of the organ from which it has been derived. Such "pseudoorgan" cultures (*36*) may be constructed about a bed of capillary tubes made from a variety of organic polymers such as polysulphone/polyacrylonitrile (*44*). The cells may be grown on the "inside" or within the lumen of the tubes or on the "outside" of the tubes. In either case it may be expected that the cells will grow across the gaps between the tubes if on the outer surface or they will completely fill the tube lacuna if on the inside. Indeed, it has been shown that cells will grow across and fill the voids in a 100 mesh stainless steel screen (*31*) (opening size 0.15 mm). By choosing an appropriate material for capillary construction, it is possible to provide cultured cells with an environment similar to that which pertains *in vivo*. The semi-permeable nature of the artificial capillaries mimics the performance of the body capillaries in that exchange of molecules can take place, and with suitable control of the cell bathing fluids toxic products of metabolism as well as potentially valuable cell excretion products may be removed from the system, thus improving the longevity of the cells and enabling product recovery operations.

It is facile to claim that under such conditions cells will perform as they do *in vivo*, yet investigators hold that it is possible to produce a wide range of hormones, blood, milk and immunoregulatory proteins in such systems (6). Perhaps surprisingly, one of the most likely products is the hepatitis A virus (36), while others describe a system used for urokinase and migration inhibition factor generation (28).

Notwithstanding the obvious advantages of pseudoorgan culture, problems arise because of the difficulties of obtaining a homogeneous environment within the reactor. This often results from cell growth occluding the space in which the medium flows (44). Thus, while this phenomenon is not serious in small reactors, such as the ones commercially available (capafusion system) (5, 25), when the system is scaled up yields do not go up in proportion. The countermeasure to this is to increase the complexity of the unit by adding thin fibre pads with special diffuser plates (micrometallic filters), which not only ensure the even distribution of liquid, but also retain the cells within the apparatus (44). Larger scale units would then have to be built of batteries of the largest successful stand-alone unit.

Such systems do offer the prospect of obtaining conditions close to those found *in vivo* in a laboratory or pilot plant environment. Should such conditions be shown to be obligatory, then the synthetic capillary systems will find a role in product manufacture. If, however, it is shown to be possible to maintain the cells in a productive state in systems whose scale-up is less complex, then the capillary systems would be used to provide opportunities for the initial examination of the productive potential of the cultured cells and the conditions which control that productivity.

2.4. Packed-bed Reactors

The previous section described a system of technology which set out to mimic the *in vivo* environment cells experience. This section describes an alternative system which has for its analogy the numerous liquid–gas contact reactors which are in common use throughout the chemical industry for gas absorption and adsorption, for liquid–liquid extraction and for heat transfer. In addition, the trickle filter system of the effluent water treatment plants constitutes another example of a widespread, large-scale packed-bed reactor system. The features of this equipment which make it so useful are the large surface areas which can be generated, the low resistance to fluid flows and the homogeneous distribution of fluid over the elements which make up the bed.

Packed-bed reactors used for cell growth are normally of the perfusion type. The cells are firmly attached to the solid substratum in the cell growth vessel while the bulk of the medium is held in a second vessel which is attached to the cell growth vessel via a circulation system (Fig. 6).

The earliest work describing vessels with internally placed materials de-

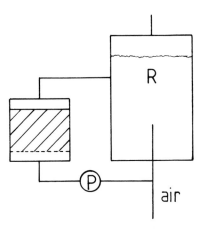

R

air

bed elements — discrete

O — spheres   Pyrex glass

@ — helices   —glass
   springs   —steel

@ —Raschig rings —glass

⬚—diatomaceous earth

⅄ — jacks — polystyrene

—fused

⬚ —sintered spheres — glass

⬚ — Koch static mixer

⬚ sponge

⬚ folded cellophane—perforated

Fig. 6.   Arrangement of packed-bed inhomogeneous unit process system.

signed to provide more surface for cells to grow on was described in 1947 by Evans and Earle, who used a system for growing cells on perforated cellophane sheets (10). This work was extended to include the use of such packing materials as cut-up glass tubing (Raschig Rings) (8), or glass helixes (9). In 1962 a recognisable perfusion system was portrayed by McCoy et al. (41).

A complete description of a cell propagation system based on circulating medium through a bed of glass spheres was given by Lorans in a French patent of 1970 (32). This was followed by descriptions of "novel culture vessels" packed with glass spheres (52, 71) and later work by the author and his colleagues, who showed that specially shaped culture vessels were unnecessary and that a bed of glass spheres could be held in parallel sided vessels and at scales of up to 100 litres of medium volume with normal productivities of BHK monolayer cells and the foot-and-mouth disease viruses which would grow in those cells (60, 69). These observations were paralleled by work in the Beecham group (1, 3, 54) with a human diploid fibroblast cell system based on either glass or polycarbonate spheres. Other workers have shown that reactor elements could be made up from polystyrene jacks (49), diatomaceous earth (66) or stainless steel springs (42). However, there is little doubt that glass sphere systems have advantages in the conventional nature of the substrate, known and calculable packing features, ready availability in a wide range of diameters and high degree of structural stability. Alternative structures for the reactor bed based on fused or continuous elements have also been tried. The use of cellophane sheets was described in 1947 (10) and in 1951 the use of a cellulosic sponge coated with collagen was introduced (29, 58). More recently, the immobilisation of cells within a calcium alginate gel has been shown to be a preferred way of exploiting the productive capabilities of monoclonal antibody–secreting cells (47). Two fused element systems have been reported; one is based on the generation of a glass matrix (27), while the other is based upon specially shaped stainless steel elements combined in such a fashion as to create conditions of maximum mixing (15, 18–20).

The packed bed–perfusion system for growing industrial quantities of anchorage-dependent cells is simple to use, robust and reliable. The apparatus can be built in-house and it is easy to monitor and control as the medium-holding vessel can provide adequate accommodation for probes and the aeration of the system in this vessel does not interfere with the cells growing on the bed elements. A significant advantage of this type of system is the ability to remove spent growth medium or chemicals which control the cellular biochemistry rapidly and efficiently and to change the culture fluid volume/cell ratio. This latter feature enables cell growth to be achieved under one set of conditions, while product generation may be obtained under conditions of higher cell concentrations.

Methods have been developed to trypsinise the cells from such culture vessels (62) and the cells so produced have been shown to be useful seed materials for the next culture. It has also been shown that the cells grow throughout the matrix of the bed and that it is possible, by increasing the volume of medium presented to the cells, to produce BHK cells at concentrations of 3×10^7 cells/ml culture vessel volume (70). The cells so produced are competent for virus production. In addition to being able to determine the optimum time to harvest virus from such cultures (61), it is also possible to automate the collection of that product material (40).

Although packed bed systems suffer from the disadvantages inherent in the inhomogeneous nature of the system (cf. Section 1.2), they can be scaled up to large volumes (15,000 litres medium volume) (see Appendix for calculation) and therefore they can be expected to serve a useful function for the manufacture of materials derived from cultured anchorage-dependent animal cells.

APPENDIX: CALCULATION OF MAXIMUM SIZE OF GLASS SPHERE–BASED STATIC BED PROPAGATOR

Assumptions:

1. Oxygen demand/cell/hr = 6×10^{-12} g.
2. At saturation medium carries 7.6×10^{-6} g oxygen/ml, of which 6.6×10^{-6} g is available for cell growth (lowest percentage saturation ca. 10%).
3. Density of cells in bed = 7.5×10^6/cm^3 bed volume (equivalent to 150×10^6 cells per 200 cm^2 of surface in a 20-cm^3 bed).
4. Linear flow rate of medium down the bed = 2 cm/min.
5. Bed depth/diameter ratio = 1.

Calculation: Maximum depth of bed

Oxygen demand/cm^2 of bed = $7.5 \times 10^6 \times 6.6 \times 10^{-12}$ g/hr
$$= 4.95 \times 10^{-5} \text{ g/hr or } 8.2 \times 10^{-7} \text{ g/min}$$

Oxygen supply/cm^2 of bed/min (2 ml of medium supplied/cm^2 of bed/min)
$$= 13.2 \times 10^{-6} \text{ g/min}$$

Taking a bed of cross section 1 cm^2, then the oxygen supplied would be depleted to less than 10% of the level at saturation after the medium had passed through.

$$\frac{13.2 \times 10^{-6}}{8.2 \times 10^{-7}} \text{ cm of bed depth} = 16.1 \text{ cm}$$

Extrapolations:

1. With a 2 cm/min linear flow rate the maximum bed volume would be 3.3 liters to give a system volume of 15 liters of medium.

2. With double the flow rate down the bed [a situation which does not significantly affect cell growth (60)] and using oxygen as the gassing medium (as opposed to air) then the bed depth could be increased to 161 cm. This would give a bed volume of 3300 liters and system volume of 15,000 liters of medium.

REFERENCES

1. Beecham Group (1976). Procède pour la culture de cellules. Belgian Patent 842,002.
2. Birch, J. R., Cartwright, T. B., Ford, J. A. (1981). Verfaren zum Zuchten von tierischen Zellen und mit einem Plattenstapel versehene Zellzuchtungsapparatur. German Patent DE 3,031,617.
3. Burbidge, C. (1979). The mass culture of human diploid fibroblasts in packed beds of glass beads. *Dev. Biol. Stand.* **46**, 169–172.
4. Burbidge, C., and Dacey, I. K. (1984). The use of plate heat exchangers in growing human fibroblasts. *Dev. Biol. Stand.* (in press).
5. CELLCO (1975). Cell culture on semi-permeable tubular membranes. British Patent 1,-395,291.
6. Community Blood Council of Greater New York (1977). Cell-containing hollow fibres and use of same for metabolic processes. British Patent 1,491,261.
7. Discussion (1982). *Dev. Biol. Stand.* **50**, 333–334.
8. Earle, W. R., Schilling, E. L., and Shannon, J. E. (1951). Growth of animal tissue cells on three-dimensional substrates. *J. Natl. Cancer Inst. (U.S.)* **12**, 179–193.
9. Earle, W. R., Bryant, J. C., and Schilling, E. J. (1953–1954). Certain factors limiting the size of the tissue culture and the development of massive cultures. *Ann. N.Y. Acad. Sci.* **58**, 1000–1011.
10. Evans, V. J., and Earle, W. R. (1947). The use of perforated cellophane for the growth of cells in tissue culture. *J. Natl. Cancer Inst. (U.S.)* **8**, 103–119.
11. Fontages, R., Beaudry, Y., and Matot, G. (1974). Tissue culture of diploid cells in a new type of rotating spiral apparatus (summary). *Biotechnol. Bioeng. Symp.* **4**, 859–860.
12. Gey, G. O. (1933). An improved technique for massive tissue culture. *Am. J. Cancer* **17**, 752–756.
13. Girard, H., and Buhler, R. (1979). Cultivating tissue cells. British Patent 2,002,814.
14. Girard, H. C., Sutcu, M., Erdem, H., and Gurhan, I. (1980). Monolayer cultures of animal cells with the Gyrogen equipped with tubes. *Biotechnol. Bioeng.* **22**, 477–493.
15. Grabner, R. W., and Paul, E. L. (1979). Emploi de la surface d'un mélangeur immonbile comme propagateur d'une culture de cellules. Belgian Patent 873,653.
16. House, W., Shearer, M., and Maroudas, N. G. (1972). Method for bulk culture of animal cells on plastic film. *Exp. Cell Res.* **71**, 293–296.
17. House, W. (1973). Bulk culture of cell monolayers. *In* "Tissue Culture: Methods and Applications" (P. F. Kruse, Jr., and M. K. Patterson, eds.), pp. 338–344. Academic Press, London.
18. Huber, M. (1974). Mixing apparatus and method. U.S. Patent 3,785,620.

19. Huber, M., and Schutz, G. (1975). Mixing apparatus and method. U.S. Patent 3,871,624.
20. Huber, M., and Schutz, G. (1975). Static mixing device. U.S. Patent 3,871,625.
21. Hurni, W. M., McAleer, W. J., and Hilleman, M. R. (1978). Apparatus for removing tissue cells from culture discs. British Patent 1,498,354.
22. Instrumentation Laboratory Inc. "A New Gas Permeable Membrane Method for Quantity Cell Cultivation in Automatically Controlled Environment." IL, Lexington, Massachusetts.
23. Izuka, M. (1979). Apparatus and method for growing cells. British Patent 1,539,263.
24. Jensen, M. D., Wallach, D. F. H., and Lin, P. S. (1974). Comparative growth characteristics of Vero cells on gas permeable and conventional supports. *Exp. Cell Res.* **84**, 271–281.
25. Knazeck, R. A., Gullino, P. M., Koller, P. O., and Dedrick, R. L. (1972). Cell culture on artificial capillaries: An approach to tissue growth in vitro. *Science* **178**, 65–67.
26. Knight, E. (1977). Multisurface glass roller bottle for growth of animal cells in culture. *Appl. Environ. Microbiol.* **33**, 666–669.
27. Kontes Instrument Group (1977). "Kontes Superculture Apparatus." Vineland, New Jersey.
28. Ku, K., Kuo, M. J., Delente, J., Wildi, B. S., and Feder, J. (1981). Development of a hollow-fibre system for large-scale culture of mammalian cells. *Biotechnol. Bioeng.* **23**, 79–95.
29. Leighton, J. (1951). A sponge matrix method for tissue culture formation of organised aggregates of cells in vitro. *J. Natl. Cancer Inst. (U.S.)* **12**, 545–561.
30. Linbro Chemical Co. "Linbro-rola Cartridge." Linbro Chem. Co., New Haven, Connecticut. pp. 49–54.
31. Litwin, J. (1976). A new type of multisurface rotating vessel for anchorage dependent cells. *Proc. Gen. Meet. Eur. Soc. Animal Cell Technol., 1st, 1976 Amsterdam* (R. E. Spier and A. L. van Wezel, eds.), pp. 49–54. Rijks Inst. Volksgezondheid, Bilthoven, Netherlands.
32. Lorans, G. (1970). Procède et appareillage permettant, notament de cultiver des cellules. French Patent 1,598,245.
33. Lowenstadt, H. (1925). Einige neue Hilfsmittel zur Angelgung von Gewebe Kulturen. *Arch Exp. Zellforsch.* **1**, 251–256.
34. Mann, G. F. (1972). A new system for cell cultivation in perfused layered cultures. I. Design and operating principles. *Bol. Of. Sanit. Panam. (Engl. Ed.)* **6**, 33–36.
35. Mann, G. F. (1974). Culture apparatus. British Patent 1,369,593.
36. Markus, H. Z., and McAleer, W. J. (1981). High titre production of hepatitis A virus. U.S. Patent 4,301,249.
37. Max-Planck-Gesellschaft zur Forderung der Wissenschaften EV (1972). Cultivation of cells and tissues. British Patent 1,389,411.
38. Max-Planck-Gesselschaft zur Forderung der Wissenschaften Ev (1972). Cultivation chambers. British Patent 1,398,412.
39. McAleer, W. J., Schlabach, A. J., and Spier, R. E. (1978). Growth surfaces for tissue cell cultures. British Patent 1,509,826.
40. McAleer, W. J., and Hurni, W. M. (1978). Moyen automatise de recolte des virus. Belgian Patent 867,759.
41. McCoy, T. A., Whittle, W., and Conway, E. (1962). A glass helix perfusion chamber for massive growth of cells in vitro. *Proc. Soc. Exp. Biol. Med.* **109**, 235–237.
42. Merck, W. A. M. (1982). Large-scale production of human fibroblast interferon in cell fermenters. *Dev. Biol. Stand.* **50**, 137–140.
43. Molin, O., and Heden, C. G. (1969). Large scale cultivation of human diploid cells on titanium discs in a special apparatus. *Prog. Immunobiol. Stand.* **3**, 106–110.

44. Monsanto Company (1978). Culture reactor and method. British Patent 1,514,906.
45. Munder, P. G., Modolell, M., and Wallach, D. F. H. (1971). Cell Propagation on films of polymeric fluorocarbon as a means to regulate pericellular pH and pO$_2$ in cultured monolayers. *FEBS Lett.* **15**, 191–195.
46. Nicklin, P. M., and House, W. (1976). Large scale production of virus. *Biotechnol. Bioeng.* **18**, 723–727.
47. Nilsson, K., and Mosback, K. (1980). Preparation of immobilised animal cells. *FEBS Lett.* **118**, 145–150.
48. Nunc "Multitray Unit" (1-64327). Kamstrup, Nunc, Denmark.
49. Pagano, L. F., and Valenta, J. R. (1973). Cell propagator. U.S. Patent 3,740,321.
50. Pakos, V., and Johansson, A. (1983). A large scale production of human fibroblast interferon in multitray battery systems. Abstr., *Proc. Gen. Meet. Eur. Soc. Animal Cell Technol., 6th, 1983, Copenhagen.*
51. Parisius, J. L. E. W., Cucakovich, N. B., and Maccorine, G. M. G. (1974). Method of tissue cell culture. British Patent 1,358,321.
52. Passage, E., Rudiger, H. W., and Wohler, W. (1974). Verfahren und Ginrichtung zum Zuchter Diploider Fibroblasten Oder ahnlicher Zellen. Federal Republic of Germany Patent 2,300,567.
53. Pharmacia Fine Chemicals (1981). "Microcarrier Cell Culture: Principles and Methods." Pharmacia Fine Chemicals, AB, Upsalla, Sweden.
54. Robinson, J. H., Butlin, P. M., and Imrie, R. C. (1980). Growth characteristics of human diploid fibroblasts in packed beds of glass beads. *Dev. Biol. Stand.* **46**, 173–181.
55. Santero, G. G. (1972). The "rotary column" method for growth of large scale quantities of cell monolayers. *Biotechnol. Bioeng.* **14**, 753–775.
56. Schratter, P. (1976). Cell culture with synthetic capillaries. *In* "Methods in Cell Biology" (D. M. Prescott, ed.), Vol. 14, pp. 95–103. Academic Press, London.
57. Skoda, R., and Pakos, W. (1977). Procède pour la culture en masse de cellules et dispositif de chambres pour son execution. Belgian Patent 854,898.
58. Sloane, N. H. (1970). Production of viral vaccines. U.S. Patent 3,493,651.
59. Spier, R. E. (1976). The production of foot-and mouth-disease viruses from BHK21C13 cells grown on the surface of DEAE Sephadex A50 beads. *Biotechnol. Bioeng.* **18**, 659–667.
60. Spier, R. E., and Whiteside, J. P. (1976). The production of foot and mouth disease virus from BHK21C13 cells grown on the surface of glass spheres. *Biotechnol. Bioeng.* **18**, 649–657.
61. Spier, R. E. (1977). Determination of the time to harvest foot and mouth disease virus cultures by measurements of the supernatant concentration of lactic dehydrogenase. *Biotechnol. Bioeng.* **19**, 929–930.
62. Spier, R. E., Whiteside, J. P., and Bolt, K. (1977). Trypsinisation of BHK21 monolayer cells grown in two large-scale unit process systems. *Biotechnol. Bioeng.* **19**, 1737–1738.
63. Spier, R. E. (1980). Recent developments in the large scale cultivation of animal cells in monolayers. *Adv. Biochem. Eng.* **14**, 119–162.
64. Sterilin "Bulk Culture Vessel." Sterilin Ltd., Teddington Middlesex.
65. Taylor, W. G., and Evans, V. J. (1975). Evaluation of the dyna-cell vessel for production of surface-substratum dependent cells. *Biotechnol. Bioeng.* **17**, 1847–1851.
66. Telling, R. C., Passingham, R. J., Kitchener, B. L., and Hopkinson, D. G. (1973). Improvements in cell and virus culture systems. British Patent 1,436,323.
67. Thilly, W. G. (1977). Compact cell unit may trim vaccine costs. *Chem. Week* Aug. 10, 34–35.
68. Weiss, R. S., and Scleicher, J. B. (1968). A multisurface tissue propagator for the mass-scale growth of cell monolayers. *Biotechnol. Bioeng.* **10**, 601–616.

69. Whiteside, J. P., and Spier, R. E. (1981). The scale-up from 0.1 to 100 litres of a unit process system based on 3mm diameter glass spheres for the production of four strains of FMDV from BHK monolayer cells. *Biotechnol. Bioeng.* **23**, 551–565.
70. Whiteside, J. P. (1983). Investigations on anchorage dependent animal cells grown in packed bed propagators. Ph.D. Thesis, University of Surrey.
71. Wohler, W., Rudiger, H. W., and Passage, E. (1972). Large-scale culturing of normal diploid cells on glass beads using a novel type culture vessel. *Exp. Cell Res.* **74**, 571–573.

11

Monolayer Growth Systems: Homogeneous Unit Processes

ANTON L. VAN WEZEL

Rijksinstituut voor Volksgezondheid en Milieuhygiëne
Bilthoven, The Netherlands

Animal Cell Biotechnology, Vol. 1

1. INTRODUCTION

Most animal cells require for their growth *in vitro* a suitable substratum to which they can attach and spread. This phenomenon has delayed the application of homogeneous culture techniques for animal cell cultivation, as generally used in standard microbiological cultivation processes. It has led to the development of various types of more or less non-homogeneous stationary culture systems for large scale as described in Chapter 10 of this volume. The disadvantages of these systems regarding control and scaling up of the cultivation process are obvious and have been extensively described in the literature (*8, 18, 21, 24*). Through the development of the microcarrier culture system the homogeneous cultivation of these so-called anchorage-dependent cells became possible (*26*). Initially the microcarrier system was not very well accepted as an alternative to stationary culture systems (*21*). However, through the development of better carriers, it has grown to be the most preferred system for the large-scale cultivation of animal cells as substrates for production of cellular components and viral vaccines.

2. CONCEPT OF THE MICROCARRIER CULTURE SYSTEM

The concept of microcarrier culture is rather simple. It comprises the cultivation of cells on small solid particles, the so-called microcarriers, suspended in the growth medium by stirring. Cells attach and spread on the microcarriers and gradually grow out to a confluent monolayer (Fig. 1). By using small particles of about 200 μm at a concentration of roughly 10^4 particles/ml of culture, a quasi-homogeneous system is achieved which resembles the traditional microbial submerged culture with all its advantages. In microcarrier culture the features of both suspension and monolayer culture are brought together in one system.

3. MICROCARRIERS

Initially standard DEAE–Sephadex A-50 (*26*) was used as the microcarrier. This product appeared to be toxic at concentrations above 1 mg/ml of culture. This was circumvented by coating the beads with celloidin (*28*) or pretreatment with serum proteins or other large polymers such as carboxymethylcellulose (*7, 22*). However, the real break-through came with the development of lower charged DEAE–Sephadex beads by Levine (*6*) and his colleagues (*7*). They found that when the positive charge of DEAE–Sephadex was reduced from an exchange capacity of approximately 4 meq/g

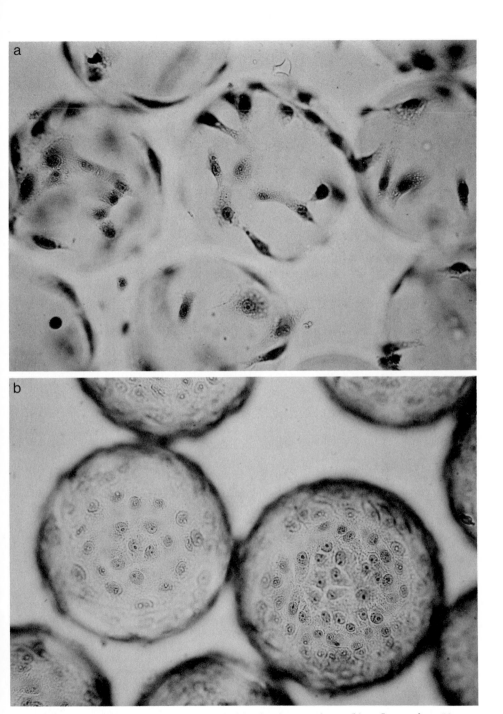

Fig. 1. Microcarrier culture of Vero cells: (a) 24 hr after inoculation; (b) confluent culture 96 hr after inoculation.

of dry dextran to 2 meq/g, the toxic effect of the beads disappeared. This development has contributed very much to a more general application and acceptance of microcarrier culture.

Besides several positively charged carriers of the DEAE–Sephadex type (1, 15, 17, 19), collagen-coated Sephadex (2, 17) and negatively charged polystyrene beads (10, 16) are now also commercially available. From these, the low-charge DEAE–Sephadex (Cytodex 1 of Pharmacia and Superbeads of Flow Laboratories), collagen-coated Sephadex (Cytodex 3 of Pharmacia) and polystyrene beads (Biosilon of Nunc) are most generally used for large-scale cultivation of animal cells. The last one has the disadvantage that the density of the beads is relatively high and special stirrer systems are required to keep them in suspension under such conditions that good attachment of cells is achieved. Recently, several other carriers have been described in the literature. For a review the reader is referred to a recent publication of Reuveny (18). One of the most interesting is the hollow glass bead carrier (34), which resembles the original substratum on which the cultivation of anchorage-dependent cells has been developed and which has proved to be the most general and optimal substratum for the cultivation of cells in monolayer systems. Recent experiments with these carriers, carried out at our laboratory, demonstrate that they are very useful in the microcarrier culture system.

4. REQUIREMENTS FOR THE OPTIMAL MICROCARRIER

The large variety of beads used for microcarrier culture may be an indication that the optimal carrier has not yet been found or in any case is not yet commercially available. The requirements for such a carrier have been defined by several groups (17, 18, 29). On the basis of our present experience, a few marginal notes may be made on the requirements formulated so far.

4.1. Charge

Although initially positively charged materials were used and advised, it is now clear that negatively charged or amphoteric materials such as proteins and amino acids polymerised to the surface may also be used. The adhesion and spreading are less dependent on the type of charge as well as on the serum proteins in the medium adsorbed to the surface. In this respect the specific adsorption of fibronectins and collagens appears to be very important (3, 35). In general it may be said that the lower the charge the better, as in this way aspecific adsorption of proteins from the medium will be avoided. This has the advantage that serum proteins can more easily be removed from the culture by washing in the final stage of the production process. As for positively charged surfaces, relatively high charges are required for optimal cell attachment, it might be better to use negatively or amphoterically charged materials for the preparation of microcarriers. An important point is that the beads have evenly arrayed surface properties, otherwise an uneven distribution of inoculum might occur.

4.2. Density

It is now clear that in order to keep the beads in suspension easily with low stirrer speeds and to avoid flotation of the beads, the specific gravity of the beads should be between 1.02 and 1.04. The optimal density is 1.030.

4.3. Diameter

To achieve an homogeneous culture system a large number of beads per unit of volume is indicated. On the other hand, the surface of one bead should be large enough to carry several hundred cells, as in most cases cells will not easily go from one bead to another. On the basis of these two points a diameter of roughly 200 μm is optimal in our opinion. Further, the size distribution should be as narrow as possible and preferably within ±25 μm. The reason for this is that great differences in size distribution give an uneven inoculum distribution on the beads even if the beads have equal surface properties. At an uneven size distribution cells attach preferably to the smaller beads, which may be explained by the fact that larger beads will move relatively more rapidly than smaller beads with regard to the cells under the influence of the hydrodynamic environment. In addition, a big advantage of an even size distribution is a higher filtration velocity during washing and trypsinization processes.

4.4. Transparency

A high transparency of the bead material is required to allow microscopic observation of cells. In some cases it might be improved by transferring the beads after staining in a liquid with a higher refractive index such as glycerol.

4.5. Porosity

The material should not be porous as this hampers the washing procedure even if medium ingredients such as serum proteins are not adsorbed to the material.

4.6. Rigidity

Although initially it was advised that the bead material should be smooth and non-rigid in order to reduce damage of cells due to collision of microcarriers during stirring, so far this problem has not been observed when rigid materials such as polystyrene and hollow glass beads are used. It might even be advantageous to make the microcarriers of more rigid materials such as polystyrene and glass as the porosity of these materials is much lower and, in addition, the filtration properties of rigid beads will be better than those of soft smooth beads.

4.7. Toxicity

The beads themselves should not be toxic to cells and the monomeric basic material should not be toxic or it should be polymerised in such a way that toxic material will not be excreted into the culture fluid. The carriers should be authorized for use in production of biologicals for human application by the appropriate regulatory agency (cf. Chapter 9, Volume 2).

5. CULTIVATION EQUIPMENT

In order to achieve optimal cultivation results it is very important that the beads can be kept in suspension by slow agitation. For small-scale experiments in magnetically stirred vessels the Techne (Cambridge Ltd.) stirrer device appears to work very well. The agitation is based on a slow circular movement of a glass rod with a bulb at the end rotating in specially designed vessels. Also, other spinner-type magnetically stirred vessels may be used (4). However, one should take care that stirrer devices are applied which keep the microcarriers in suspension at slow agitation and that the beads are not pulverised between the stirrer and the culture vessel wall.

For the large scale, standard commercially available bioreactors may be applied as long as they are equipped with a stirrer device by which the carriers can be kept in suspension at slow agitation and they do not have slack points where microcarriers may accumulate (31). In view of the last point it is advisable to use bioreactors with a rounded bottom and smooth walls. At our laboratory the "Bilthoven unit" bioreactors developed by Van Hemert (25) for bacterial vaccine production have been used successfully. The advantage of these bioreactors is that all piping and probes are mounted in the top plate and the bioreactors themselves consist of glass or stainless steel vessels with rounded bottoms and smooth walls. For monitoring and control of the environmental culture conditions the bioreactor should be equipped with a control system for at least temperature, pH, dissolved oxygen and stirrer speed. Extension of these features to the monitoring and control of CO_2, glucose etc. would be very useful.

6. INITIATION OF MICROCARRIER CULTURES

6.1. Cell Seed

In experimental work at the small scale, research workers generally use cells from standard bottle monolayer cultures for the inoculation of microcarrier cultures. Generally, cells are continuously subcultivated for successive experiments. When cells are used for production of biologicals or as a substrate for virus vaccine production this approach is not acceptable. In view of license requirements one has to work according to the "cell seed lot system", which means that each production run should be started with cells of the same passage level, or better, of the same generation level, and that the cells should be used for production after an equal number of passages or generations. In this case one is obliged to start each production run from cells frozen in liquid nitrogen, the "working cell bank". The background of this approach is that in this way the danger of cell transformation due to successive subcultivations is reduced. Also, for research work it may be advisable to apply the same principle as it certainly will improve the reproducibility of the experiments. In practice, cells may be frozen in medium plus 5% dimethyl sulphoxide (DMSO) at a concentration of about 20×10^6 cells/ml, which means that in ampoules of 10–30 ml enough cells may be frozen to start a 1- to 5-litre microcarrier culture at a standard inoculation level of about 10^5 cells/ml. In order to obtain cells with a high viability and a short lag phase at the start of a culture, it is very important that they are harvested and frozen in liquid nitrogen when they are still in the logarithmic growth phase.

6.2. Concentration of Microcarriers

Beside the medium, the available cell culture area is one of the primary limiting factors for final cell yield of cultures of anchorage-dependent cells because of the phenomenon of density inhibition. Confluent monolayer cultures carry $1–2 \times 10^5$ cells/cm^2, depending on the cell type. From this it may be calculated that for a final cell concentration of 10^6 cells/ml a culture surface area of 5–10 cm^2 is required. This means that at an optimal diameter of the microcarriers of 2×10^{-4} cm (200 μm), 4000–8000 microcarriers are required. As the diameter of some microcarriers such as the Sephadex types increases approximately 4–5 times, giving an increase in surface area of roughly 20-fold, a factor of 20 fewer microcarriers are needed on the basis of dry weight in comparison to carriers which do not swell, such as polystyrene and hollow glass beads. Therefore the concentration of microcarriers must always be expressed on the basis of the number and diameter of the hydrated beads. An approximate guide may be that per millilitre of medium in micro-carrier culture about 10^6 cells can be generated, which means that if no replenishment of medium is performed as in batch cultures, roughly 7500 carriers with a diameter of 200 μm should be added. When the final cell yield is increased by medium replenishment either batchwise or by perfu-sion systems, as will be discussed below, the concentration of microcarriers should be increased proportionally.

6.3. Inoculation

As the transfer of cells from one microcarrier to another rarely occurs with anchorage-dependent cells such as human diploid and Vero cells, it is impor-tant that at least each carrier has one cell attached to it. As the cells will attach at random to the microcarriers the culture should be inoculated with at least five times as many cells as microcarriers (17). At a bead concentration of 10,000 per millilitre, which corresponds to about 2 mg DEAE–Sephadex/ml, an inoculation of 0.5×10^5 cells is required. However, for most cells higher inoculation levels are advisable as otherwise the lag phase will be too long. In practice, the optimal inoculation level will strongly depend on the viability, plating efficiency and medium, but in general cultures should be started at least with 10^5 cells/ml. For cultures with more than 2×10^4 microcarriers/ml the inoculation level should be increased in proportion. Microcarrier cultures may be started by adding the cells while stirring into the bioreactors containing conditioned medium with carriers. It is very important for good cell attachment that the agitation is kept as slow as possible and just enough to keep the beads in suspension. Several groups have claimed that better cell attachment and growth will be obtained if the inoculation is performed at a reduced initial volume and/or with intermittent stirring (17). This may be true when too high an agitation level is applied at the start. In our experience under the right stirring conditions this will not be observed. Further, in our opinion this procedure will not work well for large-scale cultivation processes where large amounts of carriers are present in the bioreactor (30).

7. CULTIVATION PROCEDURES

7.1. Control of Culture Conditions

Beside the stirring speed, the maintenance of optimal culture conditions such as temperature, pH and oxygen is very important in order to obtain maximum cell yields per unit volume of culture. A temperature of $36°-37°C$, a pH between 7.2 and 7.5 and a pO_2 of 10–50% air saturation are generally accepted as optimal for cell growth, although under certain conditions and for certain cell types a higher pH and lower oxygen tension are indicated to be optimal. For a detailed review of this the reader is referred to the micro-carrier manual of Pharmacia Fine Chemicals (17).

When using small-scale spinner vessels, only temperature control is really possible as these culture vessels are not normally equipped with probes for pH and pO_2 control. In this case the head-space of the vessel should be briefly gassed with a mixture of 95% air and 5% CO_2 before inoculation in order to bring the pH to the desired level of 7.2–7.4. Another possibility is the use of media without bicarbonate buffer. However, it is generally accept-ed that a low level of HCO_3^- is required for optimal cell growth (11). During cultivation the pH may be adjusted by manual addition of sodium bicarbo-nate. Sufficient oxygen will be present in the head-space of the vessel as long as the ratio of culture volume to total vessel volume does not exceed 0.5 (4). However, particularly for larger cultures, it might be necessary to increase the stirrer speed after the initial attachment phase of about 24 hr in order to obtain sufficient transfer of oxygen from gas phase to culture fluid.

The bioreactors for large-scale cultivation processes (generally above 5 litres working volume) are generally equipped sufficiently to control tem-perature, pH, pO_2 and stirrer speed. For cell culture it is also important that they are equipped with a nitrogen and oxygen gas supply for pO_2 control and a CO_2 gas supply for pH control. The most common procedure for pH control above the set-point is the addition of more or less CO_2 to the gas mixture passed over the culture surface. Below the set-point the pH can be controlled by addition of a solution of bicarbonate or sodium hydroxide. As the main cause of changes in the pH is the conversion of glucose to lactate, minimizing accumulation of lactate will facilitate the pH control. Various possibilities are described in the literature [for review, see Pharmacia (17)]. Oxygen control is generally the major problem of large-scale cultivation. The main reasons for this are: (1) transfer of oxygen from the gas headspace to the culture fluid is low as only slow agitation is used, since otherwise cells may be removed from the microcarriers, and (2) sparging of air or oxygen should be avoided as cells may be damaged by air bubbles and, in addition, foam formation may result in the flotation of carriers. Therefore, oxygen is nor-mally controlled by increasing the O_2 tension in the gas mixture passed over the culture surface (30). Other possibilities are aeration of the culture fluid outside the fermenter in a recycling system as shown in Fig. 2, or even more directly by sparging air through the culture fluid inside a cylindrical filter mounted on the stirrer shaft (Fig. 3) (23). Both systems appear to work quite well in our experience. In the previous chapter it has already been men-tioned that at inoculation of the culture the agitation should be as low as

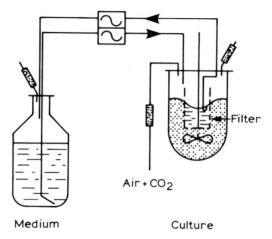

Medium   Culture

Fig. 2. Recycling system for replenishment of culture medium. The medium storage vessel may also be used for control of oxygen and pH.

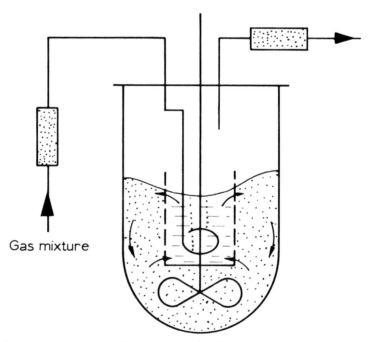

Gas mixture

Fig. 3. Aeration system for cultivation of animal cells in microcarrier culture. Air is passed through the culture fluid inside the cylindrical stainless steel sieve mounted on the stirrer shaft.

possible, but just enough to keep the beads in suspension. However, the agitation level required to keep the beads in suspension is critically dependent on the stirrer type. In the literature, special stirrer systems have been described. In our experience propeller-type stirrers with relatively large blades are generally suitable. In order to increase the oxygen transfer the stirrer speed may be increased by three to four times 24–48 hr after inoculation. At higher stirrer speeds cells may be removed from the carriers by shear force.

8. CELL MONITORING AND CONTROL

One of the major advantages of the homogeneous microcarrier culture system is that at any time during the cultivation process samples may be taken for the monitoring of cell growth, microscopic control of cells on the carriers or for a more extensive control on the absence of extraneous agents after subcultivation in standard monolayer cultures. Although monitoring of cell growth may be performed by removing the cells from the carriers with a standard trypsinization procedure and subsequent cell counting, more reliable results are obtained by counting released nuclei, as described by Sanford et al. (20). In this method as slightly modified by us (26), microcarriers with cells are, after removal of the supernatant, incubated in a 0.1% (w/v) crystal violet solution in 0.1 M citric acid for 30–60 min at 37°C. Nuclei are released by lysis of the cell wall into the solution and simultaneously stained by the crystal violet. The released stained nuclei are subsequently counted using a standard haemocytometer.

Cells on the microcarriers may be directly analysed microscopically or after staining with haematoxylin. If the refractive index between the staining solution and microcarrier material differs too much, the microscopic view may be improved by adding a fluid of higher refractive index, such as glycerol. Also, it is possible to fix the cells before staining with common fixatives and to stain them with other staining fluids such as Giemsa stain (17). For this purpose the more rigid non-swollen polystyrene or hollow glass bead microcarriers may be easier to handle than Sephadex-type carriers, whose diameter might decrease upon dehydration. For extensive control of the presence of extraneous agents, such as viruses or mycoplasmas, cells may be trypsinized from the carriers and subcultivated in standard monolayer culture. This technique is now routinely applied for the control of our production cell cultures for virus vaccine production. It has the advantage that, at the time of virus inoculation, a representative sample of the whole production culture can be taken for this purpose.

9. REPLENISHMENT OF CULTURE MEDIUM

Through replenishment of culture medium the final cell yield may be increased considerably. In all instances the replenishment should be performed before the cells are in the stationary growth phase. Further, at the start of the culture enough microcarriers have to be added to carry the expected amount of cells, which means that roughly 7500 microcarriers have to be added per 10^6 cells expected. The simplest way to replenish the medium is the batchwise procedure. This is achieved by stopping the stirrer, allowing the beads to settle, withdrawing the supernatant culture fluid and adding fresh medium. A disadvantage of this procedure is that the cells are suddenly exposed to totally non-adapted medium. Even if the medium is conditioned at the same temperature and pH, it will always introduce a short lag phase. Also, detachment of cells from the carriers is observed. These problems may be overcome by partial replenishment. Good results are obtained if only 75% of the medium is replaced. More ideal systems for medium replenishment are the continuous-flow systems with either recycling or perfusion of medium (Figs 2 and 4). With these systems the replenishment should be gradually increased in relation to the cell concentration in the bioreactors. In this way sudden changes in culture fluid composition and conditions are avoided. Microcarriers with cells have to be kept in the bioreactor. This may be achieved by a sieve mounted on the outlet tube or by a draft tube with a large diameter around the outlet tube so that the microcarriers and culture fluid are separated by gravity. The first system has the disadvantage that it may block during the process, while in the second system microcarriers and culture fluid may not be totally separated at high flow rates. The most reliable system in our experience is the application of a

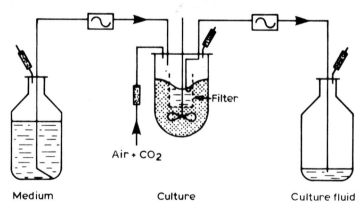

Fig. 4. Continuous perfusion system for replenishment of medium or continuous production of cellular components.

cylindrical sieve mounted on the stirrer shaft as shown in Figs 2 and 4. Even at high recycling or perfusion rates beads and cells will be dislodged from the filter by shear and centrifugal forces. Similar procedures appear to be applicable for cells growing in suspension. In addition to the supply of nutrients the recycling and perfusion system may also assist in control of pH and pO_2, while with the perfusion system removal of metabolic inhibitors for growth and survival may occur. As will be discussed below, the perfusion system appears to be particularly useful for large-scale production of biologicals excreted by cells (5).

9.1. Subcultivation and Harvesting of Cells

As real anchorage-dependent cells such as human diploid, primary kidney and Vero cells will not or will only rarely move from one carrier to another, these cells have to be removed from the carriers for subcultivation in scaling up processes. Also, for harvesting of cells from confluent cultures removal and separation of cells from the carriers will be necessary. The cells may be removed from microcarriers by methods similar to those used for removal of cells from standard monolayer cultures, such as treatment with proteolytic enzymes either with or without chelating agents. For most microcarriers the standard treatment with trypsin in combination with a chelating agent is preferred. However, for collagen-coated beads like Cytodex 3 and gelatin beads collagenase might be preferable (17). For the separation of cells from microcarriers at the small-scale differential sedimentation, filtration or density gradient centrifugation may be used. However, for large-scale subcultivation and harvesting of cells one should be able to perform the trypsinization and separation of cells under aseptic conditions. For this purpose a special trypsinization apparatus was developed (Fig. 5; 32). The procedure is performed as follows. The trypsinization apparatus and the bioreactor are aseptically connected to each other and microcarriers with cells are collected in the apparatus by removing the culture fluid through the 60-μm stainless steel screen at the bottom. The microcarriers with cells are subsequently washed with phosphate-buffered saline (PBS) to remove serum proteins and treated with the tryspin solution by passing the solution through the packed beads. After an incubation period of 10–30 min—the time depends on cell type and the composition of the trypsin solution—the cells are removed from the beads and suspended in tissue culture medium by mixing with the Vibromixer. For subcultivation, cells and microcarriers may be directly transferred to the larger bioreactor filled with fresh medium and additional microcarriers. If the cells have to be harvested, in order to freeze them in liquid nitrogen, the cells may be separated from the carriers by draining off through the screen at the bottom. As some cells may stick between the beads

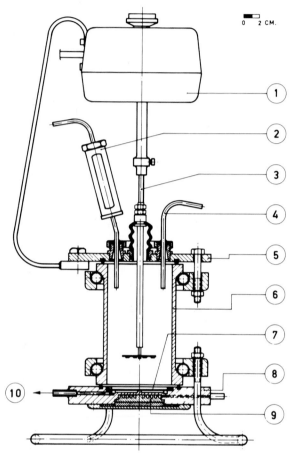

Fig. 5. Apparatus for the trypsinization of cells from microcarriers. 1, Vibromixer (model E1, Chemapec Inc.); 2, air filter; 3, Vibromixer shaft with impeller; 5, upper lid with six standardized stainless steel inlet or outlet tubes (4); 6, standard Jena glass pipe 80 × 300 mm; 8, bottom lid with stainless steel screen of 60 μm (7) and outlet (9).

a second washing with tissue culture medium will be necessary. This whole treatment may also be carried out in the bioreactor itself by using a sieve at the bottom of the bioreactor for removal of spent culture medium, washing and exposing the cells to trypsin and mixing with the stirrer. However, in many cases the sieve might be partially blocked during the process so that it is very time-consuming, which results in a large number of cells being damaged and lost. By using this special apparatus viable cell yields of 80–90% of the total amount of cells may be obtained. In addition to the smaller apparatus (suitable for cultures of 10–15 litres), a larger trypsinization apparatus for culture volumes up to 50 litres has been developed.

10. APPLICATIONS OF MICROCARRIER CULTURE

In research, a large variety of cell types from animals including bird and fish cells have been cultivated on microcarriers. For a recent review the reader is referred to the publication of Reuveny (18). Although the microcarrier culture system was originally developed for the large-scale cultivation of animal cells for the production of viral vaccines (28), its present application in the production of cellular components and for studies in cell biology has been found to be of great value. Particularly now that genes of various viral and cellular components are being cloned in mammalian cells by recombinant DNA techniques, its application in the field of large-scale production of cellular components may increase significantly. When the desired product is excreted in the culture fluid the continuous perfusion system as described for replenishment of medium (Fig. 4) has proved to be very practical for large-scale production of cellular biologicals, such as the production of tissue plasminogen activator (TPA) by Bowes melanoma cells (5). At the 40-litre scale and a perfusion rate of one culture volume per day, 750 to 1000 litres of TPA-containing culture fluid could be generated within 25 days.

However, thus far, the large-scale application of microcarrier culture is mainly used for the production of virus vaccines such as those for polio and rabies (13, 33) and foot-and-mouth disease virus (12). In this field production is already performed at the 1000-litre scale (14).

11. CONCLUSIONS

By the continuous improvement of microcarriers, the microcarrier culture system has been developed to one of the most useful large-scale cultivation

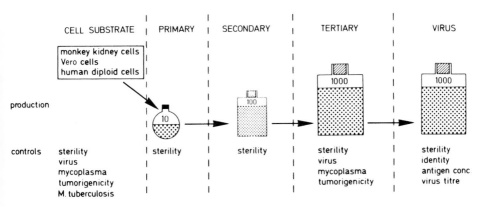

Fig. 6. Unit process system for the batchwise production of virus vaccines in microcarrier culture.

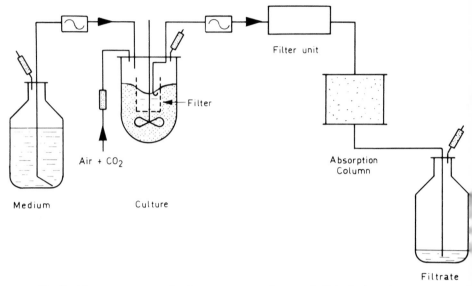

Fig. 7.   Unit process system for the continuous production of cellular biologicals in continuous perfusion systems of microcarrier culture.

systems for anchorage-dependent mammalian cells. The system is particularly suited to the development of unit processes for production of virus vaccines (Fig. 6) and cellular biologicals (Fig. 7). However, the system still needs further research on the improvement of the microcarriers and cultivation procedures. In our opinion, for the improvement of microcarriers, one should concentrate on a suitable non-porous particle of rather rigid material. They should be produced at a low cost in view of the large amounts needed for large-scale production. Research in the field of cultivation procedures should be centred upon the development of stirrer systems with optimal gas exchange between the gas and liquid phases at slow agitation levels and the control of nutrient requirements in order to achieve higher cell yields per unit culture volume. With regard to the media, the development of serum-free medium would be of great interest as this would facilitate the manipulation of the cultures; for instance it would be unnecessary to change to a serum-free medium when virus is added to a culture, and the downstream processing of cellular components would be greatly facilitated. The last point is of major interest in view of the increased application of mammalian cells as a host for recombinant DNA genes for the production of cellular biologicals. It is expected, therefore, that in future the major application of animal cell cultivation will be in this field.

REFERENCES

1. Flow Laboratories (1978). "Superbeads Microcarriers." Flow Lab.
2. Gebb, C., Clark, J. M., Hirtenstein, M. D., Lindgren, G., Lindgren, D., Lundgren, B., and Vertblad, P. (1982). Alternative surfaces for microcarrier culture of animal cells. *Dev. Biol. Stand.* **50**, 93–102.
3. Grinnell, F. (1978). Cellular adhesiveness and extracellular substrata. *Int. Rev. Cytol.* **53**, 65–163.
4. Hirtenstein, M., Clark, J. M., and Gebb, C. (1982). A comparison of various laboratory scale culture configurations for microcarrier culture of animal cells. *Dev. Biol. Stand.* **50**, 73–80.
5. Kluft, C., van Wezel, A. L., Van der Velden, C. A. M., Emeis, J. J., Verheijen, J. H., and Wijnands, G. (1983). Large scale production of extrinsic (tissue-type) plasminogen activator from human melanoma cells. *Adv. Biotechnol. Processes* **2**, 97–110.
6. Levine, D. W. (1976). Production of anchorage-dependent cells on microcarriers. Ph.D. Thesis, Massachusetts Institute of Technology, Cambridge.
7. Levine, D. W., Wang, D. I. C., and Thilly, W. G. (1977). Optimizing parameters for growth of anchorage-dependent mammalian cells in microcarrier culture. *In* "Cell Culture and Its Application" (R. T. Acton and J. D. Lynn, eds.), p. 19. Academic Press, New York.
8. Levine, D. W., Wang, D. I. C., and Thilly, W. G. (1979). Optimization of growth surface parameters in microcarrier cell culture. *Biotechnol. Bioeng.* **21**, 821–845.
9. Levine, D. W., Wong, J. S., Wang, D. I. C., and Thilly, W. G. (1977). Microcarrier cell culture: New methods for research scale application. *Somatic Cell Genet.* **3**, 149.
10. Lux (1980). "Cytopheres." Lux.
11. McLimans, W. F. (1972). The gaseous environment of the mammalian cell in culture. *In* "Growth, Nutrition, and Metabolism of Cells in Culture" (G. H. Rothblat and V. J. Cristofalo, eds.), Vol. 1, pp. 137–170. Academic Press, New York.
12. Meignier, B., Mangeot, A., and Favre, H. (1981). Foot and mouth disease virus production on microcarrier grown cells. *Dev. Biol. Stand.* **46**, 249–256.
13. Montagnon, B. J., Fanget, B., and Nicolas, B. J. (1981). The large scale cultivation of Vero cells in microcarrier culture for virus vaccine production: Preliminary results for killed poliovirus vaccine. *Dev. Biol. Stand.* **47**, 55–64.
14. Montagnon, B. J., Vincent-Falquet, J. C., and Janget, B. (1983). Thousand liters scale microcarrier culture of Vero cells for killed poliovirus vaccine: Promising results. *Dev. Biol. Stand.* **47**.
15. Monthany, J. F., Schwartz, M. D., Holis, D. F., and Polastri, G. D. (1980). U.S. Patent 4,237,218.
16. Nunc (1981). "Cultivation Principles and Working Procedures," Biosilon Bull. No. 1. Nunc, Kamstrup, Denmark.
17. Pharmacia Fine Chemicals (1982). "Microcarriers Cell Culture: Principles and Methods." Pharmacia Fine Chemicals AB, Uppsala, Sweden.
18. Reuveny, S. (1983). Microcarriers for culturing mammalian cells and their applications. *Adv. Biotechnol. Processes* **2**, 1–32.
19. Reuveny, S., Silberstein, L., Shahor, A., Freeman, E. and Mizrahi, A. (1982). DE-52 and DE-53 cellulose microcarriers. I. Growth of primary and established anchorage dependent cells. *In Vitro* **18**, 92.
20. Sanford, K. K., Earle, W. R., Evans, V. J. *et al.* (1951). The measurement of proliferation in tissue culture by enumeration of cell nuclei. *J. Natl. Cancer Inst. (U.S.)* **11**, 773–795.
21. Spier, R. E. (1982). Animal cell technology: An overview. *J. Chem. Technol. Biotechnol.* **23**, 304.

22. Spier, R. E., and Whiteside, J. P. (1976). The production of foot and mouth disease virus from BHK21 C13 cells grown on the surface of DEAE–Sephadex A-50 beads. *Biotechnol. Bioeng.* **19**, 659.
23. Spier, R. E., and Whiteside, J. P. (1984). The description of a device which facilitates the oxygenation of microcarrier cultures. *Dev. Biol. Stand.* (in press).
24. Van Hemert, P., Kilburn, D. G., and van Wezel, A. L. (1969). Homogeneous cultivation of animal cells for the production of virus and virus products. *Biotechnol. Bioeng.* **11**, 875–885.
25. Van Hemert, P. A. (1971). Vaccine production as a unit process. Thesis, Technical University, Delft, The Netherlands.
26. van Wezel, A. L. (1967). Growth of cell-strains and primary cells on microcarriers in homogeneous culture. *Nature (London)* **216**, 64–65.
27. van Wezel, A. L. (1973). Microcarrier culture of animal cells. *In* "Tissue Culture: Methods and Their Application" (P. F. Kruse, Jr. and M. K. Patterson, eds.), p. 372. Academic Press, New York.
28. van Wezel, A. L. (1972). New trends in the preparation of cell substrates for the production of virus vaccines. *Prog. Immunobiol. Stand.* **5**, 187.
29. van Wezel, A. L. (1977). The large scale cultivation of diploid cell strains in microcarrier culture. Improvement of microcarriers. *Dev. Biol. Stand.* **37**, 143–147.
30. van Wezel, A. L. (1982). Cultivation of anchorage-dependent cells and their applications. *J. Chem. Technol. Biotechnol.* **32**, 318–323.
31. van Wezel, A. L., and Van der Velden-de Groot, C. A. M. (1978). Large scale cultivation of animal cells in microcarrier culture. *Process Biochem.* **13**, 6–8.
32. van Wezel, A. L., Van der Velden-de Groot, C. A. M., and Van Herwaarden, J. A. M. (1980). The production of inactivated poliovaccine on serially cultivated kidney cells from captive-bred monkeys. *Dev. Biol. Stand.* **46**, 151–158.
33. van Wezel, A. L., Van Steenis, G., Hannik, C. A., and Cohen, H. (1978). New approach to the production of concentrated and purified inactivated polio and rabies tissue culture vaccines. *Dev. Biol. Stand.* **41**, 159–168.
34. Varani, J., Dame, M., Beals, T. F., and Wass, J. A. (1983). Growth of three established cell lines on glass microcarriers. *Biotechnol. Bioeng.* **25**, 1359–1372.
35. Yamada, K. M., and Olden, K. (1978). Fibronectins. Adhesive glycoproteins of cell surface and blood. *Nature (London)* **275**, 179–184.

12

Physical and Chemical Parameters: Measurement and Control

J. L. HARRIS

L. H. Fermentation
Stoke Poges, Buckinghamshire
United Kingdom

R. E. SPIER

Department of Microbiology
University of Surrey
Guildford, Surrey, United Kingdom

Animal Cell Biotechnology, Vol. 1

1. INTRODUCTION

The measurement and control of the parameters that affect microbial growth is a very important part of any fermentation (*11, 14*). For the researcher, the greater the number of parameters that can be measured, the greater is the information gained about the process. If the effect of varying just one variable over a series of fermentations is to be investigated, then it is important that all the other environmental conditions must be kept as constant as possible to eliminate their effect on the experiment. At the production scale, where the greatest rate of growth or product formation is required, accurate control systems are needed to maintain conditions as near optimum as possible.

This chapter examines the various physical and chemical parameters currently used in fermentation technology, and discusses the means available for their accurate measurement and the systems that can be employed to control them. Special emphasis is given to those conditions which prevail in the unit process cultivation of animal cells.

1.1. The General Control System

Figure 1 shows a simple method for controlling a parameter within a fermentation vessel. Measurement is performed using a suitable probe or sensor which produces a signal related to the magnitude of the parameter. A conditioning amplifier then brings this signal up to a standard level. The parameter value is compared with the desired value (the set-point) to produce an error signal, which is fed to an electronic controller. The controller output is used, via an actuator, to vary the physical or chemical conditions in

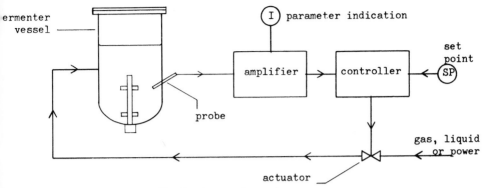

Fig. 1.   A general control system.

the vessel. This in turn varies the parameter value and hence the control loop is closed. The controller attempts to maintain the parameter at the desired set-point despite influences from outside the vessel (e.g. changes in ambient temperature) or from within (e.g. changes in oxygen demand).

The following sections deal with the elements of Fig. 1 and consider the differing requirements of different parameters. Section 2 is concerned with the various methods of measurement and, in some cases, their effects upon the conditioning amplifier. Section 3 discusses the different controllers that are available, their suitability and relative advantages for parameter control.

Finally, Section 4 lists the various actuators for varying physical quantities, such as temperature and fluid flow, and examines the possible interactions between the parameters that result.

2. PARAMETER MEASUREMENT

2.1. pH

The degree of acidity of a solution can be estimated by using a coloured indicator, measuring its electrical conductivity or by taking a sample for titration. The most widespread means of measuring pH in fermentations, though, is the pH electrode. The principle of operation of modern pH electrodes is based on the electro-chemical properties of the hydrogen electrode.

2.1.1. The Hydrogen Electrode

Although rarely used unless an extremely accurate measurement is required, the hydrogen electrode is the absolute standard against which all

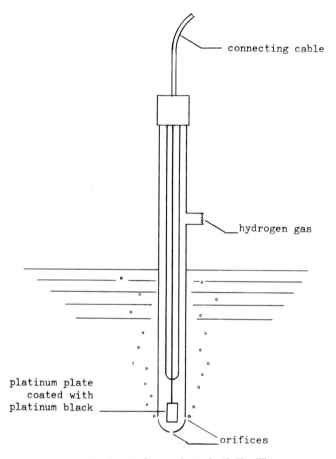

Fig. 2.   Hydrogen electrode "half-cell".

other electrodes are referenced. Figure 2 shows the construction of a hydrogen electrode "half-cell", while Fig. 3 shows a practical measurement system using two such cells. Referring to Fig. 3, the electrical potential E,
developed between the two connecting wires, when in equilibrium, is given
by:

$$E = 2.303 \frac{RT}{F} \log_{10} \frac{(H_a^+)}{(H_b^+)} \qquad (1)$$

Since pH is defined as:

$$pH = -\log_{10}(H^+) \qquad (2)$$

combining Eqs (1) and (2) gives:

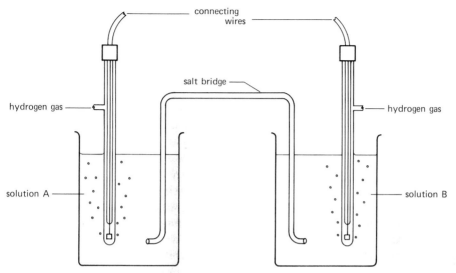

Fig. 3. Practical measurement system using two "half-cells".

$$E = 2.303 \frac{RT}{F} (\text{pH}_b - \text{pH}_a) \tag{3}$$

where R is the universal gas constant, 8.32 J/g/mol/°C, F the Faraday constant, 96,500 C, T the absolute temperature in degrees kelvin and E the generated emf in volts.

Equation (3) shows us that the developed voltage is proportional to the absolute temperature of the solutions and also to their pH difference, hence by knowing T and measuring E with a voltmeter, $(\text{pH}_b - \text{pH}_a)$ can be calculated. To measure pH_b directly we would have to set pH_a to zero, which can be done as follows:

If solution A had a hydrogen ion activity of one, and the electrode in contact with it used hydrogen gas at 1 atm, then the potential of electrode A is defined as zero at any temperature. To avoid the practical difficulties of using hydrogen electrodes the glass electrode was developed.

2.1.2. The Glass Electrode

The glass electrode relies on the pH-dependent potential generated at the surface of a very thin glass membrane in contact with the measured solution. A similar potential occurs inside the electrode at the inner surface of the membrane, but this is kept constant by filling the electrode with a buffer of constant pH (Fig. 4). The electrical signal is brought out of the electrode through a silver/silver chloride system so that the potentials generated by the contact of the wire with the buffer are also kept constant. As all pH

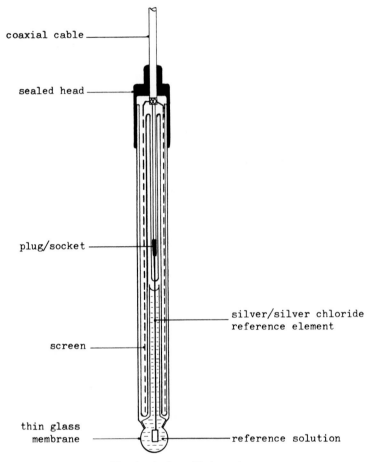

coaxial cable

sealed head

plug/socket

silver/silver chloride
reference element

screen

thin glass
membrane

reference solution

Fig. 4.   A glass pH electrode.

measurement systems require two half-cells, the electrical circuit is completed with a reference electrode, usually made from calomel.

2.1.3. *The Calomel Electrode*

A common reference electrode is that of the calomel (Hg_2Cl_2) type, shown in Fig. 5. This provides the second electrical connection to the system via a platinum wire, with a complex electrode system of calomel in saturated potassium chloride ensuring that any potentials generated are kept as constant as possible. Free exchange of ions between the electrolyte and the culture is allowed through a liquid junction made out of a porous plug set into the wall of the electrode.

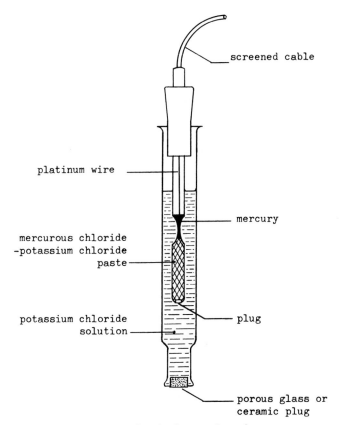

screened cable

platinum wire

mercury

mercurous chloride
-potassium chloride
paste

potassium chloride
solution

plug

porous glass or
ceramic plug

Fig. 5.   Calomel reference electrode.

2.1.4. The Combined Electrode

As pH electrodes improved, the reference half-cell was brought inside the body of the measuring cell to produce the combined electrode (Fig. 6). Within one glass body this contains the complete pH measurement system with a plug or socket provided at one end for connection to an amplifier.

2.1.5. The pH Amplifier

The use of glass as the pH-sensitive element of the electrode has several advantages. It is resistant to attack from the chemicals used in most fermentations and is easily cleaned if contaminated. Glass electrodes can also be used at much higher temperatures than previously available types (70°C continuous and 100°C intermittent use), and they can withstand sterilising temperatures, 121°C. Unfortunately, the use of glass results in an excep-

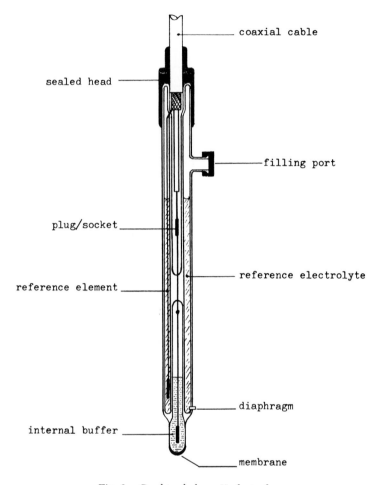

Fig. 6.   Combined glass pH electrode.

tionally high impedance electrode, typically greater than 10^8 ohms. The pH amplifier must therefore have an input impedance several orders of magnitude greater than this as current may not be drawn from the probe. The interconnecting lead and connectors used must also have a very high integrity to prevent a loss of signal between the electrode and amplifier. Coaxial cable, plugs and sockets of the type used on television aerials are commonly used, with connections at the electrode being sealed to prevent the ingress of moisture. The output from a combined glass electrode can be expressed as:

$$E = k + (2.303RT) \cdot \text{pH} \tag{4}$$

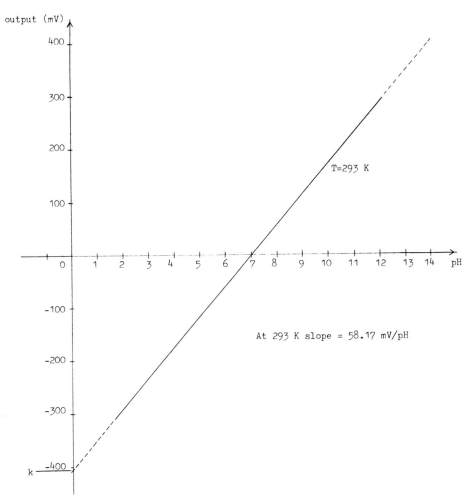

Fig. 7. Graph of pH electrode output versus pH.

Equation (4) has been plotted in Fig. 7. The point E^0, at which the output from the probe is zero, is usually at either pH 2 or 7, the latter being more widespread. The output from the probe is proportional to the absolute temperature, so pH amplifiers are fitted with a slope control, calibrated in degrees Celsius, to compensate for this. Alternatively, automatic compensation is possible using a temperature sensor (usually a platinum resistance thermometer) either built into the probe or mounted separately in the vessel. $2.303RT$ is called the "slope" of the electrode.

Calibration of the system consists of placing the probe in a buffer solution close to the E^0 value (e.g., 7) and adjusting the buffer (or zero) control of the amplifier. A second solution (e.g., 4 or 9) is then used to set the slope control of the amplifier. If automatic temperature compensation is not in use then care must be taken to ensure that the buffers are at the expected fermentation temperature or that the manual temperature compensation dial is correctly set. Calibration should be performed before each run as the characteristics of the probe change with age and use. Some drift of the E^0 value may also be expected with sterilisation.

Common problems experienced with pH measuring systems are clogging of the porous plug, preventing ion exchange, or loss of signal in the interconnecting lead. The latter can be caused by too great a distance between probe and amplifier, or contamination of the connections by oils during handling. Systematic substitution of probe, lead and amplifier can identify the source of the problem.

2.2. Temperature

Temperature can be measured in many ways, usually by converting the expansion of gases or liquids into movement across a calibrated scale. Two properties that are exploited in the construction of electrical probes are the generation of a temperature-related voltage, as in the thermocouple, and resistance changes, as in thermistors and resistance thermometers.

2.2.1. The Thermocouple

The principle behind the operation of the thermocouple is the Seebeck effect, namely that when two dissimilar metals are joined to form a closed circuit, and the two junctions are held at different temperatures, then a current flows around that circuit. The emf produced by a thermocouple is the sum of the emf's generated at the junctions of the two metals (Peltier emf's) and the emf's induced in each of the two wires due to the temperature gradient along their length (Thomson emf's). The net emf, E, is given by:

$$E = a(t_1 - t_2) + b(t_1^2 - t_2^2) \qquad (1)$$

where a and b are constants determined by the metals used. To measure the fermentation temperature it is necessary to immerse the hot junction, held in a suitable jacket, in the medium and to hold the reference junction at a constant temperature. Most thermocouple amplifiers allow the reference junction to remain at ambient and include electronic compensation for any change, a technique known as "cold junction compensation".

Advantages of the thermocouple include: linearity of output, interchangeability of probes and sensitivity (typically 40–50 μV/°C with the reference junction at 0°C). The main disadvantage is that junctions and breaks in the lead should be avoided as they introduce spurious emf's.

2.2.2. The Thermistor

A thermistor is a device, usually made from metal oxides, whose electrical resistance changes with temperature. The majority have a negative temperature coefficient, i.e. their resistance falls as temperature rises. In general, the resistance is given by:

$$R = ae^{b/t} \tag{2}$$

where t is the absolute temperature and a and b are constants. The resistance R at any temperature t can be found knowing the resistance R^0 at a temperature t^0 by using Eq. (3):

$$R = R^0 e^{(b/t - b/t)} \tag{3}$$

Many types of thermistors are available having characteristic resistances ranging from 1 kilohm to 1 megohm at 25°C, but the poor tolerance of R^0 and b ($\pm2\%$) makes them unsuitable for accurate measurement. Their non-linearity restricts their useful temperature range and poor repeatability means that the amplifier must be calibrated to each thermistor.

2.2.3. Semiconductors

The strong temperature dependence of semiconductors makes them a very accurate means of measurement. Amplifiers for semiconductor probes exploit either the change in diode current or transistor gain with temperature, both normally undesirable quantities. Their small mass also gives them fast response times. Unfortunately, the comments regarding non-linearity and the need to match each device to its amplifier apply to semiconductors more than to other thermistors. Although single semiconductor probes are rarely used, modern microchip technology has enabled complete circuits to be manufactured in which non-linear effects are made to cancel. The development of such devices as the Radio Spares 308–809 may lead to a wider use of semiconductors in temperature measurement.

2.2.4. Platinum Resistance Thermometers

Of the three metals commonly used in resistance thermometers, platinum, nickel and copper, platinum is almost universally used in the fermentation industry. Over the range 0° to 200°C the resistance of a length of platinum wire can be expressed as:

$$R = R^0(1 + At + Bt^2) \tag{4}$$

where R^0 is the resistance at 0°C. Table I shows the resistance/temperature characteristics as defined by BS1904 and DIN43760. The commonly used "PT 100" resistance thermometer conforms to this standard. From the table it can be seen that if the thermometer is assumed linear over the range 0°–100°C an error of less than 0.2°C results at 50°C. If the fermentation working

TABLE I Resistance Values in Ohms for Resistance Thermometers Based on BS1904

°C	+0°	+2°	+4°	+6°	+8°
0°	100.00	100.78	101.56	102.34	103.12
10°	103.90	104.68	105.46	106.24	107.02
20°	107.79	108.57	109.35	110.12	110.90
30°	111.67	112.45	113.22	114.00	114.77
40°	115.54	116.31	117.09	117.86	118.63
50°	119.40	120.17	120.94	121.71	122.47
60°	123.24	124.01	124.78	125.54	126.31
70°	127.07	127.84	128.60	129.37	130.13
80°	130.89	131.66	132.42	133.18	133.94
90°	134.70	135.46	136.22	136.98	137.74
100°	138.50	139.26	140.02	140.77	141.53
110°	142.29	143.04	143.80	144.55	145.31
120°	146.06	146.81	147.57	148.32	149.07
130°	149.82	150.57	151.32	152.07	152.82

temperature is restricted to a narrower range, then the amplifier can be calibrated accordingly, affording a higher accuracy over that range. Alternatively, linearisers are available, e.g. the Radio Spares components 158–418.

Measurement of resistance does not require specialised circuitry and hence amplifiers for platinum resistance thermometers are simple to construct. The amplifier passes a constant current through the probe and measures the developed voltage, which is proportional to the resistance in circuit. Care has to be taken on two points. Firstly, the current should not be so great that self-heating of the sensor alters its resistance, and secondly, the resistance of the leads to the thermometer must be taken into account, if it is significant. Three-wire and four-wire resistance thermometers use different wires to sense the voltage and to supply the current, and hence can compensate for lead length.

To isolate the sensing element from the fermentation medium, temperature sensors are enclosed in stainless steel sheaths. To minimise response time the walls of these sheaths should be thin and the inside filled with heat-conducting compound.

2.3. Dissolved Oxygen

The dissolved oxygen concentration of a fermentation can be measured directly using available dissolved oxygen electrodes. These electrodes are of the amperometric type, i.e. a current is produced in the presence of oxygen, and they can be divided into two types: the polarographic electrode, which requires a polarising voltage, and the galvanic electrode, which does not (5,8).

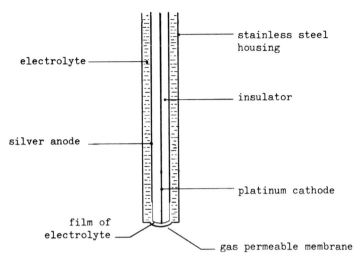

electrolyte —

silver anode —

film of
electrolyte —

— stainless steel
   housing

— insulator

— platinum cathode

— gas permeable membrane

Fig. 8.   Ingold polarographic oxygen electrode.

2.3.1. *The Polarographic Oxygen Electrode*

In 1956, Clark (3) described an electrode system enclosed within a permeable membrane which, when polarised with an applied voltage, produced a current in response to oxygen that diffused through the membrane. Figure 8 shows a schematic of the Ingold electrode, which is based on this design. The probe consists of a silver cylinder anode and a platinum wire cathode and uses a silver chloride–based electrolyte. When a polarising voltage between 600 and 700 mV is applied the following reactions take place:

$$O_2 + 2H_2O + 4e \rightarrow 4OH^- \quad \text{(at the cathode)}$$

$$4Ag + 4Cl \rightarrow 4AgCl + 4e^- \quad \text{(at the anode)}$$

The electrons circulate from anode to cathode in the external circuit and this current is measured by the amplifier. Because oxygen is consumed within the cell the reaction can only proceed at the rate at which oxygen diffuses through the membrane. Further, because the oxygen concentration within the cell is very small compared with that in the external solution, this rate is proportional to the external oxygen concentration. The disadvantage of the diffusion-limited system is that the geometry of the probe affects the output current. The exact relationship is given by:

$$I = k \cdot D \cdot a \cdot A \cdot \frac{dO_2}{x} \tag{1}$$

where I is the electrode current, k is a constant, D is the diffusion coefficient of oxygen through the membrane, a is the solubility of oxygen in the membrane, A is the area of the cathode, x is the membrane thickness and dO_2 is the oxygen concentration in the external solution. It can be seen from Eq. (1) that a large cathode area and small membrane thickness are desirable as they lead to a high output current. Thin membranes also improve the response time (which is proportional to x^2), which is typically 60 sec to reach 98% of the final value. Too large an electrode current is undesirable as polarisation of the anode and a dependence on solution flow past the membrane may result.

2.3.2. The Galvanic Oxygen Electrode

The electrode described by Mackereth (9) generates a current spontaneously in the presence of oxygen. This is the basis of the galvanic oxygen electrode. The steam-sterilisable version described by Johnson *et al.* (5) is in common use today. The electrode sold by Uniprobe Instruments Ltd. is of this type (Fig. 9). This probe consists of a lead spiral anode, silver disc cathode and an acetate buffer as electrolyte. Oxygen diffusing into the cell reacts with the anode, the overall equation being:

$$\tfrac{1}{2}O_2 + Pb + H_2O \rightarrow Pb(OH)_2$$

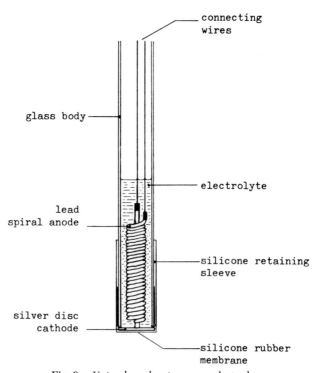

Fig. 9.   Uniprobe galvanic oxygen electrode.

Conversion of the lead anode is the primary disadvantage of this type of probe as it means the probe must be replaced when all of the anode has been consumed. The current produced, typically 10 μA in air, is passed through an external resistor, usually 1 kilohm, to produce a milli-volt signal for amplification. This is significantly larger than the output from a polarographic electrode, which is approximately 10^{-7} A in air. Both kinds of probe show good linearity and a low residual current (output in the absence of oxygen) of approximately 0.1% of the air reading. The temperature dependence of oxygen permeability through the membrane gives the probes an undesirable temperature coefficient of approximately 3% per degree Celsius. In fermentations where temperature may vary widely, temperature compensation is necessary. Automatic compensation is usually achieved with a thermistor whose temperature coefficient closely matches that of the membrane. Some amplifiers use a resistor network and a thermistor mounted separately in the vessel; others use probes which have a built-in thermistor.

Calibration consists chiefly of setting the span of the amplifier so that under a given set of conditions the solution is defined as being 100% saturated. Temperature, flow rate past the membrane and fermenter head-space pressure all affect the dO_2 reading so care should be taken to reproduce the actual measurement conditions while calibrating. Zero point checking, if required, should be performed before setting the span, by sparging the vessel with nitrogen and adjusting the amplifier "zero" control for zero output.

It should be emphasized that these electrodes measure oxygen partial pressure, and since the oxygen concentration in milligrams per litre depends also on the solubility, the oxygen concentration cannot be obtained unless the solubility is also known. Spier and Griffiths have reviewed the factors which relate to the optimum oxygen levels for animal cell growth and product generation. (12).

2.4. Redox

A reduction–oxidation (redox) reaction is a reversible reaction in which one component is reduced by the addition of electrons and another is oxidised by the loss of those electrons. This can be expressed generally as:

$$A + ne^- \rightleftarrows B \tag{1}$$

where A is the oxidised form and B is the reduced form of the atom, ion or molecule involved. In an electrolytic cell, the "redox potential" is defined as the voltage that has to be applied to cause oxidation to take place at the anode and reduction to occur at the cathode. This potential is referred to that of a standard hydrogen electrode (defined as zero). Conversely, when a metal electrode is placed in an aqueous solution a reaction will proceed until the potential acquired by the electrode prevents any further addition or removal of electrons. This potential is a function of the true redox potential, Eh, which is given by the Nernst equation:

$$Eh = E^0 + 2.303 \frac{RT}{nF} \log_{10} \frac{(A)}{(B)} \qquad (2)$$

where Eh is in milli-volts, R is the universal gas constant, F is the Faraday constant, T the absolute temperature, n the number of electrons involved and (A) the activity of A. Compare with Eq. (1) in Section 2.1.

The potential acquired by the redox electrode depends on whether the components of the reaction prefer to occupy their oxidised state, A, by donating electrons to the probe, or by removing electrons from the probe and becoming reduced, B. The potential relative to the solution will become negative or positive respectively. Just as the pH electrode measures hydrogen ion activity, the redox electrode can be thought of as measuring electron activity.

The Redox Electrode

The combined redox electrode is very similar in design to the pH electrode except that the potential-sensitive element is a metal such as platinum, gold or silver. The reference electrode is usually a silver/silver chloride mixture used as electrolyte. The same comments regarding pH amplifiers can be made for redox amplifiers. Some buffer adjustment is required to accommodate the different E^0 values of the probes and temperature compensation must be built in as the electrode slope varies with temperature. Calibration consists of placing the probe in two solutions of known redox potential and alternately adjusting the zero and slope controls until agreement is reached. Suitable solutions can be prepared with saturated solutions of quinhydrone at two different pH (7) values. The potential of these solutions is given by:

$$Eh = 699 - 2.303 \frac{RT}{F} \cdot pH \qquad (3)$$

where pH is the pH value of the solution and Eh is in milli-volts.

When measuring the redox potential of biological systems it must be remembered that the value obtained is an aggregate of the potentials of all the redox pairs in the system. Even so, redox is still a useful parameter in the investigation of the performance of animal cell systems (4). Indeed, it has been shown recently that the production of herpes suplic virus in cell culture is best followed or controlled by redox measurements.

2.5. Pressure

It is useful to measure the head-space pressure during a fermentation for several reasons. Firstly, it provides early warning of a blockage in the off-gas line such as a wet filter. Secondly, the pressure in the vessel affects the partial pressure of gases in the medium and hence the reading of dissolved oxygen probes. Finally, it may be desirable to run the fermentation at elevated pressures and so measurement of the pressure is required for the control system.

The Strain Gauge

The principle underlying the strain gauge is that when a material is subjected to mechanical stress its electrical resistance changes. Fine wire, foil and semiconductors are all used in modern transducers, which usually consist of four gauges connected together to form a Wheatstone bridge (Fig. 10). An excitation voltage is applied across opposite sides of the bridge, the output being measured from the other two corners. When the elements are strained their resistance changes, thus changing the output voltage. By suit-

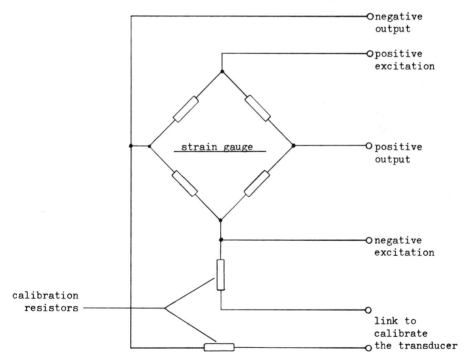

Fig. 10.   Wheatstone bridge pressure transducer.

able arrangement of the gauges, the Wheatstone bridge provides a greater change in output than would be attainable with only one gauge.

When measuring the pressure above a fermentation it is important that the transducer has no crevices where micro-organisms may collect and also that suitable materials are used, such as stainless steel. For these reasons the diaphragm type of transducer is commonly used. This consists of a force-summing diaphragm linked to the strain gauges by a push-rod. The space behind the diaphragm can either be vented to the atmosphere, if it is required to measure gauge pressures, or evacuated for absolute pressure measurement. Apart from isolating the active elements of the transducer from the fermentation environment, the diaphragm also provides a degree of protection from temperature changes in the vessel.

Modern techniques in printed circuit board manufacture have made available very small transducers using silicon or foil tracks on a suitable backing material. Although semiconductors such as silicon have a much greater sensitivity to strain than metals, they also exhibit a strong temperature coefficient, which is undesirable. With the inclusion of special "dummy" gauges in the circuit, temperature compensation over the range 0°–100°C is available to typically 0.03% of full range output. Other advantages of such transducers are that they are linear, may contain calibration resistors and can be sterilised.

A further use of pressure transducers is to determine the volume of fluid in the culture. This generally involves two sensors, one at the bottom of the vessel and the other in the gas space; the differential pressure is that caused by the weight of fluid in the vessel.

2.6. Gas Flow

The most common reason for adding gas to a fermentation is to oxygenate the medium. Air is frequently used, with other gases being added if required. For example, extra oxygen may be used if the demands of the culture are particularly high, or carbon dioxide may be added for pH control. These gases may be passed across the head-space or sparged directly into the medium. It is important that the flow rates of gases into the fermenter are measured to ensure the correct aeration rate and blend of gases. If gases are to be used for dissolved oxygen or pH control, then their flow rates must also be controlled. In situations where gas analysers are used to obtain mass balance equations or to calculate the respiratory quotient (RQ), the gas flow rate must be held constant if accurate results are to be obtained. Four kinds of flowmeters are discussed here: the inferential meter, the differential pressure meter, the variable aperture meter and the thermal mass flowmeter. A comparison of the first three types is given in Fig. 11.

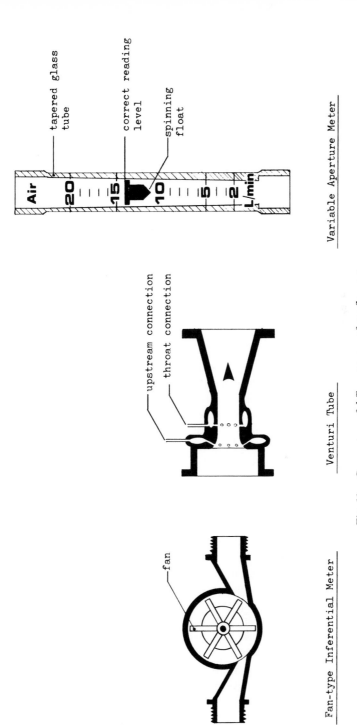

Fig. 11. Comparison of different types of gas flowmeter.

Fan-type Inferential Meter

fan

Venturi Tube

upstream connection

throat connection

Variable Aperture Meter

tapered glass
tube

correct reading
level

spinning
float

Air

20

15

10

5

2

L/min

2.6.1. The Inferential Meter

Meters of this type are so called because the velocity of gas flow is measured and the volumetric flow is "inferred", knowing the cross-sectional area of flow. The gas velocity is measured by placing a fan, turbine or similar rotating propeller in the line carrying the gas. The speed of rotation of the propeller varies with the gas velocity in the line and hence the volumetric flow can be calculated by multiplying by the area across which the gas has to flow. In modern types of meter, the fan either drives a small generator or its rotation is electro-magnetically sensed by coils mounted around the pipe. A suitable amplifier then scales the signal so that a direct read-out in litres per minute is obtained.

2.6.2. The Differential Pressure Meter

These meters force the gas through a nozzle, venturi tube, orifice plate or similar obstruction and measure the pressure upstream and downstream of the constriction. As the gas flows through the narrowing portion of the pipe its velocity increases and its pressure decreases. The difference in pressure is related to the difference in velocity and, knowing the specific gravity, the volumetric flow can be calculated. Pressure transducers such as those described in Section 2.5 can be used to provide an electrical signal from which the flow rate may be calculated. The main disadvantage of this kind of meter is its inherent non-linearity, which makes the electronics complicated. Also, the flow element itself must be very accurately made if compressibility of the gas and eddy currents around the constriction are not to cause significant errors.

2.6.3. The Variable Aperture Meter

A very common means of measuring gas flow in fermentation systems is the tapered tube and float arrangement such as the "rotameter". The gas to be measured is introduced into the bottom of a glass tube whose internal diameter increases towards the top of the tube. A float in the tube is pushed upwards by the gas stream until the aperture around the float is large enough to allow the gas to escape at such a velocity that the force on the float just balances its weight. Inclined grooves on the float cause it to rotate, thus producing visual indication that the float is not stuck and helping to prevent any contamination in the gas from sticking to its surface. By calibrating the glass tube a very simple means of measuring gas flow is obtained, which no doubt accounts for its popularity. The one drawback is that a continuous electrical signal proportional to gas flow is very difficult to obtain.

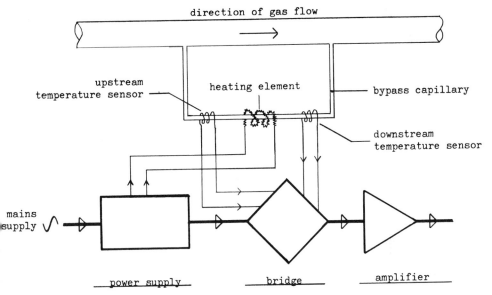

direction of gas flow

upstream
temperature sensor

heating element

bypass capillary

downstream
temperature sensor

mains
supply

power supply

bridge

amplifier

Fig. 12. Principle of operation of Brooks mass flowmeter.

2.6.4. The Thermal Mass Flowmeter

In applications requiring continuous and accurate measurement of gas flow rates the thermal mass flowmeter is widely used. The previously described systems all measure volumetric flow rates and hence their readings will vary with the temperature and pressure of the gas in the line even though the mass flow is constant. In fermentation, it is important that the mass of gas added to the medium is known and, in situations where the pressure and temperature can both vary, the thermal mass flowmeter has a distinct advantage.

The meter consists of a tube, through which the majority of the gas flows, bypassed by a small capillary (Fig. 12). A constant proportion of the gas flows through this capillary, which is uniformly heated by an electrical coil. Temperature sensors upstream and downstream of the coil detect the heating effect on the gas flow. The output of the system is balanced to be zero under no-flow conditions. When gas flows, the temperature difference between the sensors is directly proportional to the mass flow rate. The final output is usually displayed in litres per minute, assuming the gas to be at NPT.

Because the heat transfer rate is dependent on the heat capacity of the gas, the thermal flow sensor is sensitive to variations in gas composition. Air, nitrogen and oxygen can be mixed with negligible error, but if other gases such as carbon dioxide are to be used, their individual flow rates should be measured prior to mixing. This unfortunately increases the cost of the system.

2.7. Liquid Flow

Liquids are added to the fermentation media to provide nutrients, or to correct pH changes. During continuous cultures, media must be added at a known rate to dilute the culture fast enough to maintain a high growth rate but not so fast that "wash-out" occurs. Automatic control of liquid addition is therefore required and hence some accurate means of measurement. The methods of measuring gas flow described in the preceding section can be applied to liquids with varying degrees of success. Turbines, flowmeters, orifice plates and rotameters are all available for liquids but invariably their ranges are orders of magnitude higher than the few millilitres per minute typical of fermentation additions. Two methods of indirectly obtaining liquid flow rates are discussed here, namely measuring pump rates and addition vessel weight.

2.7.1. Pumping Rates

If liquid is to be pumped into a fermenter, then it is sometimes feasible to measure the speed of the addition pump, having previously obtained a flow rate against pump speed calibration curve. The main drawback of this method is that the flow rate varies even though pump speed is constant. Peristaltic pumps, which are cheap and easy to take apart for autoclaving, are notorious for the falling flow rates that result from tube wear. The flow rate also varies with the pressure in the fermenter. As a result, such pumps must be calibrated before and after each run and cannot be relied upon during long fermentations. A more reliable method is to use a positive displacement pump or even a metering pump so that a known volume of liquid may be added with each stroke. Such pumps are expensive and, unless the head can be detached from the motor, the addition line cannot be heat-sterilised.

2.7.2. Weighing Addition Vessels

With the advent of microcomputers it has become possible to obtain a measurement of flow rate and a running total of the liquid added by weighing the addition/collection vessels. The computer regularly samples the output of an electronic balance and from the changes in weight and density of the liquid, calculates the flow rate into the fermenter. If the computer is itself deciding when additions are to be made then it need only measure the weight of the vessel before and after the addition. Such a sophisticated system is only likely to be practical if very high accuracy is required or the fermentation must run for long periods unattended.

2.8. Agitator Speed

In order that the contents of the fermenter be thoroughly mixed, some form of agitation is required. Air-lift fermenters rely on the action of sparged air while, in other systems, mechanical impellers are used. The range of impeller types and speeds is very wide, ranging from marine impellers turning at only a few revolutions per minute (for animal cell culture), to flat blade turbines turning at speeds in excess of 400 rpm for bacterial fermentations.

The stirrer speed affects not only the mixing but also the level of broth in the vessel and, in some cases, the dissolved oxygen tension. Agitation should be fast enough to prevent the heavier elements of the medium from settling out but not so fast that the organisms, particularly cells, are damaged by shear forces. Indication of stirrer speed is usually achieved by connecting a small tachogenerator to the motor or vessel shaft and feeding this signal to an electrical meter. The display is calibrated in revolutions per minute. The signal from the tachogenerator can also be used by the motor controller to keep the agitation rate constant despite fluctuations in load caused by changes in aeration rate or medium density.

2.9. Agitator Power

Measurement of power required to stir a culture could be used in the future as one method of estimating biomass, particularly where this causes a change in the viscosity of the medium (1, 2). The power transmitted by a rotating shaft is equal to the product of its angular speed and the torque, or twisting force, it experiences. Measurement of shaft torque is mechanically difficult, so motor power is often measured instead, using ac or dc watt-meters. The cost, and poor accuracy at low power, means that such systems are unlikely to be used on fermenters used to cultivate animal cells.

2.10. Weight

If a fermentation is being run at constant volume, then measuring the weight of the vessel contents immediately provides its density. Alter-natively, it may be preferable to run a continuous fermentation at constant weight rather than constant volume. In either case, the fermenter can be weighed by supporting the vessel on load cells. These are strain gauges connected in a bridge arrangement similar to the pressure transducer discussed in Section 2.5. Three or four cells may be used, to compensate for uneven loads, with their signals averaged. The amplifier will display the load on the cells in kilograms force and is likely to have a large zero offset or "tare" so that the vessel weight may be subtracted from the signal. The display therefore reads only the weight of vessel contents. Aeration and agitation can cause large errors when weighing small vessels. The ratio of the vessel contents' weight to vessel weight also improves with vessel size, so load cells are usually found on vessels upwards of 50 litres in volume.

2.11. Optical Density

A method of estimating the density or concentration of organisms in the medium is to measure the ease with which a beam of light can be passed through the culture. Optical density meters have been used on samples taken from fermentation over many years. First, a calibration curve is drawn of optical transmittance against a series of samples of known cell concentration. A sample from the vessel is then tested and the cell concentration looked up on the graph. Several difficulties make on-line measurement of cell density inaccurate. One is a tendency for the organisms to grow on the window of a probe, and also changes in ambient light levels can affect the sensor. Care must be taken in selecting the frequency of light used. The transmittance of white light will be affected by colour changes in the medium that have nothing to do with cell density. Ideally, a frequency should be used which is absorbed or scattered only by the organisms, or the particular product, of interest. This method, while widely used in the culture of bacteria and yeast, has not been much used in animal cell systems as the variable optical properties of the medium during the course of a culture preclude consistent and useful measurements. Also, the change in the optical properties of the system resulting from a two- or fourfold increase in cell number is relatively quite small (in a bacterial culture growing from 10^5 to 10^9 cells/ml, a factor of 10^4!).

2.12. Foam

Foam presents a problem in fermentations with a high concentration of proteins, or where there are high aeration and agitation rates. As well as reducing oxygen transfer rates, culture fluid can be carried out of the fermenter and clog the off-gas filter. In some systems, cells and microcarriers trapped in the foam can seriously deplete the culture, cause undesirable reactions and result in a large number of dead organisms. It is important, therefore, that foaming be detected and controlled.

In some fermentations it may suffice to add an antifoam reagent to the medium once only, before inoculation. In other cases a regular addition of antifoam by automatic timer is sufficient. If the antifoam is expensive, or excessive use may have a detrimental effect on the culture, then it should only be added when required and a probe capable of detecting the presence of foam is used. Antifoams particularly useful in animal cell culture are the Midland Silicone variety. Three kinds of probe are discussed here, the capacitive, conductive and thermal probes.

2.12.1. Capacitance Probe

These probes consist of two electrodes, either side by side or in the form of concentric cylinders, separated by an air gap. As foam rises between the electrodes it decreases their mutual capacitance. A small alternating current applied to the probes detects this and triggers the addition of the antifoam reagent.

2.12.2. Conductive Probes

Two conducting probes, or a single probe with the vessel wall forming the second electrode, have a voltage applied across them. Foam touching these probes shorts out the voltage and triggers the addition. These systems are usually designed so that the resistance between the probes that causes triggering can be pre-set from a few thousand to many millions of ohms, so the sensitivity can be adjusted. Unlike the capacitance probe, which can be covered in an insulator, such as PTFE, the conductivity probe must have an exposed metal part. Passing a small current through the foam in this way is sometimes considered undesirable.

2.12.3. The Thermal Probe

This is a single probe, completely enclosed in a stainless steel sheath, of similar construction to the resistance thermometer described in Section 2.2. Sufficient current is passed through the resistance element to cause self-heating. When foam touches the probe, the sudden cooling changes its electrical resistance and this is detected as before. The advantages of this probe are that no current is passed through the medium and permanent shorting cannot occur. All foam systems employing probes should have a few seconds delay before addition is made to avoid nuisance triggering by splashing. Alternative mechanical foam disintegrators either work (no need of foam sensing)—or they do not—in which case the mechanisms described above could be of value.

3. PARAMETER CONTROLLERS

The function of the controller in a system is to accept signals from the probes and amplifiers previously described, compare them with the desired level of the parameter and decide how much corrective action should be taken. The output from the controller is then passed to a suitable actuator, which can affect the conditions within the fermenter so that the parameter is brought back towards the desired level.

Controllers can have different numbers of set-points, differing types of control action and different types of output signal. They are usually, but not always, electronic in nature, and this section will describe the common types of electronic, analogue controller. With the advent of microchip technology, a wide range of digital controllers has become available; but discussion of these is postponed until the next chapter.

3.1. On/Off Controllers

As their name implies, on/off controllers have only two states. They either apply, or do not apply, corrective action, depending solely on the current value of the measured parameter. As an example, consider a simple temperature control system. The desired temperature, or set-point, is dialed up on the controller and the controller is arranged so that while the measured temperature lies below the set-point, heat will be applied to the fermentation. The temperature will rise until it equals the set-point, and then the controller will turn off, allowing the medium to cool by natural losses. As soon as the temperature falls below the set-point, heat will again be applied.

Controllers of this type are frequently used to control pH. An upper set-point controls the acid addition and a lower, the alkali addition. The parameter may vary freely within the "dead-band" between the set-points, but if it is driven outside the band, by chemicals produced by the fermentation, then the appropriate reagent is added to return the pH to within the band. A controller with two on/off set-points must obviously have two separate outputs: one for acid and one for alkali, in this example. Care should be taken to ensure that the band is sufficiently wide that overshot cycling does not occur as this could alter the volume and salt concentration of a culture very quickly.

On/off controllers perform a very simple function and, to make them more suitable to a particular task, sophistication may be added in several ways. Three such methods are described here; they are hysteresis, the addition of a time constant and percentage derating of the output. Their effect can be seen in Fig. 13.

3.1.1. Hysteresis

This is a feature whereby the point at which the controller turns on is different from the point at which it turns off. In the temperature example, the medium temperature may have to rise by 0.1° above the set-point before

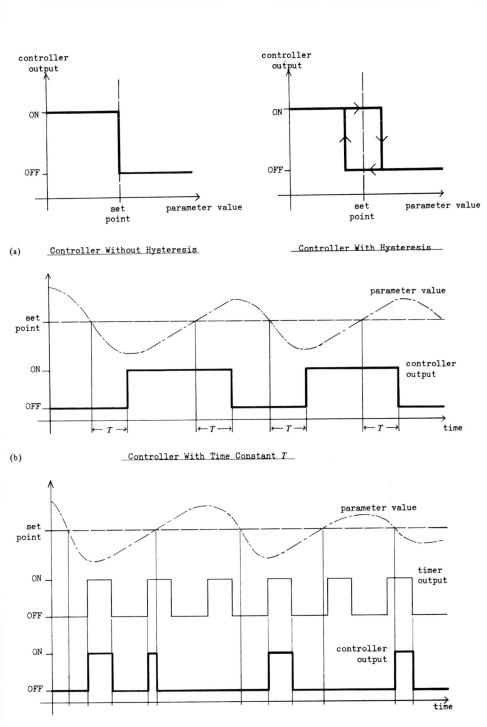

Fig. 13. The effects of (a) hysteresis, (b) time constant and (c) percentage derating on an on/off controller.

heating is turned off, but fall to 0.1° below the set-point before heating is turned on. Hysteresis is used to prevent the rapid switching that may occur if the medium cools quickly. If the controller switched at the set-point, a small amount of heat would raise the temperature enough to turn off the heating, but a slight temperature fall would turn it on again. Hysteresis also prevents noise, in the signal from the amplifier, from switching the controller incorrectly when the parameter is very close to the set-point.

3.1.2. Time Constant

The addition of a time constant to the controller causes it to delay before changing state, even though the parameter may have crossed the set-point. Such a pause prevents nuisance tripping, by splashes in a foam controller, or by a noisy signal. It ensures that corrective action is taken only if the parameter deliberately crosses the set-point, and conversely, that enough control action is applied to return the parameter within the set-point.

3.1.3. Percentage Derating

This method requires the addition of a cycling, or percentage, timer between the output from the controller and an actuator, such as a pump. The pump operates only when the controller demands it and the timer is in the "on" part of its cycle. Sometimes, therefore, the controller will demand an addition but nothing will happen until the timer reaches its "on" state. This has the same effect as the time constant previously described. At other times, the timer may "time out" and stop addition even though the controller still requires it. This allows time for the reagents already added to change the parameter, which could then turn off the controller. This has the advantage that less reagent is added than if the controller alone were in charge of the pump.

If a parameter is continually driven in the same direction by the fermentation it will reach an on/off set-point and "bounce" along it. The relative strengths of the driving force and corrective action will determine the widths of the excursions about either side of the set-point. Such oscillation of the parameter may be undesirable. If it is required to control the parameter stably at a set-point, then a more sophisticated control action is necessary, such as that provided by a proportional controller.

3.2. The Proportional Controller

A proportional controller consists of a set-point surrounded by a proportional band, the width of which may be varied (10). Unlike the on/off controller, the proportional controller can vary the amount of applied control action linearly with the position of the parameter within the proportional band. For

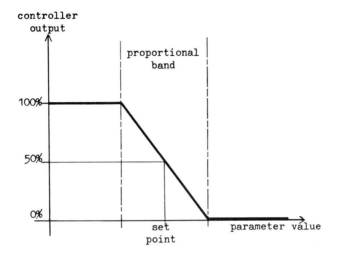

Single Output

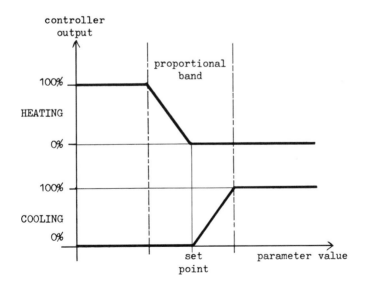

Split Proportional Band

Fig. 14.   Control action of two types of proportional controllers.

example, consider a controller with a set-point at 40°C and a proportional band width of 20°C. Below the proportional band full heating is applied. The fermentation temperature will rise, but as it passes 30°C less heat will be applied until, at the set-point, exactly 50% heating is being used. If the temperature continues to rise then less and less heat will be added until, at 50°C and above, no heating at all will be used. This is shown plotted as a graph in Fig. 14.

Alternatively, a split proportional band system may be used with two outputs, one for heating and one for cooling. If we use the same set-point as above, but have independent proportional bands each 10°C wide, then the controller behaves as follows. Below the lower band (30°C) full heat is applied, falling to 50% at 35°C, and no heating at all at and above the 40°C set-point. If the temperature exceeds the set-point, cooling is progressively applied with 50% at 45°C and full cooling above 50°C. In both examples a point will exist, provided the heating and cooling methods are strong enough, where the heat applied by the proportional function will be equal to the losses at that temperature. The proportional controller will then apply just the right amount of heat to keep the temperature stable.

The drawback of a proportional controller is that this stable point may be far away from the set-point. Also, the stable point depends on heat produced by the fermentation and ambient temperature, so off-setting the set-point will work for one set of conditions only. Decreasing the width of the proportional band increases the gain of the controller as it increases the change in output for a similar change in input. This will bring the temperature closer to the set-point but may cause instability as the medium has a finite response time. If the proportional band is made too narrow, then heat already applied may cause the temperature to overshoot. The system then applies too little heat, and the temperature falls through the stable point because the medium temperature cannot change quickly enough as more heat is applied. The system therefore oscillates unless the gain is reduced.

Purely proportional controllers always have a steady-state error because, like on/off controllers, they only calculate the amount of control action based on the current parameter value. If this error is unacceptable, then a yet more sophisticated controller is required such as the proportional and integral controller.

3.3. The Proportional and Integral (P & I) Controller

In addition to the proportional term described above, the P & I controller has an integral or reset term that is capable of removing any steady-state error. It does this by accumulating the error between the actual parameter value and the set-point and using the total to modify the amount of control action. If, for example, temperature is being controlled below the set-point by the proportional term, the integral term will steadily increase the heating as long as the error remains. As the temperature rises, the contribution of the proportional term will fall and the integral term will accumulate more slowly. At the set-point, the integral term remains constant at the total accumulated so far, because the error is zero. If the sum of the P & I terms is too great, then the temperature will overshoot and the integral term will now begin to decrease because the error is negative. In this way, the integral term modifies the applied heat depending on past errors and brings the temperature to the desired set-point.

The degree of applied integral action is described by a time constant in either seconds or minutes. An integral term of 30 sec means that, for any error, it will take 30 sec for the integrator output to change by the magnitude of the proportional term. Alternatively, integral terms can be quoted as a reset rate. This is the number of times per minute that the integrator output changes by the magnitude of the proportional term. An integral time of 30 sec therefore corresponds to a reset rate of 2.

For stability, the rate at which the integral term accumulates should not be too fast, and its weighting compared to that of the proportional term should not be too large. The integral term takes time to remove the error and, therefore, P & I controllers have a poor response if disturbed by, say, the addition of cold reagents to the medium. A large overshoot of the parameter before the system settles down may also be undesirable. To improve the controller, a third term, the differential or derivative, may be added.

3.4. The Proportional, Integral and Differential (PID) Controller

The differential term of a PID controller differentiates, or calculates, the rate of change of the error (the difference between the set-point and the value of the measured variable). The result of the calculation is then added to the P & I term, the total deciding how much control action is applied. If the temperature of the medium should suddenly fall, or the set-point be increased, the rapid rate of change of the error will cause the differential term to immediately increase the heating. This counteracts the disturbance quickly and thus improves the transient response of the controller. Such effects are of particular value in rapidly changing systems with low response time; in other systems they can accentuate instabilities.

The differential term also decreases the initial overshoot while the controller is settling down. Before the parameter reaches the set-point, the integral term will accumulate rapidly because of the high error. This may cause too much control action to be applied and hence overshoot. The differential term, in response to the rapidly decreasing error, will attempt to decrease the control action and hence it opposes the integral term. By resisting rapid changes, the differential term decreases the speed with which the parameter crosses the set-point and hence reduces overshoot. The level of derivative action is described by a time constant which, like the integral term, is quoted in seconds or minutes. If the error changes at a constant rate then the contribution of the derivative term will be constant. The proportional term, though, will be constantly changing. A derivative time of 20 sec means that, under the above conditions, it will take 20 sec for the proportional term to vary by the magnitude of the derivative term.

3.5. Controller Outputs

Once the controller has determined the amount of control action to be applied, this must be passed to the actuators. Two common types of output signal are called proportional and time-proportional. A truly proportional signal can take any value between defined limits; 0–5 V or 4–20 mA are commonly used. These are fed to motor controllers or positioning valves so that the speed of a pump or flow of a gas can be continually varied by the controller.

Time-proportional signals are used to control actuators that have only two states, either on or off, such as a fixed-speed pump. The device is alternately switched on and off; the ratio of its "on time" to its "off time" is varied by the controller between the extremes of always off and always on. Electrical cartridge heaters can be controlled like this as the response time of the media is long enough to average out the pulsed heat input. Controllers with time-proportional outputs employ relays or solid-state switches to which the load can be connected. In this way, mains-operated devices can be safely controlled. The various ways in which the parameters may be controlled are described in the next section.

4. ACTUATORS AND PARAMETER CONTROL

The preceding sections have described how various fermentation parameters may be measured and their values processed by a controller, which provides suitable outputs for taking corrective action. To "close the loop" these outputs must drive actuators which affect the physical and chemical conditions in the fermenter, and this section reviews some of the systems currently employed.

4.1. Temperature

Laboratory-scale fermenters frequently use electrical cartridge heaters, in a stainless steel pocket, as a heat source and a "cold finger", or cooling coil, through which mains or chilled water may be passed. Time-proportional control of the electrical supply to the heater is adequate. Similarly, the cooling water can be controlled with a solenoid valve in the supply line, or left running continually.

In a gently agitated system, the surface temperature of a cartridge heater can rise significantly above that of the surrounding medium. If the medium contains sugars which may burn onto the heater, or cells that are sensitive to higher temperatures, then an alternative means of heating must be found. The fermentation can be indirectly heated by using a thermocirculator to pump temperature-controlled water through a vessel coil or jacket. The thermocirculator heats water in its reservoir using cartridge heaters or steam, and cools it down either with a cooling coil or by direct injection of mains water. The advantages of this system are that only one coil or pocket need be used in the vessel and the medium temperature can never rise above that of the process fluid in the thermocirculator. Disadvantages are that the response time is slower and the cost is higher than for the conventional system. The medium temperature will never equal that of the process fluid, due to losses, unless the thermocirculator is operated by a controller that can measure medium temperature directly. One variation of this system is to stand glass vessels in a temperature-controlled water bath or hot room.

On pilot-scale and production vessels, heating is achieved with steam, either through a thermocirculator or by direct injection into the vessel jacket. *In situ* sterilisation of all vessels is most efficiently performed using steam. The sterilising temperature can be pre-set by regulating the supply pressure of the steam (15 psig corresponds to 121°C) or, alternatively, the steam supply can be switched on and off with a solenoid valve. In laboratories where steam is not available, electrical thermocirculators are used, although the process water must be pressurised to prevent boiling above 100°C.

4.2. pH

The most common means of controlling pH is by the addition of liquid acid and alkali to the medium. Fixed-speed peristaltic pumps operated on a time-proportional basis are suitable for most fermentations, although for fine control, variable-speed pumps operated from a truly proportional 0–5 V or 4–20 mA are better. Large vessels use solenoid or air-actuated valves to regulate the flow of reagents from pressurised addition vessels. Hydrochloric acid and sodium hydroxide are the two most common reagents, although animal cell cultures frequently have a 5% carbon dioxide in air mixture as the sparge gas. By varying the carbon dioxide content, in conjunction with a bicarbonate buffer in the medium, the pH of the culture may be prevented from becoming too alkaline.

4.3. Dissolved Oxygen Concentration

The dissolved oxygen concentration of a fermentation can be controlled in three ways: stirrer speed, sparge gas concentration and substrate addition rate in a continuous culture. The characteristics of the fermentation determine the most suitable method.

If the sparge gas rate and composition are constant, then varying the stirrer speed varies the dissolved oxygen level. An increased stirrer speed produces smaller gas bubbles and increases their residence time in the medium. A greater area for diffusion is available and hence the oxygen transfer rate rises. If this exceeds the rate of oxygen consumption, the dissolved oxygen level will rise. Oxygen control by stirrer speed is effective in the range of a few hundred to a few thousand revolutions per minute. A proportional signal from the dissolved oxygen controller controls the agitator motor within set limits. The lower limit is high enough to provide adequate mixing and the higher limit prevents damage to the organisms by shear stresses. If more oxygen is required than safe stirrer speeds can provide, then gas control can be used above the upper speed limit.

For those cell cultures that are easily damaged, gas control is nearly always used. Extra oxygen can be provided to a culture by either increasing the sparge rate or increasing the proportion of oxygen in the gas. In both cases solenoid valves can be used, but if turning the sparge gas on and off is too coarse, then proportional valves must be used instead. An inert "ballast" gas, such as nitrogen, can be added to the sparge mixture. This has the advantages that if the oxygen-rich gas is completely cut off, sparging does not stop, and it decreases the dissolved oxygen level by replacing the oxygen in solution. Although air and nitrogen are commonly used as sparge gases, pure oxygen may be added to enrich the mixture. Care must be taken to ensure that all gas lines passing oxygen are completely grease-free. Some cultures may be damaged by the shear stresses caused by gas bubbles in the medium, in which case special care must be taken when sparging (12, 13).

The dissolved oxygen concentration can also be controlled by limiting the supply of a substrate feed to the culture. By depriving the organisms of raw material, their oxygen uptake rate can be limited. Adding more substrate increases their activity and therefore decreases the dissolved oxygen level.

As with pH reagents, gaseous and liquid substrates can be controlled using valves and pumps respectively.

4.4. Redox Potential

The redox potential of a culture is often measured when the dissolved oxygen concentration is very low. It provides an indication of activity in anaerobic or near-anaerobic conditions, when dissolved oxygen electrodes are useless. Redox control systems are comparatively rare but they employ gas control methods similar to those for dissolved oxygen. Small percentages of air are added to a carrier gas, causing the redox potential to rise. The air can either be switched in using a solenoid valve, or blended using a proportional valve.

4.5. Gas Flow

Control of the sparge gas flow rate and composition is important for many other control systems. In all cases, the final control element is a valve mounted in the gas line operated by one of the parameter controllers. Simple pH or dissolved oxygen systems employ time-proportional solenoid valves operated by mains voltage or 24 V dc. These valves switch air, carbon dioxide or extra oxygen into the main gas line with a resultant change in the total flow rate. More sophisticated systems use proportional valves that continuously vary the flow rates. Up to about 100 litres/min modulating solenoid valves may be used, but above that motorised butterfly valves or needle valves provide better control. Large butterfly valves consist of a servo-control mechanism which positions the valve between fully shut and fully open in response to a proportional control signal. Air-operated needle valves also provide an elegant means of controlling pilot-scale flow rates. They may be positioned by applying a 3–15-psi pneumatic control signal, derived from a current/pressure converter, which responds to a 4–20-mA signal from the controller. If the flow rate of sparge gas must be held constant, then a mass flow system may be used to measure and control the mixed gases before they enter the vessel.

4.6. Pressure

Vessel head-space pressure can be controlled by placing a modulating valve in the off-gas line from the fermenter. By varying the restriction in the line the desired pressure may be built up in the vessel. Care should be taken not to overpressurise glass vessels. If it is required to vary the pressure during a run, then a control system on the gas inlet is desirable, otherwise the sparge flow rate will vary.

4.7. Agitation

Mechanically stirred vessels consist of a shaft mounted through the top or bottom plate on which impellers are mounted. Gas sparged in under the lower impeller is efficiently mixed with the liquid. Rushton-type fixed blade turbines are commonly used, although many designs are available, ranging from variable pitch impellers to high-intensity aeration systems employing marine-type propellers and downdraft tubes.

Tissue culture systems which require very gentle mixing use paddles or marine impellers turning at a few revolutions per minute to keep shear damage as low as possible. Suspended PTFE-covered impellers are also used, magnetically coupled to a stirrer unit, which removes the need for passing the impeller shaft through the vessel wall.

Air-lift fermenters rely on the circulation of medium around a central baffle driven by the sparged gas alone (6). The rate of gas flow affects the mixing in the medium and it is desirable to keep this constant. Gas blending systems are therefore used to add extra gases for pH and dissolved oxygen control without changing the overall flow rate. As air lifts can be used for cell cultures where oxygen demand is low, the poor oxygen transfer rate caused by the gentle mixing is not a problem.

4.8. Weight

Continuous fermentations may be very simply controlled by arranging an overflow weir in the vessel and pumping in medium at a suitable dilution rate. If it is required to run the fermentation at constant weight, then a weight control system is needed. Two configurations are possible.

If medium is to be continually added to the fermentation, then a simple weight controller must open the drain or harvest valve whenever the vessel exceeds a pre-set weight. The vessel contents can then flow under gravity, or the vessel may be pressurised, into a receiving tank. Alternatively, the product can be continuously pumped from the tank and fresh medium added whenever the weight falls below the set-point. If accurate weight control is required flow rate in or out of the vessel should be regulated. Variable-speed pumps or modulating valves operated by the weight controller provide the best solution.

REFERENCES

1. Brauer, H. (1979). Power consumption in aerated stirred tank reactor systems. *Adv. Biochem. Eng.* **13**, 87–119.
2. Brown, D. E. (1977). The measurement of fermenter power input. *Chem. Ind. (London)* 684–688.

3. Clark, L. C. (1956). Artificial internal organs. *Trans. Am. Soc. Artif. Intern. Organs* **2**, 41.
4. Griffiths, B. (1984). The use of oxidation–reduction potential to predict the state of an animal cell culture. *Dev. Biol. Stand.* (in press).
5. Johnson, M. J., Borkowski, J., and Englom, C. (1964). *Biotechnol. Bioeng.* **6**, 457–468.
6. Katinger, H. W. D., and Schierer, W. (1982). Status and developments of animal cell technology using suspension culture techniques. *Acta Biotechnol.* **2**, 3–41.
7. Kjaergaard, L. (1977). The redox potential: Its use and control in biotechnology. *Adv. Biochem. Eng.* **7**, 131–147.
8. Lee, Y. H., and Tsoa, G. T. (1979). Dissolved oxygen electrodes. *Adv. Biochem. Eng.* **13**, 36–86.
9. Makereth, F. J. H. (1964). *J. Sci. Instrum.* **41**, 38–41.
10. Miller, J. T. (1971). "The Instrument Manual." United Trade Press, London.
11. Nyiri, L. K. (1972). "Equipment Considerations in Animal Cell Suspension Cultures Fermentation Design (New Brunswick Scientific) Internal Paper."
12. Spier, R. E., and Griffiths, B. (1984). An examination of the data and concepts germane to the oxygenation of cultured animal cells. *Dev. Biol. Stand.* (in press).
13. Spier, R. E., and Whiteside, J. P. (1983). The description of a device which facilitates the oxygenation of microcarrier cultures. *Eur. Soc. Anim. Cell Technol. Newsl.* (prelim. note).
14. Tanne, L. P., and Nyiri, L. K. (1979). Instrumentation of fermentation systems. *In* "Microbial Technology (H. J. Peppler and D. Perlman, eds.), 2nd ed., Vol. 2, pp. 331–373. Academic Press, New York.

13

Computer Applications in Animal Cell Biotechnology

J. L. HARRIS

L. H. Fermentation
Stoke Poges, Buckinghamshire
United Kingdom

R. E. SPIER

Department of Microbiology
University of Surrey
Guildford, Surrey, United Kingdom

Animal Cell Biotechnology, Vol. 1

1. INTRODUCTION

Computers have been used to study or improve fermentations for many years (9,2). In the 1960s they were used to replace the conventional analogue control loops of fermenter instrumentation and to provide a process log, a cumbersome and expensive operation then (6). With the recent, rapid developments in microprocessor technology, both the size and the cost of computing power have fallen dramatically. The widespread availability of computers and their peripherals has made possible the application of this new technology to solve some old fermentation problems.

This chapter covers the ways in which computers may be connected to fermentations, the advantages that are to be gained thereby and the wider role of computers in overseeing data handling and plant operations.

2. INTERFACING COMPUTERS TO FERMENTATIONS

There are many ways in which computers can help to monitor and control fermentation processes (12). The choice of computer and configuration depends on many things: the number of fermenters, cost of running each fermentation and resources available to install and program the computer system. Obviously, the needs of a small research group with a few laboratory fermenters are going to be very different from those of a large production facility handling many thousands of litres per day. The former will probably require an inexpensive system which can be easily reprogrammed and which is flexible enough to follow the changes in direction of their research. The latter will put an emphasis on reliability and ease of operation by untrained staff, with alarms and back-up systems to protect against possible failure.

The size of computers that are used can vary widely, as can their function, from dataloggers to plant controllers. There are essentially three different system configurations: supervised analogue control, direct digital control and hierarchical control systems. They are discussed below.

2.1. Supervised Analogue Controllers

In this configuration the parameters are controlled by analogue control loops (see previous chapter). Signals from the instrumentation amplifiers are passed through a discrete analogue-to-digital (A/D) converter, so that the actual parameter values may be logged by the computer. The computer may vary the set-points of the analogue controllers by returning a suitable voltage via a digital-to-analogue (D/A) controller. It should be noted that analogue controllers with *remote set-point control* are necessary for this capability (Fig. 1a) (3,1).

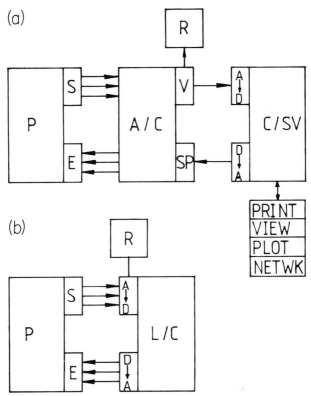

Fig. 1. (a) A supervised analogue control system. P, process; S, sensors; E, effectors; A/C, analogue control loops; V, output voltage; R, recorders; SP, set-point; A → D, analogue-to-digital converter; D → A, digital-to-analogue converter; and C/SV, computer/supervisor. (b) A direct digital control system. L/C, local computer; the other abbreviations are the same as those for Fig. 1a.

Digital set-point control (DSC) has the following advantages: (1) it is inexpensive to set up using a desk-top microcomputer, (2) the computer can be programmed in a high-level language and (3) depending on the power of the microcomputer, several fermenters can be controlled at once. The features that are available are automatic datalogging, time profiled or conditional set-point changes and control of valves, pumps etc. via relays. The main disadvantages of this system are: (1) it is expensive to expand, (2) if too many parameters are controlled the microcomputer may be too slow to handle calculations and data storage and (3) the characteristics of the control loops are determined by the analogue controllers.

2.2. Direct Digital Control

In this configuration the fermentation parameters are directly controlled by the program within the computer. Many versions of this system are possible, ranging from dedicated single-board microprocessors to mainframes controlling many fermenters and also plant operations. As with DSC (Section 2.1), the parameter values are obtained via A/D converters, usually built into the computer itself. The actuators, pumps and valves are controlled directly through the computer program deciding how much acid, alkali etc. is to be added. Digital-to-analogue converters may be used to control analogue devices such as the stirrer motor (Fig. 1b) (*4,13*).

Advantages of this method are as follows: the control loops may be varied in any way by the computer, the system may be configured to handle many fermenters, very large speed improvements are possible if the control loops and A/D routines are written in machine code and supervisory programs may be written in high-level languages to control the overall strategy by interacting with the "background" control loops. The disadvantages are: large systems are relatively expensive, initial programming is a difficult and specialised task, reprogramming difficulty impairs the system's flexibility and, if the computer is located away from the fermenters, large numbers of cables carrying the analogue and digital signals are required.

2.3. Hierarchical Control

This system consists of a number of dedicated microprocessors each controlling one or two fermenters by DDC as described above. These are located close to each fermenter but are connected to a supervisory computer, which is in overall control via a standard digital link such as the RS232 protocol (Fig. 2) (*13*).

Such a system is extremely flexible. Extra microprocessors may be added to the system with new fermenters. The fermenters may be used "manually," as they do not rely on the supervisor for their control algorithms. The use of RS232 lines permits the supervisor to be located well away from the hostile environment of the plant, and keeps the number of interconnecting leads to a minimum. The microprocessors execute each control algorithm in machine code, which leaves the supervisor free to run its programs in a high-level language. As data acquisition is performed by the microprocessors and transmitted in scientific units, the supervisor is left to perform the data handling and storage functions to which it is most suited. The distributed nature of the system also minimises the effect of failures. Should one control loop, or even the whole microprocessor, fail, the other fermenters are not affected, and the failure can be detected by the supervisor, which then

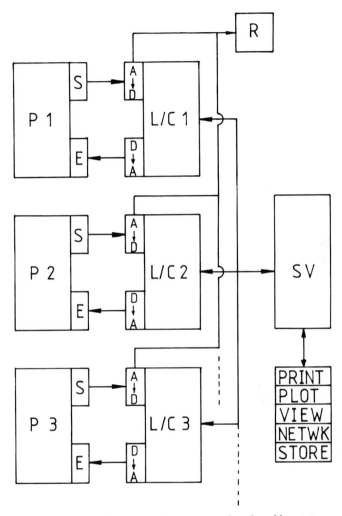

Fig. 2. A hierarchical control system. SV, supervisor; the other abbreviations are defined in the legends of Figs 1a and b.

sounds an alarm. Indeed, the supervisor itself can fail without the loss of the fermentations, although data storage and overall control will be lost. Unfortunately, the redundancy of the system makes it expensive for small numbers of fermenters. Duplication of the microprocessors, memory and RS232 transmission systems increases the cost over a centralised system. Initial programming of the microprocessors to take full advantage of the system's flexibility is a complex and difficult exercise.

3. THE BENEFITS OF USING COMPUTERS

Once a system has been decided upon, the computer can be used to advantage in several areas (6,7,10). The speed and efficiency of data acquisition and storage can be improved, as can the accuracy and flexibility of the control loops. These and other features are described in the following sections. To emphasise the advantages, the hierarchical system previously described will be assumed.

3.1. Monitoring

The signals from probes in the fermentation vessel are conditioned by instrumentation amplifiers and passed to A/D converters for translation into a digital form which the microprocessor can handle. Further conditioning can be performed at this stage, which improves the overall performance. For example, it is possible to linearise outputs from the probes using calibration tables so that the accuracy of measurement may be improved. The output from a resistance thermometer is often assumed to be linear over the range from ambient to sterilising temperature, when in fact the characteristic is a slight curve. Using a look-up table removes this error and provides the controller with a true value of vessel temperature. Compensation for temperature effects can also be performed digitally, particularly the variation in slope of pH and Eh electrodes which obey well-known laws. Removing the membrane-induced temperature coefficient of oxygen electrodes is more difficult and is only partially successful using hardware. Digitally, the oxygen electrode would have to be placed in an oxygen-saturated solution at a sufficient number of different temperatures for the microprocessor to define the probe's characteristic. This could then be stored in a table for future reference during the fermentation. Similarly, compensation for head-space pressure variations can also be made.

Arithmetic functions, which are difficult to perform in hardware, are more readily handled by the microprocessor at this stage. The extraction of the pCO_2 value from the pH of a bicarbonate buffer, for example, involves finding the logarithm of a signal. If this is implemented by the microprocessor, then the supervisor can receive the actual pCO_2 in scientific units with a saving in complexity of its own program.

Probe characteristics may be stored by the micro, thus allowing buffer and slope adjustments to be made in software. Calibration would then consist only of typing in the value of, say, a pH buffer solution. The arithmetic is then handled by the software, which simplifies considerably the design of the amplifiers.

Extra parameters may be monitored by the system by introducing the outputs of gas analysers or chemical balances into the A/D as if they had come from instrumentation amplifiers. Any of the above processing may be carried out before these values are passed on to the supervisor.

3.2. Datalogging

Automation of the storage of data by a computer confers several advantages (14). Data can be collected when the plant is left unsupervised, e.g. at night, and the sampling rate may be automatically changed with events. Samples can be taken more frequently if a parameter changes rapidly, thus providing better resolution, or taken less frequently if a parameter is stable, thus saving storage space on the disc.

The data stored on a disc are obviously easier to manipulate than those on chart-recorder rolls and extra space can be allowed for the addition of off-line results, such as cell count measurements, which would be entered later. It should also be recognised that strip chart display can be more valuable for a rapid assessment of trends and deviations from designed conditions. However, the printout from the computer can be set to print only values deviating by a predetermined amount from the set-point. Perhaps the biggest advantage is the ability of the supervisor to calculate derived parameters, such as motor power or respiratory quotient (RQ), from directly measured parameters (5). These can be saved on disc alongside simple parameters, or converted to an analogue output via a D/A converter, so that the plant supervisor has a chart-recorder trend of the fermentation's progress.

3.3. Control

Traditional analogue control loops are dependent upon component values for their function. The components in a controller have tolerances, and their value can change with temperature. In contrast, the digital controller depends solely upon the steps in its program and, by removing the dependence on components, should give a more predictable and stable performance. For optimum results, analogue proportional, integral differential (PID) loops require fine tuning to conditions within the fermenter. As these conditions can change greatly during a fermentation, the PID loop should be continually tuned, which is virtually impossible during a run. In particular, the exponential nature of some parameters, such as oxygen demand, can outrun a PID controller if it grows too fast. A digital control loop can be continually changed, if the supervisor deems it necessary, or can be made adaptive so that it adjusts its own characteristics with changing conditions. Additionally, different types of control algorithm can be available to the microprocessor and selected as required by the supervisor (13).

In this way pH, say, can be initially controlled with a two-point on/off algorithm with a wide "dead-band" that allows the pH freedom to move, and later a PID or iterative loop may be switched in that holds the pH at some optimum value. Numerical or iterative algorithms can be used which do not have the disadvantages of PID loops, such as overshoot, or a settling down period.

Changing set-points is readily achieved as both supervisor and controller handle digital information, thus removing the need to interface via a D/A converter to an analogue loop.

3.4. Process Parameters

In addition to the improvements on existing systems outlined above, extra features can be provided that are not available with analogue controllers. The microprocessor directly controls the addition of corrective action and can therefore totalise the amount of air to control the dissolved oxygen level (8) or acid and alkali that has been added in controlling the pH. This can provide information on the state of the fermentation, the quantity of salt formed in the fermenter or the cost of running the fermentation. The microprocessor also measures the rate of acid or alkali addition that is required to hold the pH at the set-point. This indication of the culture's "resistance" to being controlled provides more information about the biochemical state of the fermentation. An obvious application is to use the stirrer speed or gassing rate to reflect the culture's oxygen demand and hence determine the end of the growth phase or any change in steady-state conditions. Derived parameters, such as RQ, may be passed to the digital control loops and controlled in the same way as directly measured parameters (5). This provides the possibility of controlling biochemical factors directly rather than working indirectly via pH, dissolved oxygen etc.

4. THE ROLE OF THE SUPERVISOR

As well as the routine operation of parameter control, the computer system can perform more general management tasks, handle the acquired data and control the overall strategy of the fermentation's progress. These tasks are best written in a high-level language on a system with many peripherals, such as printers, disc drives and visual display units (VDUs). It is likely that this software will vary frequently with the configuration and needs of the process under investigation, whereas the primary parameter control loops in the microprocessors will be changed less often. The purpose of this section is to describe some of the likely processes that such a supervisory computer will have to perform.

4.1. Modelling

In order that the course of a fermentation may be accurately predicted or controlled, the supervisor should have some form of mathematical model capable of mirroring the behaviour of the culture (6,8). Changes in conditions can be simulated using the model, and, when an optimum response has been found, those conditions can be imposed on the culture itself. Unfortunately, it is extremely difficult to develop a theoretical model of a culture and for this reason the technique is rarely used. There is usually insufficient data about the fermentation itself—heat transfer rates, oxygen transfer rates or which species are contributing to the redox potential—to construct a comprehensive model. Furthermore, factors may be influencing the culture which the model does not take into account, such as the history of the cells used as the inoculum or slight differences in the complex and often biologically derived components of the medium.

The lack of success of the analytical method does not mean that modelling is impossible, simply that an empirical approach may prove more suitable (4). In this technique the culture is represented as a set of transfer functions (temperature change with applied heat, pH change with acid addition, oxygen demand with growth rate etc.) in matrix form. The input to the culture is represented by a matrix of possible stimuli, and the predicted response is the product of these two matrices. During the fermentation, the computer applies different stimuli to the culture, sets up a corresponding matrix and predicts the results. The transfer function is then updated so that the predicted results agree with what actually happened. Rather than have one model representing the entire course of the fermentation, the computer is creating an adaptive model that predicts one or two steps ahead at a time. The time-varying transfer functions so produced can provide valuable information about the different processes involved.

Even if the computer does not use a model, it can present the collected information in a variety of ways that enable the user to better understand the fermentation. Data can be presented in graph form on a VDU with either all the parameters from one run shown simultaneously, or one parameter from many runs shown superimposed so that the effect of changing other parameters can be clearly seen. Sections of the graphs can be expanded down to the resolution of the original sampling rate, or one parameter may be plotted against another to see if a correlation exists between the two. Changes in applied stimuli can be plotted time-shifted alongside their effects to discover the time constants involved, and a variety of statistical tools may be applied to the raw data to test the validity of new theories. In this way the operator may evolve a model of the culture that can be rapidly checked against past records of fermentation runs.

4.2. Control of Strategy

The computer is a very powerful tool for guiding the course of the fermentation. It can operate round the clock, is accurate and can make decisions based on pre-defined algorithms. All of the parameters under the computer's control may have their set-points progressively changed according to a time-defined profile (6), or they may be varied when other parameters reach critical values (4). In addition, the type of control algorithm used may be varied during a run. A simple on/off algorithm could be used during the growth phase, and then changed to a PID function for accurate control during the steady-state continuous culture.

Simulations of conditions in large vessels can be tested prior to scale-up in laboratory fermenters, where smaller quantities of medium and cells are involved. For example, the pressure within the vessel can be cycled periodically to simulate the circulation of cells in a tall air-lift fermenter, where large changes in hydrostatic head are experienced by the organisms.

4.3. Plant Operations

In a large production plant there are many other functions that a computer can perform in addition to process control. Tanks may be automatically filled and emptied, or their contents transferred temporarily whilst washing or sterilisation takes place. Provided the supervisor has control of the valves in the process and service lines, complex cleaning in place, filtering and mixing of media operations may be performed under its direct control. As well as the obvious savings in time and manpower, the supervisor can also provide more information about the processes involved. For example, if holding vessels are being continuously weighed, the computer can totalise the amount of acid, alkali or medium added to a culture, or use the change in weight to accurately meter the flow rate of additions, thus enabling the calculation of a process mass balance. During a sterilisation cycle, the supervisor could integrate the temperature while it was above 100°C and adjust the time spent at 121°C, so that each part of the plant is exposed to sterilising conditions (7,1,10).

4.4. Optimisation

Perhaps the most important feature of the supervisor is its ability to op-timise the fermentation to produce the maximum yield of a desired sub-stance, be it biomass or a by-product. As long as a quantity relevant to the optimisation can be measured or derived from other measured variables, the computer can vary conditions so that its yield is maximised (14).

Optimisation can be performed either by using a model as previously described or by using "hill climbing" techniques. In the latter approach, one parameter is varied slightly to see if the yield changes, and if so in which direction. A small step is taken in the direction that increases yield and then another parameter is tested. The technique is analogous to climbing a hill by alternately taking steps in a north–south and then an east–west direction, always moving so as to climb. The fermentation situation is greatly compli-cated by the number of dimensions available for movement and the great interdependence of the variables. Mathematical techniques are available which ensure that a situation is reached whereby any change in conditions decreases yield, and hence the top of the hill has been found. This system only discovers local maxima, however, and several different starting points should be tried to ensure that there is not a taller hill somewhere else.

4.5. Reducing Running Costs

Maximising yield does not necessarily give the best financial return when the biomass or final product is sold. On an industrial scale, the raw materials and energy input that a process requires can be very expensive and if a large computing system is to be justified then it should pay for itself by reducing the cost of the fermentation. For example, is it desirable to run the fermen-tation at a high dissolved oxygen level when the cost of the electricity to drive impellers or gas compressors is taken into account? By providing the computer with the relative cost of the media, corrective chemicals and elec-tricity for mixing and heating, it could control a well-understood process at the most cost-effective point rather than the most productive point. A con-tinuous culture could be run with the supervisor monitoring the motor power, the flow rates of all additions being made and the heat input (either electrical or steam). The cost of these would be accumulated and the process parameters adjusted to maximise the yield-to-cost ratio. Similarly, a batch culture could be monitored so that as soon as the rate of production of the desired product fell below optimum, the fermentation would be halted. The product could then be automatically harvested even if there were no plant personnel available, e.g. at night or during the week-end.

5. FUTURE DEVELOPMENTS

As well as the improvements to the process that faster, smaller and cheaper computers will bring, it is likely that they will soon encompass other aspects of cell culture, such as upstream and downstream processing and also administration.

On the upstream side, computers could be used to keep stock control of raw materials, to keep records of what chemicals each culture uses for costing purposes and also for the automatic weighing, mixing and preparation of media (A. Telling, personal communication). Not only would batches of media be produced consistently and faster but poor yield in cultures could be cross-correlated with sources of serum, conditions of storage etc., which would help to trace sources of contamination.

Downstream, more sophisticated equipment such as gas analysers and mass spectrographs will yield more information about the culture. Automatic sampling and analysis will provide faster measurements of parameters that, currently, cannot be measured on-line. Computer control of harvesting, processing and freezing will also make extraction of the final product quicker and cheaper.

A new generation of intelligent probes may also develop with built-in microprocessors to linearise and temperature-compensate signals prior to transmitting them digitally to the instrumentation. Such probes could incorporate automatic buffer and span controls and even provide a warning if the probe is coming near to the end of its useful working life.

Record-keeping and data analysis are two areas in which computers can be used to great effect. As more data are collected, large archives of results could be established. Eventually, networks may be established across the country, enabling researchers to exchange results and programs. Computer libraries of graphs, data and the conditions for optimum growth of well-known organisms would provide greater access to this information and prevent duplication of work. In addition to the scientific aspects of record-keeping, there are those which require data to be kept for presentation to a *regulatory authority* to characterise a batch of a biological product such as a vaccine. Such records also enable managers and supervisors to be sure that the process has been maintained reliably and provide an additional check on whatever manual system is normally operative.

REFERENCES

1. Arminger, W. B., and Humphrey, A. E. (1979). Computer applications in fermentation technology. *In* "Microbial Technology" (H. J. Peppler and D. Perlman, eds.), Vol. 1, pp. 375–401. Academic Press, New York.

2. Arminger, W. B. (1979). Computer applications in fermentation technology. *Biotechnol. Bioeng. Symp.* **9**, 1–398.

3. Breame, A. J., and Spier, R. E. (1977). A low cost microcomputer for the control of microbiological systems. *Lab. Pract.* **26**, 957–961.

4. Breame, A. J., and Spier, R. E. (1980). Down market computers for fermentation research. *Int. Conf. Comput. Appl. Ferment. Technol. Soc. Chem. Ind.*, *3rd, 1980*, pp. 13–22.

5. Bushell, M. E., and Fryday, A. (1983). The application of materials balancing to the characterisation of sequential secondary metabolite formation in *Streptomyces cattleya* NRRL 8057. *J. Gen Microbiol.* **129**, 1733–1741.

6. Constantinides, A., Spencer, J. L., and Gaden, E. L. (1970). Optimisation of batch fermentation processes. II. Optimum temperature profiles for batch penicillin fermentation. *Biotechnol. Bioeng.* **12**, 1081–1098.

7. Hampel, W. A. (1979). Application of microcomputers in the study of microbial processes. *Adv. Biochem. Eng.* **13**, 132–33.

8. Koga, S., Burg, C. R., and Humphrey, A. E. (1967). Computer simulation of fermentation systems. *Appl. Microbiol.* **15**, 683–689.

9. Nyiri, L. K. (1972). Applications of computers in biochemical engineering. *Adv. Biochem. Eng.* **2**, 49–95.

10. Nyiri, L. K., and Kirishnaswami, C. S. (1974). Fermentation process analysis, modeling and optimisation. *Am. Soc. Microbiol. Meet.*, Sess. 56.

11. Grayson, P. (1969). Computer control of batch fermentation. *Process Biochem.* March, 43–44.

12. Unden, A., Rindone, W. P., and Heder, C.-G. (1979). The Computer in Fermentation Process Research. *Process Biochem.* March, 8–12.

13. Rolf, M. J., Hennigan, P. J., Mohler, R. D., Weigand, W. A., and Lim, H. C. (1982). Development of a direct digital-controlled fermentor using a micromini computer hierarchical system. *Biotechnol. Bioeng.* **24**, 1191–1210.

14. Saguy, I. (1982). Utilisation of the "Complex Method" to optimise a Fermentation Process. *Biotechnol. Bioeng.* **24**, 1519–1825.

Index